AF615212

Microstructural characterisation of fibre-reinforced composites

Microstructural characterisation of fibre-reinforced composites

Edited by

John Summerscales

CRC Press
Boca Raton Boston New York Washington, DC

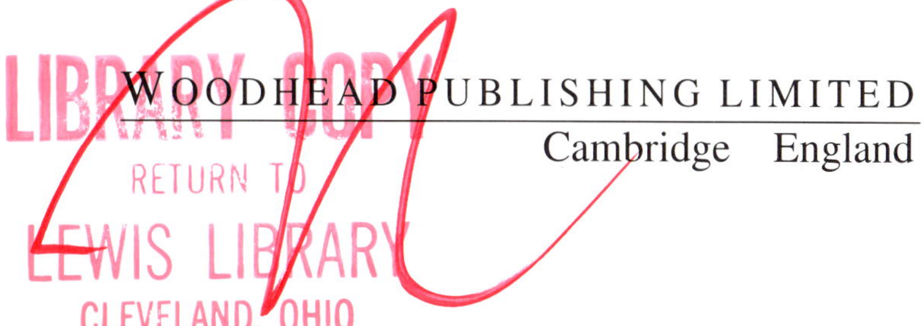

WOODHEAD PUBLISHING LIMITED
Cambridge England

Published by Woodhead Publishing Limited, Abington Hall, Abington
Cambridge CB1 6AH, England

Published in North and South America by CRC Press LLC, 2000 Corporate Blvd, NW
Boca Raton FL 33431, USA

First published 1998, Woodhead Publishing Ltd and CRC Press LLC

British Library Cataloguing in Publication Data
A catalogue record for this book is available from the British Library.

Library of Congress Cataloging in Publication Data
A catalog record for this book is available from the Library of Congress.

Woodhead Publishing ISBN 1 85573 240 8
CRC Press ISBN 0-8493-3882-4
CRC Press order number: WP3882

Cover design by The ColourStudio
Typeset by Techset Composition Ltd, Wiltshire, England
Printed by TJ International, Cornwall, England

Contents

Preface

That same day Pharaoh commanded the taskmasters of the people as well as their supervisors "You shall no longer give the people straw to make bricks, as before: let them go and gather straw for themselves. But you shall require of them the same quantity of bricks as they have made previously"

Genesis 5: 6–8

The use of natural composite materials by human kind is almost as old as the species itself, notably through animal and plant fibres adapted to perform each task. The last fifty years have seen a significant advance in the development of artificial fibre-reinforced composite materials for structural applications, notably with continuous glass or carbon fibres at high fibre volume fractions. The effective development of this class of materials involves a close understanding of the micromechanical features and the macroscopic performance. The use of microstructural characterisation techniques has been applied widely in extending this understanding. However, the literature has proliferated primarily in conference proceedings and journals across a wide range of subjects, throughout science, technology and engineering, with marginal coverage in a few books. The purpose of this book is to provide a contemporary review of the application of advanced microstructural characterisation techniques to fibre-reinforced composites. It is aimed at senior undergraduates, research degree candidates and engineers entering, or already practising, in the field of composite materials. It should also prove useful to microscopists who are given the task of examining these novel materials. The chapters are intended to provide a stand-alone introduction to each topic with comprehensive referencing of appropriate basic texts and of the original works for those readers wishing to pursue the subject further.

The introductory chapter presents an overview of the need for appropriate microscopical techniques, together with a brief consideration of textures and of those advanced techniques which are not covered elsewhere in the book and are current (image processing and three-dimensional visualisation) or are likely to become important in the near future (optical coherence tomography, microradiography, magnetic resonance). Chapter 2 considers the microscopy of flexible textile composites: this important area has much to offer those of us who have traditionally been associated with 'structural' rigid composites. Chapter 3 describes the novel 3D confocal laser scanning microscopes at the University of Leeds. To date, these have been primarily employed in the characterisation of short fibre composites: the systems have high potential for detailed characterisation of continuous unidirectional and fabric-reinforced composites. Chapter 4 describes the work of the University of Delaware in the geometric modelling of yarn and fibre assemblies, whilst Chapter 5 extends this by considering the yarn shape in woven fabric composites. Chapter 6 deals with the quantitative microstructural definition of fabric-reinforced composite materials and the potential to develop process–property–microstructure relationships. The final three chapters each consider the use of a specific microstructural characterisation technique: Chapter 7 deals with electron microscopy in the study of interfacial bonding, Chapter 8 deals with the measurement of strain or stress by Raman microscopy and last (but by no means least) Chapter 9 looks at the acoustic microscope in the context of ceramic-matrix and metal-matrix composites.

The authors selected for each chapter were most often the first choice of the editor, but a couple were either too busy or otherwise could not give their commitment to the project. The editor thus accepts any blame for imbalance in the coverage of the book. Extreme care has been taken in creating this volume, but inevitably there will be faults which escape proofreading. I should be most grateful to receive notice of any such problems which readers might detect, and especially to receive reprints of appropriate key papers which might point the way to any future volume.

John Summerscales

Acknowledgements

The editor would like to acknowledge the help of colleagues in the School of Manufacturing, Materials and Mechanical Engineering for their assistance in the preparation of this book, either through direct advice on points of detail or simply for being there to share the work of the School equitably. Further, I should like to express personal gratitude to Mr David Short (now Head of SMMME) for his encouragement and guidance throughout the twenty years which have elapsed since I adopted composites as a career.

Thanks are also due to Patricia Morrison at Woodhead Publishing Limited for supporting the idea of this book and for undertaking much of the routine administrative work in a cheerful and professional manner.

Last, but by no means least, I should like to thank each of the chapter authors for producing manuscripts in a timely manner which required minimal editing and who thereby made a pleasure of the task of putting the book together.

Where figures have not been originated by the chapter authors, we are grateful to the originators and publishers for permission to reproduce the figures included in the book. They are acknowledged in the individual chapters. The authors and publishers have attempted to trace the copyright holders of all figures reproduced in this publication and apologise to any copyright holders if permission to publish has not been obtained through error or omission.

List of contributors

PATRICIA A ANNIS *Department of Textiles, Merchandising and Interiors, The University of Georgia, Athens, GA 30602, USA*

GEOFF ARCHENHOLD *Molecular Physics and Instrumentation Group, Department of Physics and Astronomy, University of Leeds, Leeds LS2 9JT*

G ANDREW D BRIGGS *Department of Materials, University of Oxford, Parks Road, Oxford, OX1 3PH*

ASHLEY R CLARKE *Molecular Physics and Instrumentation Group, Department of Physics and Astronomy, University of Leeds, Leeds LS2 9JT*

NIC C DAVIDSON *Molecular Physics and Instrumentation Group, Department of Physics and Astronomy, University of Leeds, Leeds LS2 9JT*

COSTAS GALIOTIS *Department of Materials, Queen Mary and Westfield College, Mile End Road, London E1 4NS and Institute of Chemical Engineering and High Temperature Processes, Foundation for Research & Technology – Hellas, Stadiou Street, Platani, PO Box 1414, GR-265 00, Patras, Greece*

MICHÈLE GUIGON *Laboratoire de Génie Mécanique pour les Matériaux et les Structures, UPRES A 6066, Université de Technologie de Compiègne, BP 20529, 60205 Compiègne, France*

FELICITY J GUILD *Department of Mechanical Engineering, University of Bristol, Queen's Building, University Walk, Bristol BS8 1TR*

JULIUS JORTNER *34360 Highway 101 South, PO Box 219, Cloverdale, OR 97112-0219, USA*

MICHAEL KEEFE *Mechanical Engineering Department, University of Delaware, Newark, Delaware DE 19716-3140, USA*

CHARLES W LAWRENCE *Department of Materials, University of Oxford, Parks Road, Oxford OX1 3PH*

THOMAS W QUIGLEY Jr *Department of Textiles, Merchandising and Interiors, The University of Georgia, Athens, GA 30602, USA*

JOHN SUMMERSCALES *Advanced Composites Manufacturing Centre, School of Manufacturing, Materials and Mechanical Engineering, University of Plymouth, Plymouth, Devon PL4 8AA*

STEVEN W YURGARTIS *Department of Mechanical Engineering, CAMP Building, Clarkson University, Potsdam, NY 13699-5729, USA*

1
Introduction

JOHN SUMMERSCALES

1.1 Introduction

Fibre-reinforced composites have risen from very small beginnings around the time of the Second World War to an industry which is now delivering in excess of one million tons of material in each of Europe and the United States every year. The vast majority of these materials are polymer-matrix composites. The percentage of metal-matrix and ceramic-matrix composites is still in single percentage figures by volume and weight, but represents a larger proportion in value.

The intention of this book is to highlight some of the current state-of-the-art in the microstructural characterisation of fibre-reinforced materials. The increasing interest in mesostructures (intermediate between micro- and macrostructures) is reflected here.

Those readers who are not familiar with the basic principles are referred to the recent literature for microscopy [1–35] and for microprobe analysis [36–43]. Flewitt and Wild [28] offer a good balance of comprehensive coverage and realistic price, albeit that polymers and composites are given only superficial treatment. A biennial annotated bibliography of chemical microscopy is produced by Cooke [44].

1.2 Microscopy of polymers and composites

Three existing books provide a very strong background for the microscopy of polymers. They are recommended sources for that broader topic. Hemsley [45] has described the more advanced techniques of polymer microscopy using light (or optical) microscopy. He outlines the principles and practice of individual methods which may be applied to various forms of specimen. His book gives only fleeting coverage of composites; sectioning of composites (pages 33/34) and ultraviolet and fluorescence microscopy for curing and permeability of thermosetting resins (pages 266–269).

Sawyer and Grubb [46] provide a slightly more extended coverage; specimen preparation (first edition pages 82/84), etching of reinforcement fibres (pages 111/112), fibre, plastics and composites fracture (pages 131–136), application of microscopy to fibres and fabrics (pages 155–176) and to composites (pages 214–230), the surface coatings on fibres (page 238) and liquid crystalline polymer composites (pages 252–254).

Roulin-Moloney [47] describes the techniques that are currently employed in fractographic investigation, including case histories of specific polymeric and composite materials to which the techniques have been applied. The book opens with six chapters on the microscopical techniques including image analysis, and closes with six chapters on fractography and failure mechanisms. Of the latter six, three are specific to carbon fibre composites, to short-fibre reinforced semi-crystalline thermoplastics and to stress corrosion cracking in glass reinforced plastics.

Engel *et al.* [48] have published a book which shows different kinds of damage in polymers that can often only be interpreted by microscopic examination combined with experience. Some 40 examples are relevant to this book, notably:

- fracture of unsaturated polyester resin matrix composites (plates 196, 197, 346–348 and 420–421),
- fracture of fibre-reinforced nylon 6 (plates 296–298, 404 and 405),
- fracture of fibre-reinforced PTFE (plate 300),
- fracture of fibre-reinforced polypropylene (plates 301, 302, 338–340, 384 and 385),
- fracture of fibre-reinforced polycarbonate (plates 305–307, 356–358 and 429–430),
- fracture of cotton-reinforced phenolic resin (plates 349 and 350),
- weathering of glass-reinforced nylon (plates 205–207),
- frictional wear of carbon fibre-reinforced epoxy resin (plates 111–114) and
- a glass fibre-reinforced PTFE bearing cage (plates 117 and 118).

A companion book exists [49], but does not include micrographs of either fibre- or whisker-reinforced composites. The Royal Microscopical Society organises a biennial International Conference on Microscopy of Composite Materials [50], although the polymer composite content is often low.

1.2.1 Defects

Adams and Cawley [51,52] reviewed defect types and non-destructive testing techniques for composites and bonded joints.

Summerscales [53] recently reviewed the manufacturing defects which are likely to occur in fibre-reinforced plastics composites. Special emphasis was laid on the ability to detect these defects by non-destructive techniques. The principal problem areas were considered to be fibre orientation, layer stacking sequence errors, fibre waviness, fibre clustering, state-of-cure of the matrix resin, voids or inclusions and retained moisture.

Piggott [54] has reviewed the effect of fibre waviness on the mechanical properties of unidirectional fibre composites. Waviness has been shown to play a major rôle in compressive strength and fatigue endurance. Shear strength and delamination resistance are also influenced by fibre waviness.

1.2.2 Damage

Heslehurst and Scott [55] reviewed both defects and damage pertaining to composite aircraft components. They identified that the structural degradation in engineering properties varied with: defect severity, defect location and orientation, frequency of defect occurrence, component load path criticality and stress state, defect idealisation, design load levels and nature, defect detection capability, local repair capability, component configuration, environmental conditions, loading history, material property variations, and acoustic vibration response. The paper includes comprehensive lists of problems likely to be encountered in composite materials.

Cantwell and Morton [56] reviewed the various failure modes which occur in long fibre composites. The significance of each fracture mechanism mode, in terms of the energy dissipation capacity and the residual load bearing properties, was considered. In general, failure modes which involve fracture of the matrix offer low fracture energies, whereas fibre dominated modes of fracture involve a greater dissipation of energy. The tensile strength of long-fibre composites is sensitive to fibre damage. The compressive properties are influenced by matrix fracture, most particularly delamination. The fractography and failure mechanisms described in Roulin-Moloney [47], which were referred to earlier, are also pertinent here.

1.3 Textures

1.3.1 Microtexture

Microtexture conjoins the study of microstructural features with (crystallographic) textural features in materials. These areas have been regarded as separate until recently because of the difficulty of applying techniques for texture determination on a microscopic level. The microstructure is usually characterised by various microscopy techniques and the texture is commonly determined by X-ray or neutron diffraction techniques [57]. However, these methods do not easily permit the correlation of the microstructure and the individual orientations of the constituent parts.

Spatially resolved microtexture analysis allows the determination of this relationship. The method is usually based on diffraction techniques in electron microscopy, which is ideally suited to this problem through the combination of high spatial resolution with the imaging capability.

One microtexture tool uses the determination of individual orientations by Kikuchi patterns (projections of the geometry of lattice planes in a crystal due to reflections which satisfy the Bragg condition) in SEM (scanning electron microscopy) reflection or in TEM (transmission electron microscopy). Kikuchi patterns arise as a result of the divergence of the electron beam in all directions as it penetrates the specimen, and are essentially maps of the angular relationships in the crystal.

The second microtexture tool obtains orientation information from selected areas by measuring local pole figures (the stereographic projections of the distribution of crystallographic planes). Schultz [58] developed the X-ray pole figure technique, which ushered in the modern era of texture measurement. The publication of a mathematical method of pole figure inversion, now used for the calculation of the orientation distribution functions, by Bunge [59] and Roe [60] initiated modern texture analysis. A practical guide to microtexture determination is available [61] and the topic is the subject of a triennial conference series [62].

1.3.2 Mesostructures

The concept of mesomechanics was introduced by Haritos *et al.* [63] for the study of the microstructure–mechanics relationship at scales intermediate between the microscopic and the macroscopic. The concept is particularly relevant to the study of assemblies of reinforcement fibres, especially in flexible composites (Chapter 2), short-fibre composites (Chapter 3) and woven fabric composites (Chapters 5 and 6). The geometric modelling of yarn and fibre assemblies is introduced in Chapter 4.

1.3.3 Stochastic geometry and tessellated space

Complicated geometric patterns occur in many areas of science and technology. The analysis of such a data set requires suitable mathematical models and appropriate statistical methods. The area of mathematical research concerned to provide such models and methods is called stochastic geometry. The basic subject considers problems concerning a finite number of geometrical objects of fixed form, where the positions are completely random and, to a certain extent, uniformly distributed. The modern theory of stochastic geometry considers (infinite) random geometrical patterns of more complicated distribution.

The book by Stoyen *et al.* [64] has become the reference work in this subject. Topics covered include the basic theories of point processes, random sets, fibre (defined as a sufficiently smooth simple curve of finite length in the plane, *not* a filament as in reinforcement!) and surface processes, random tessellations, stereology and the statistical theory of shape. Other recent key texts include Diggle [65], Hermann [66] and Okabe *et al.* [67].

Dirichlet [68] and Voronoi [69] considered regular tessellations of planes and higher dimensional spaces, motivated by problems in number theory. The Voronoi cell associated with any single point within a planar point process is the polygonal region of the plane which is closer to that point than to any other point in the process. Applications of Dirichlet and Voronoi tessellations in meteorology, metallurgy and crystallography, and ecology appear to have arisen independently [64]. When the locally finite system of points in n-dimensional Euclidean space is a lattice (as was the case for Dirichlet and for Voronoi), then the cells are called Wigner–Seitz zones by physicists and metallurgists. If the Voronoi tessellation has almost every node touched by exactly three cells (in the planar case) or exactly four cells (in the spatial case), then the Delaunay tessellation is produced. The construction of Voronoi and Delaunay tessellations is discussed in Aurenhammer [70].

Stereology is the branch of stochastic geometry which considers the problem of reconstructing a three-dimensional structure from planar (two-dimensional) or linear (one-dimensional) sections.

1.3.4 Fractal geometry

Mandelbrot [71] claims 'that many patterns of Nature are so irregular and fragmented, that, compared with Euclid [standard geometry] Nature exhibits not simply a higher degree but an altogether different level of complexity. The number of distinct scales of length of natural patterns is for all practical purposes infinite. ... Responding to this challenge, [Mandelbrot] conceived and developed a new geometry of nature and implemented its use in a number of diverse fields'. This concept of fractal (Latin for irregular) geometry has been widely adopted as evidenced by the thousands of citations of the above work (Science Citation Index Search, Bath Information and Data Services, 27 August 1996).

Fractal analysis may provide a way forward for the quantitative evaluation of microstructures that are difficult to accommodate by more traditional methods. Hornbogen [72] reviewed the use of fractals for the characterisation of the microstructure of metals. He concluded that fractal microstructure–property relationships was a 'hardly explored field'. Promising relationships were found for surface structure–catalysis, for martensite formation–shape memory and for mode of fracture–fracture energy.

In a review of the applications of fractal geometric analysis to microscopic images, Cross [73] discusses the theory using classic examples of the van Koch curve, the Cantor set and the Sierpinski gasket. The concept of fractal dimension is introduced to describe these objects. The availability of microcomputer-aided image analysis systems allows the iterative analytical processes to be implemented. Kaye [74] considers random walk modelling of fractal systems and demonstrates the descriptive power of fractal dimensions applied to a diverse range of materials (but not to fibre-reinforced plastics).

Worrall and Wells [75] used fractal variance analysis to characterise differences in filamentisation between bundled and filamentised press-moulded long discontinuous glass-fibre/polyester resin composites. Changes in the slope of Richardson plots ('measured length of coastline' plotted against the 'size of the measuring stick' on a log–log scale) were used to identify changes in the composite structure using optical microscope images of polished sections. A higher Izod impact strength was achieved by the bundled material ($50.8 \pm 15.4\,\mathrm{kJ\,m^{-2}}$) than by the filamentised material ($35.1 \pm 7.17\,\mathrm{kJ\,m^{-2}}$). This is qualitatively in agreement with theories which suggest that bundles act as large diameter fibres during pull-out.

Flook [76] has described an alternative strategy to Richardson's method for estimating the fractal dimension, which is more suitable for computer evaluation. The approach is based on the method used by Cantor to 'tame' non-differentiable curves. The curve is considered to be made of a series of closely spaced points. A circle of known radius is drawn on each point of the curve. The circles then describe a path of width equal to the diameter of each of the circles covering the curve. The area of the path divided by its width gives an estimate of the length of the curve. As the circle radius r increases, the circles have a greater degree of overlap and increasingly obscure the fine details of the curve, hence reducing the length estimate of the curve, $L(r)$. The fractal dimension may be evaluated from the plot of log $L(r)$ against r. This boundary dilation method provides a rapid and accurate method of estimating the fractal dimension.

Dzenis [77] undertook atomic force microscopy studies of the surfaces of several reinforcement fibres and revealed their self-similar fractal properties over a scale range from nanometres to micrometres. The fractal nature of the fibre surfaces implies a fractal morphology for the fibre/matrix interface. Simple models for the effective properties of an unidirectional composite, using a general scaling relationship with the thickness of the interlayer as a scale parameter, suggests that the fractal dimensions of the interface strongly affect the mechanical properties.

1.4 Advanced techniques

1.4.1 Image analysis

Image analysis and image processing are used for two distinct purposes:

- preparing images for the measurement of features and structures present therein
- improving the visual appearance of images for the human viewer.

The use of quantitative image analysis in the characterisation of material microstructures has not been held back by a lack of suitable techniques or equipment. The real obstacle has been the time required to select an appropriate and efficient technique from the vast array of possible methods. It is not the intention of this book to provide a comprehensive course on quantitative image analysis. There are several excellent texts which fulfil this rôle [78–81] and a number which are specific to microscopy of materials [82–84].

1.4.2 Three-dimensional visualisation

The visualisation of data from three-dimensional (3D) volumes has undergone significant development in recent years. In particular, the development of computerised tomography (CT) has become an important means of interpreting medical non-invasive diagnostic techniques. The use of computer tomography for the macroscopic non-destructive examination of fibre-reinforced plastics composites has been comprehensively reviewed by Bossi *et al.* [85].

McNab and Cornwell [86] have demonstrated that ray-tracing and surface rendering may be extremely useful techniques for the interactive assessment of 3D data sets. For the rendering of the entire 3D volume in one view, ray-casting is an excellent tool which permits the representation of symbolic depth. In ray-tracing a set of rays emanate from an arbitrary plane and propagate through the volume. Certain parameters along the ray are used to determine the amplitude value of each pixel in the resulting image which has one pixel per ray. In ray-casting, only primary ray paths are traced, with no calculation of reflected ray paths. Surface rendering represents the external surfaces of the 3D volume. It cannot render the entire volume in a single view, but does permit an understanding of 3D features in an image. Estimation of positions and sizes may be achieved using locations in excavation planes: the interior of the object is seen by removing a portion of the image and recalculating the visible surfaces. The two techniques are complementary: the former permits a quick aggregate view of the volume and the latter provides quantitative positional information on image features.

1.4.3 Optical coherence tomography

Optical coherence tomography (OCT) is a novel imaging technique which permits the acquisition of tomographic images with high resolution ($\sim$15 μm in three dimensions) and a high dynamic range ($>$100 dB). It has found extensive use in biomedicine, notably for ophthalmic procedures. In low coherence reflectometry (LCR), the coherence property of light reflected from a sample provides information on the time-of-flight delay from the reflective boundaries and backscattering sites in the sample. The delay information is then used to determine the longitudinal location of the reflection sites. LCR can be performed with continuous wave light using diode light sources and fibre optics.

Huang *et al.* [87] extended the LCR technique for tomographic imaging of biological systems, using a fibre optic Michelson interferometer illuminated by an 870 nm wavelength superluminescent diode. Their OCT system performs multiple longitudinal scans at a series of lateral locations to provide a two-dimensional map of reflection sites in the sample. This mode of operation is similar to ultrasonic B-scan pulse-echo imaging. The resolution of the OCT system is limited only by the coherence length of the light sources. The longitudinal resolution (FWHM; full-width half-maximum) was 17 μm in air, with repeatable location of the origin of a sample reflection at a spatial resolution of <2 μm in the absence of 'strong nearby reflections'. Image reconstruction is far less computation intensive than for X-ray computer tomography (X-ray CT) or magnetic resonance imaging (MRI).

Kulkarni *et al.* [88] have enhanced the sharpness of OCT images through a new linear shift invariant system model. The use of a constrained iterative restoration algorithm produced an enhancement of the longitudinal image sharpness by a factor of >2.5, with a decrease of only 2 dB in the dynamic range. The lateral resolution was improved, from 14 μm before, to 7 μm after deconvolution through 10–30 iterations.

The National Institute for Standards and Technology (USA) has recently used OCT to obtain 'some beautiful pictures of a glass/vinyl ester composite that showed in great detail the porous structure' [89]. These pictures are not yet in the public domain.

1.4.4 Microradiography

The characterisation of composites can be difficult using traditional non-destructive techniques. Hentschel and Lange [90] developed 'X-ray diffraction scanning microscopy' to overcome this problem. Scanning microscopy is performed by Thompson elastic X-ray scattering to reveal a two-dimensional image of a section through a thick laminate. The selection of the diffraction angle provides a chemical contrast. The system described consisted of 5000 pixels with a resolution of 50 μm × 200 μm × 1.5 mm. Demonstration images were obtained from a 25 mm thick 31 layer carbon fibre composite and a 4.5 mm thick carbon fibre ceramic plate.

X-ray tomographic microscopy (XTM) is a high resolution variant of computed tomography (CT). The X-ray attenuation coefficient at any point in a material can be determined from a finite set of X-ray attenuation measurements (projection data) taken at different angles. Kinney *et al.* [91] have developed an XTM system specifically to address many fundamental questions in composites materials science. The spatial resolution is 15 μm using conventional X-ray generators and better than 5 μm using 25–50 keV synchrotron radiation. This permits the study of fracture, cyclic fatigue and creep crack growth as well as composite fabrication technologies. Experiments with 8-ply SiC fibre in an aluminium matrix clearly resolved the 142 μm diameter fibres and their 33 μm graphite cores. Variations in the SiC stoichiometry adjacent to the graphite core were also detected, as were small cracks in the outer plies attributed to the CTE (coefficient of thermal expansion) mismatch between the fibres and matrix causing cracking during post-consolidation cool-down.

Kinney *et al.* [92] have reported further studies with a pixel sampling size of 5.6 μm, using imaging of three examples of fibre composite structures:

- unidirectional SCS8 SiC fibres in an aluminium matrix,
- $0^\circ/90^\circ$ cross-plied SCS6 fibres in calcium aluminium silicate (CAS) matrix, and
- woven 500 fibre tow amorphous SiC in (10–20 μm diameter Nicalon)/SiC matrix.

A crack, revealed by SEM to be less than 1.5 μm wide along its entire length, was revealed in the SCS6/CAS composite demonstrating that detection can be achieved for features which are smaller than the system spatial resolution. The XTM image of the fully reacted Nicalon/SiC woven fabric distinguished between the fibre and the matrix, identified individual fibres at the tow periphery and revealed both macroporosity (lying outside the tow) and microporosity (within the tow) consequent upon the chemical vapour infiltration (CVI) process.

In a subsequent paper [93], it was reported that the microporosity could have much smaller pore sizes than the spatial resolution of the instrument. Material consolidation was measured by equating the average attenuation coefficient in a supervoxel (an assemblage of contiguous voxels that average over several micropores) to the amount of SiC that has been CVI deposited. Porosity can thus be measured as a function of infiltration time referenced to the initial value of porosity. Volume fraction values obtained by the supervoxel approach were indistinguishable from those obtained by conventional destructive sectioning techniques.

Erre *et al.* [94] chose to develop X-ray projection microscopy based on a simple modified scanning electron microscope. The system is claimed to permit X-ray images to be obtained in <10 s with a lateral resolution of 2 μm. The system was expected to be enhanced by incorporating a capability for X-ray microtomographic reconstruction. The microscope has been demonstrated using the spherical interfaces of zirconia spheres with a 100 μm thick SiC outer coating and by the detection of 0.1 μm thin gold film in a 150 μm thick carbon matrix.

1.4.5 Magnetic resonance

Nuclear magnetic resonance (NMR) in a complex molecule is generally achieved by placing the material in a strong constant magnetic field and then applying a perpendicular radiofrequency (RF) alternating magnetic field. The nuclei absorb or emit energy at characteristic frequencies in the RF field, and the magnitude and frequency of this energy can indicate the chemical structure of the material under test.

Similarly, the magnetic properties inherent in free radicals, due to the presence of unpaired electrons, can be utilised using electron paramagnetic/spin resonance (EPR/ESR) techniques with microwave frequencies instead of radiofrequencies. The use of NMR and ESR techniques for the macroscopic non-destructive testing of fibre-reinforced polymer matrix composites has been reviewed by Summerscales and Short [95].

Such techniques may be used to produce line spectra of chemical species within materials, but the data may also be transformed to present spatial information about species distributions within the sample. Perhaps the best known example of this is MRI used in health care. Information is presented using CT. The magnetic resonance techniques have greater potential as they may be used to follow metabolic processes, whereas X-ray CT simply records density variations. Note that the NMR technique does not involve radioactivity, but the N is omitted from MRI for medical applications to avoid undue concern in the patients.

Possibly the first use of NMR imaging for polymer composites was a study of the distribution of water in unidirectional glass fibre/epoxy resin samples by Rothwell *et al.* [96]. The results presented a clear indication of the heterogeneous macroscopic absorption of water by the material.

Magnetic resonance techniques for the microstructural characterisation of materials have immense potential, and are the subject of one key book [97]. This book does not include a chapter on this new technique: a brief survey of the literature follows here.

The concept of microscopic NMR imaging was introduced at the same time as that of macroscopic NMR imaging [98]. Magnetic resonance microscopy is a rapidly developing field of research with a wide variety of applications in different areas. The technique combines the wealth of information provided by NMR spectroscopy with high spatial resolution. Spatial selectivity is achieved primarily by the application of magnetic field gradients, although surface coils may be successfully used. The bulk of the work is undertaken using NMR for experimental simplicity. At the turn of the decade, the best reported resolution was a voxel (volume element: the 3D equivalent of a pixel) of 260 μm^3 using a 300 MHz NMR system. The inherent limitations of the achievable resolution are on a purely practical level: the spatial information is frequency encoded such that a higher spatial resolution requires a more accurate frequency measurement or a larger frequency spread of the encoding process.

Comprehensive coverage of magnetic resonance techniques for microstructural characterisation may be found in Blümich and Kuhn [99]. They include chapters on NMR imaging of polymers, *in situ* polymerisation reactions and fluid displacement in porous media (of potential interest to those working with resin transfer moulding processes). Figures presented therein show the water distribution in a glass/polyamide pultruded rod, a section of a natural rubber tyre, a carbon–carbon composite rod implanted in a rabbit knee femur and the progress of cure in a carbon-fibre prepreg composite.

1.4.6 Other techniques

Various chapters of this book consider novel microscopical techniques which have been applied to the study of composite micro- and mesostructures. Chapter 3 looks at the use of confocal laser scanning microscopy primarily in the context of short-fibre composites. Electron microscopy is widely used for fractography of composite failure surfaces: in Chapter 7 the use of this technique to study interfacial phenomena is described. Chapter 8 describes the use of laser Raman microscopy to determine the stress transfer profiles and interfacial shear stress distributions. Finally, Chapter 9 considers scanning acoustic microscopy for the study of ceramic fibres in high modulus matrix systems.

References

1. S Bradbury, *Introduction to the Optical Microscope*, RMS Microscopy Handbook 01, Royal Microscopical Society, Oxford, 1989. ISBN 0-19-856419-8.
2. P J Goodhew, *Specimen Preparation for Transmission Electron Microscopy of Materials*, RMS Microscopy Handbook 03, Royal Microscopical Society, Oxford, 1984. ISBN 0-19-856403-1.
3. S K Chapman, *Maintaining and Monitoring the Transmission Electron Microscope*, RMS Microscopy Handbook 08, Royal Microscopical Society, Oxford, 1986. ISBN 0-19-856407-4.
4. P C Robinson and S Bradbury, *Qualitative Polarized-Light Microscopy*, RMS Microscopy Handbook 09, Royal Microscopical Society, Oxford, 1992. ISBN 0-19-856410-4.
5. Nomenclature Committee of the Royal Microscopical Society, *RMS Dictionary of Light Microscopy*, RMS Microscopy Handbook 15, Royal Microscopical Society, Oxford, 1989. ISBN 0-19-856421-x.
6. P M Budd and P J Goodhew, *Light-Element Analysis in the Transmission Electron Microscope*, RMS Microscopy Handbook 16, Royal Microscopical Society, Oxford, 1988. 0-19-856417-1.
7. D Chescoe and P J Goodhew, *The Operation of Transmission and Scanning Electron Microscopes*, RMS Microscopy Handbook 20, Royal Microscopical Society, Oxford, 1990. ISBN 0-19-856420-1.
8. J F Watts, *Introduction to Surface Analysis by Electron Spectroscopy*, RMS Microscopy Handbook 22, Royal Microscopical Society, Oxford, 1990. ISBN 0-19-856425-2

9. S Bradbury, *Basic Measurement Techniques for Light Microscopy*, RMS Microscopy Handbook 23, Royal Microscopical Society, Oxford, 1991. ISBN 0-19-856426-0.
10. D W Humphries, *Preparation of Thin Sections of Rocks, Minerals and Ceramics*, RMS Microscopy Handbook 24, Royal Microscopical Society, Oxford, 1992. ISBN 0-19-856431-7.
11. B Bracegirdle, *Scientific PhotoMACROgraphy*, RMS Microscopy Handbook 31, Royal Microscopical Society, Oxford, 1994. ISBN 1-8727-4849-x.
12. P H Greaves and B P Saville, *Microscopy of Textile Fibres*, RMS Microscopy Handbook 32, Royal Microscopical Society, Oxford, 1995. ISBN 1-8727-4824-4.
13. B Bracegirdle and S Bradbury, *Modern PhotoMICROgraphy*, RMS Microscopy Handbook 33, Royal Microscopical Society, Oxford, 1995. ISBN 1-8599-6090-8.
14. S Bradbury and P J Evennett, *Contrast Techniques in Light Microscopy*, RMS Microscopy Handbook 34, Royal Microscopical Society, Oxford, 1996. ISBN 1-8599-6085-5.
15. C Sheppard and D Shotton, *Confocal Laser Scanning Microscopy*, RMS Microscopy Handbook 35, Royal Microscopical Society, Oxford, 1997. ISBN 1-8727-4872-4.
16. P J Goodhew and F J Humphreys, *Electron Microscopy and Analysis*, 2nd edn, Taylor & Francis, London, 1988. ISBN 0-85066-415-2 (hb). ISBN 0-85066-414-4 (pb).
17. E Metcalfe, *Characterisation of High-Temperature Materials: Microstructural Characterisation*, Institute of Materials, London, 1988. ISBN 0-901462-47-0.
18. P W Hawkes and E Kasper, *Principles of Electron Optics, Volume 1: Basic Geometrical Optics*, Academic Press, London, 1988. ISBN 0-12-333351-2. *Volume 2: Applied Geometrical Optics*, Academic Press, London, 1989. ISBN 0-12-333352-0. *Volume 3: Wave Optics*, Academic Press, London, 1994. ISBN 0-12-333354-7.
19. N Barakat and A A Hamza, *Interferometry of Fibrous Materials*, Adam Hilger/Institute of Physics Publishing, 1990. ISBN 0-85274-100-6.
20. T Wilson, *Confocal Microscopy*, Academic Press, London, 1990. ISBN 0-12-757270-8.
21. F W D Rost, *Quantitative Fluorescence Microscopy*, Cambridge University Press, Cambridge, 1991. ISBN 0-521-39422-8.
22. J J Bozzola and L D Russell, *Electron Microscopy*, Jones & Bartlett, Boston MA, 1992. ISBN 0-86720-126-6.
23. P C Cheng, T H Lin, W L Wu and J L Wu (eds), *Multidimensional Microscopy*, Springer-Verlag, New York, 1993. ISBN 0-387-94118-5. ISBN 3-540-94118-5.
24. W G Hartley, *The Light Microscope: its Use and Development*, Senecio, Witney, Oxon, 1993. ISBN 0-906831-05-9.
25. E Hunter, P Maloney and M Bendayan, *Practical Electron Microscopy: a Beginners Illustrated Guide*, Cambridge University Press, Cambridge, 1993. ISBN 0-521-38539-3.
26. M H Loretto, *Electron Beam Analysis*, Chapman & Hall, London, 1993. ISBN 0-412-47790-4.
27. E M Slayter and H S Slayter, *Light and Electron Microscopy*, Cambridge University Press, Cambridge, 1993. ISBN 0-521-32714-8.
28. P E J Flewitt and R K Wild, *Physical Methods for Materials Characterisation*, Institute of Physics, Bristol, 1994.

29. S J Spells, *Characterisation of Solid Polymers: New Techniques and Developments*, Chapman & Hall, London, 1994. ISBN 0-412-58490-5.
30. B Bousfield, *Surface Preparation and Microscopy of Materials*, John Wiley, Chichester, 1992 (republished with corrections, July 1994). ISBN 0-471-93181-0.
31. B D Dunn, *Metallurgical Assessment of Spacecraft Parts, Materials and Processes* (2nd revised edn), Wiley-Praxis, 1996. ISBN 0-471-96428-x.
32. R Egerton, *Electron Energy-Loss Spectroscopy in the Electron Microscope*, 2nd edn, Plenum Press, New York, 1996. ISBN 0-306-45223-5.
33. A W Robards and A J Wilson, *Procedures in Electron Microscopy*, John Wiley, Chichester, 1996. ISBN 0-471-92853-4.
34. D B Williams and C B Carter, *Transmission Electron Microscopy: a Textbook for Materials Science*, Plenum Publishing, New York, 1996. ISBN 0-306-45324-x.
35. S Amelinckx, D van Dyck, J van Landuyt and G van Tendeloo, *Handbook of Microscopy: Applications in Materials Science, Solid-State Physics and Chemistry*, Three Volumes: *Methods I*, ISBN 3-527-29280-2. *Methods II*, ISBN 3-527-29473-2. *Applications*, ISBN 3-527-29293-4. VCH, Weinheim, Germany, 1997.
36. M K Miller and G D W Smith, *Atom Probe Microanalysis: Principles and Applications to Materials Problems*, Materials Research Society, Pittsburgh PA, 1989. ISBN 0-931837-99-5.
37. Tien T Tsong, *Atom-Probe-Field Ion Microscopy: Field Ion Emission and Surfaces and Interfaces at Atomic Resolution*, Cambridge University Press, Cambridge, 1990. ISBN 0-521-36379-9.
38. D A Bonnell, *Scanning Tunnelling Microscopy and Spectroscopy*, VCH, Weinheim, 1993. ISBN 3-527-27920-2.
39. C J Chen, *Introduction to Scanning Tunnelling Microscopy*, Oxford University Press, Oxford, 1993. ISBN 0-19-507150-6.
40. S J B Reed, *Electron Microprobe Analysis*, 2nd edn, Cambridge University Press, Cambridge, 1993. ISBN 0-521-41956-5.
41. R Weisendanger, *Scanning Probe Microscopy and Spectroscopy*, Cambridge University Press, Cambridge, 1994. ISBN 0-521-42847-5.
42. S N Manganov and M-H Whangbo, *Surface Analysis with STM and AFM*, VCH, Weinheim, 1996. ISBN 3-527-29313-2.
43. S J B Reed, *Electron Microprobe Analysis and Scanning Electron Microscopy in Geology*, Cambridge University Press, Cambridge, 1996. ISBN 0-521-48350-6.
44. P M Cooke, Chemical microscopy, *Anal. Chem.* 1986 **58**(9) 1926–1937; *Anal. Chem.* 1988 **60**(12) 212R–226R; *Anal. Chem.* 1990 **62**(12) 423R–441R; *Anal. Chem.* 1992 **64**(12) 219R–243R; *Anal. Chem.* 1994 **66**(12) 558R–594R; *Anal. Chem.* 1996 **68**(12) 333R–378R.
45. D A Hemsley, *Applied Polymer Light Microscopy*, Elsevier Applied Science, London, 1989. ISBN 1-85166-335-5.
46. L C Sawyer and D T Grubb, *Polymer Microscopy*, 2nd edn, Chapman & Hall, London, 1995. ISBN 0-412-60490-6.
47. A C Roulin-Moloney, (ed) *Fractography and Failure Mechanisms of Polymers and Composites*, Elsevier Applied Science, London and New York, 1989. ISBN 1-85166-296-0.
48. L Engel, H Klingele, G W Ehrenstein and H Schaper (translated by M S Welling), *An Atlas of Polymer Damage: Surface Examination by Scanning Electron*

Microscope, Wolfe Science Books/Carl Hanser Verlag, London, 1981. ISBN 0-7234-0751-7.

49. L Engel and H Klingele (translated by S Murray), *An Atlas of Metal Damage: Surface Examination by Scanning Electron Microscope*, Wolfe Science Books/ Carl Hanser Verlag, London, 1981. ISBN 0-7234-0750-9.
50. Proceedings of the International Conferences on Microscopy of Composite Materials, The Royal Microscopical Society, First conference; *J. Microscopy*, February 1993, **169**(2); Second conference; *J. Microscopy*, March 1995, **177**(3); Third conference; *J. Microscopy*, February 1997, **185**(2).
51. R D Adams and P Cawley, A review of defect types and non-destructive testing techniques for composites and bonded joints, *NDT Internat.* 1988 **21**(4) 208–222.
52. P Cawley and R D Adams, Defect types and non-destructive testing techniques for composites and bonded joints, *Mater. Sci. Technol.* 1989 **5**(5) 413–425.
53. J Summerscales, Manufacturing defects in fibre-reinforced plastics composites, *Insight*, 1994 **36**(12) 936–942.
54. M R Piggott, The effect of fibre waviness on the mechanical properties of unidirectional fibre composites: a review, *Composites Sci. Technol.* 1995 **53**(2) 201–205.
55. R B Heslehurst and M Scott, Review of defects and damage pertaining to composite aircraft components, *Composite Polym.* 1990 **3**(2) 103–133.
56. W J Cantwell and J Morton, The significance of damage and defects and their detection in composite materials: a review, *J. Strain Anal.* 1992 **27**(1) 29–42.
57. H Weiland, Microtexture determination and its application to materials science, *JOM: Journal of Minerals, Metals and Materials*, 1994 **46**(9) 37–41.
58. L G Schultz, A direct method for determining preferred orientation of a flat reflection specimen using a Geiger counter x-ray spectrometer, *J. Appl. Phys.* 1949 **20** 1030–1036.
59. H J Bunge, Zur Darstellung Allgemeiner Texturen, *Z. Metall.* 1965 **56** 872–874.
60. R J Roe, Description of crystallite orientation in polycrystalline materials III. General solution to pole figure inversion, *J. Appl. Phys.* 1965 **36** 2024–2031.
61. V Randle, *Microtexture Determination and its Applications*, Institute of Materials, London 1992. ISBN 0-901716-35-9.
62. *The International Conference series on Textures of Materials (ICOTOM)*, 1st, Clasthal-Zellerfeld, 1968; 2nd, Krakow, 1971; 3rd, Pont-a-Mousson, 1973; 4th, Cambridge, 1975; 5th, Aachen, 1978; 6th, Tokyo, 1981; 7th, Noordwijkerhout, 1984; 8th, Santa Fe, 1987; 9th, Avignon, 1990; 10th, Clausthal, 1993.
63. G K Haritos, J W Hager, A K Amos, M J Salkind and A S D Wang, Mesomechanics: the microstructure-mechanics connection, *Internat. J. Solids Structures* 1988 **24**(11) 1081–1096.
64. D Stoyen, W S Kendall and J Mecke, *Stochastic Geometry and its Applications*, 2nd edn, John Wiley, Chichester, 1995. ISBN 0-471-95099-8.
65. P J Diggle, *Statistical Analysis of Spatial Point Processes*, Academic Press, London, 1983. ISBN 0-12-215850-4.
66. H Hermann, *Stochastic Models of Heterogeneous Materials*, Materials Science Forum **78**, Trans Tech Publishers, Zurich, 1991.
67. A Okabe, B N Boots and K Sugihara, *Spatial Tessellations – Concepts and Applications of Voronoi Diagrams*, John Wiley, Chichester, 1992. ISBN 0-471-93430-5.

2
Flexible textile composite microscopy

PATRICIA A ANNIS
AND THOMAS W QUIGLEY JR

2.1 Introduction

Flexible textile composites are defined as modified textile materials that have the textile structure as the major component or matrix. Unlike rigid composites, used for applications requiring extreme strength and high modulus, flexible textile composites find use in innumerable industrial and consumer applications where flexibility is a necessary requirement. In fact, flexible textile composites probably are the most numerous category within the broad range of materials called composites. In this chapter, the microscope analysis of flexible textile composites is discussed. The composition and architecture of flexible textile composites engineered at the fiber, yarn and fabric levels are investigated and illustrated using optical and non-optical microscopy.

Microscopy is a useful tool for analyzing the microstructure of flexible textile composites because the various components of these materials are within the magnification range of most microscopes. Microscopy also lends itself well to analysis of flexible textile composites because sample preparation is straightforward and data usually are easily collected and interpreted. Light microscopes, in particular, offer many advantages for analysis of flexible textile composites because these instruments provide high enough magnification to provide good images without the effort and expense associated with non-optical microscopes.

Specimen preparation for optical microscopy also is easily accomplished and is, in many cases, non-destructive. The coupling of electronic imaging to optics has improved the performance of optical microscopes and introduced new capabilities that enhance and greatly expand the applicability of these microscopes to the analysis of flexible textile composites. Photodocumentation of the structure and composition of flexible textile composites is easily acquired with an optical microscope when the optics of the system are interfaced with video and image processing capabilities.

2.1.1 Textile composites

Textiles are fibrous materials and products made from fibers. Under this broad definition, textiles include fibers, filaments, yarns, threads, cords, ropes, fabrics, cloths, nets, braids, felts and tarpaulins. Carpets and rugs also are textiles as are clothes and home furnishings. Materials used for industrial, aerospace, transportation, medical, agricultural and engineering applications qualify as textiles if they are composed primarily of fibrous materials.

Fibers are the basic component of textiles and are characterized by flexibility, fineness and a length to width ratio of at least 100 : 1. The mechanical properties of fibers are determined by polymer morphology, that is crystallinity and degree of molecular orientation, as well as by geometric characteristics such as fiber length, diameter, cross-sectional shape and three-dimensional configuration (crimp). Typical commercial textile fibers include elongated single-cell seed hairs such as cotton and kapok; elongated multicellular structures such as wool or hair; aggregates of elongated cells such as flax, jute or sisal; continuous filaments or short lengths of organic materials such as silk, rayon, nylon, acrylic, polyester and carbon; and inorganic materials such as silicon, ceramic and metal [1]. Short textile fibers are called staples and fibers of continuous length are called filaments.

Staple or filament fibers are twisted or combined into flexible strands called yarns for use as threads or cables or made into fabrics by weaving, knitting, braiding or otherwise intertwining. The properties of yarns depend upon the properties of constituent fibers and upon the structural characteristics of the yarn. Yarns may be composed of one or more continuous filaments, discontinuous staple fibers, or a combination of filament and staple fibers. Individual yarns can be twisted together to form plied yarns and plied yarns can be twisted to form cords or cables.

Fabrics are manufactured assemblies of fibers or yarns which have substantial surface area in relation to their thickness and sufficient mechanical strength to give the assembly inherent cohesion [2]. Fabrics retain more or less completely the strength, flexibility and other properties of their original fibrous nature. Fabric properties also are determined by yarn properties and the many variables of fabrication.

2.1.2 Flexible textile composites

Traditionally, composites have been used for load bearing, high strength and heat resistant applications. Examples of rigid composites include construction materials, automotive and aerospace components, maritime structures, military armor, medical implants and recreational items such as canoes, golf clubs and fishing rods.

Rigid composites consist of a fiber-reinforced matrix. The function of the matrix is to provide stability and to hold the reinforcement in a prescribed suspension, position or orientation [3]. The matrix may be polymeric, ceramic, metallic, glass or carbon. The fibrous reinforcement in rigid composites usually is distributed uniformly throughout the matrix as inclusions (whiskers) or embedded in the form of staple or continuous fibers, yarns and fabrics, or any combination of these. The purpose of the fibrous reinforcement in rigid composites is to augment one or more high performance properties such as strength, stiffness or thermal stability.

Production of rigid composites requires a densification step in which the fibrous reinforcement is surrounded by a matrix material which is caused to harden and take on a solid shape. Densification is accomplished in a variety of ways, all of which require a heating or curing step [4–12]. After densification, a rigid composite is ready for use or it may undergo additional processing and/or shaping to meet various specifications.

Rigid composites with fibrous reinforcement are referred to as fibre- or textile-reinforced composites, textile structural composites or, simply, fibrous or textile composites. Regardless of their label, all of these materials are characterized by a high modulus and tenacity. Technically, rigid composites comprise one subgroup of a large and economically important class of materials called textile composites. Textile preforms, fibers, yarns, or fabrics impregnated or saturated with rigid matrix materials, are one type of rigid textile composite.

Flexible textile composites, the focus of this chapter, are defined as modified textile materials that have the textile structure as the major component or matrix. The primary components of flexible textile composites are fibrous, that is, fibers, yarns and/or fabrics with fibers being the most elemental component of yarns and fabrics. Fibers used in flexible textile composites may be made from high modulus fibers and/or yarns which resist elongation and exhibit high tensile strength or fibers with moderate to low tensile properties such as those used in consumer applications. Fiber length in flexible textile components may be discontinuous staple or continuous filament or a combination of both of these lengths. Cross-sectional shape is variable and linear density runs from the extremes of ultra microfibers to coarse filaments. The three-dimensional configuration of fibers in flexible textile components may be flat or textured (crimped). Textile fibers, crystalline, high molecular weight macromolecules, differ from whiskers which are single filamentous crystals of very short length and a fine diameter [13]. Textile fibers usually need to be at least 5 mm long to be processed by most textile manufacturing equipment. Textile fibers are the primary component of flexible textile composites whereas whiskers are used as reinforcement in rigid fibrous composites.

Secondary components or additives in flexible textile composites consist of organic or inorganic particulates or adhesives (thermoplastic, thermosetting or elastomeric) [9]. The secondary components in flexible textile composites contribute to the flexibility of the composite or, at least, do not hinder it. Both the primary and secondary components of flexible textile composites are chosen to optimize mechanical properties and performance characteristics suitable for an intended end use. In practice, the selection of materials and construction techniques is matched with product performance, productivity and cost requirements [11].

Textile composites need not be rigid to be useful in a vast array of industrial and consumer products. In fact, flexible textile composites can be used in almost every industry in some form or fashion. A compilation of a few of the many uses for flexible textile composites is presented in Table 2.1. The non-rigid character of flexible textile composites is a decided advantage in many of these end uses.

The term flexible, when used to describe textile composites, does not mean flabby or unsubstantial. On the contrary, flexible textile composites frequently have better impact resistance, elongation and tensile properties at low loads than rigid composites. They certainly are more porous and absorbent. Flexible textile composites usually are light in weight and low in density and when primary and secondary components are wisely chosen, they can provide resistance to wear, heat, flame, corrosion, photochemical degradation or dimensional change.

Flexible textile composites also possess a decidedly better drape, hand and cover than rigid composites. Elastomeric textile composites have excellent elastic recovery but not all flexible textile composites are elastomeric. Flexible textile composites may be biostable or biodegradable depending upon their intended end use. Compared with many of the traditional materials that they are replacing, flexible textile composites are inexpensive, relatively easy to fabricate, and require little, if any, maintenance.

As with all composites, flexible textile composites are composed of two or more components that exist as separate and distinct phases combined without chemical reaction [7]. If each of the phases has significantly different physical properties, the properties of the resulting composite will be substantially different from the individual components [8]. In fact, one of the unique characteristics of composites is that their material, structure and form lead to synergistic effects in which the behavior of the whole is much greater than the sum of the parts, from a performance point of view [11].

At the macroscopic level, flexible textile composites appear homogeneous. However, the boundaries between the individual components are physically distinguishable and have a clear interface which makes the components of flexible textile composites easily resolvable at the microscopic level. Because textiles are formed directionally, flexible textile composites are naturally anisotropic. Thus, most flexible textile composites have biaxial strength and stiffness characteristics. However, materials with isotropic properties are desirable for some end uses and fabrication of flexible textile composites with uniaxial properties usually is readily achieved. Fiber orientation and type, yarn structure and properties and the

Table 2.1 Flexible textile composites: representative examples

Industry	Application
Geotextiles/ geomembranes	Flexible pond/reservoir/dam liners, coastal protection, roadbed/embankment stabilization, subsurface concrete reinforcement, drainage/filtration systems, waste containment, landfill covers, inflatable dams.
Architecture/construction	Reinforced domes and covers, air houses, translucent roofs, chimney covers, insulation, ventilation tubing/ducts, pipe liners, asphalt roof tiles, awnings/marquees.
Biomedical/healthcare	Surgical gowns/masks/bonnets/footwear, drapes, surgical dressings/bandages, dissolvable sutures, vascular grafts, artificial ligaments/skin, casts/trusses/stockings, superabsorbent pads and sponges.
Military/safety/ protective clothing	Bullet proof vests/body armor, camouflage tarpaulins/nets, parachutes, life rafts/life jackets/life lines, inflatable radomes, wet suits, firefighter turnout coats, race car drivers' suits, pesticide/chemical/vapor protection suits, antistatic fabrics/carpets, chain saw chaps, cut resistant gloves/aprons, clean room garments, aprons.
Automotive/transportation	Tires, flame retardant seat covers, belts/hoses, air bags, upholstery/seat covers, headliners/wall/trunk liners, inflatable boats/catamarans.
Aerospace	Lift-off/in-flight space suits, ceramic tile cushions, reflective insulation.
General industrial/agricultural	Conveyor belts, tarpaulins/covers, ropes/cords, lifting bags, hoses, scrims, lattice fabrics/webbing, packaging materials, collapsible containers, grain silos, inflatable green houses, snow fence, thermal screens, oil absorbing materials, gas/liquid/dust/chemical filters, thermoplastic binder fibers, lightening strike protection.
Electronics	Insulated cables, computer disk liners, insulation, conductive fibers/yarns, electromagnetic shields, electronic clean room fabrics.
Interiors	Fabric/vinyl coated wall coverings, carpet and rugs, tufted fabrics, wall panels/flexible space dividers, cushions/pillows, drapery/curtains, upholstery/slip covers, mattress pads, comforters/coverlets, vacuum cleaner bags, bandage springs, window screen.
Consumer products (Recreation/apparel/food)	Sleeping bags/tents, luggage, book bags/totes/backpacks/handbags, boat covers, sail cloth/sail covers, tennis nets/racket strings, gymnasium mats, stadium cushions, indoor/outdoor pool covers/linings, umbrellas/sunshades, outdoor furniture, laminated table covers, infant carriers/swings, folding strollers, sling-type furniture, artificial leather, baseballs/footballs/tennisballs/basketballs, breathable/waterproof outerwear, high tech outerwear, shoes/slippers/shoe liners, gloves/glove liners, interfacing/interlinings, bicomponent manmade fibers, fiber blended yarns/fabrics, compound/metallic/coated yarns, delustered/pigment colored manmade fibers, sausage skins, stretch yarns, carpet yarns.

many variables of fabrication all influence the performance and properties of flexible textile composites.

Flexible textile composites are constructed from fibers, yarns and/or fabrics that are stacked in layers, machined to specified contours and dimensions, or molded to flexible three-dimensional shapes. Polymer modified textile materials and coated or laminated textiles and yarns also are flexible textile composites. Fibers, yarns and fabrics with two or more physically distinct components are flexible textile composites. Because each component of a flexible textile component interacts with all the other components to influence properties of the whole, flexible textile composites are systems. Therefore, flexible textile composites are best analyzed using a systematic approach that considers not only individual fiber, yarns and fabric properties but also the interaction of these components. The microscope is the instrument of choice to analyze the components of flexible textile composites and to understand how structure-/property relationships influence performance.

The purpose of this chapter is to discuss the microscope analysis of flexible textile composites. The composition and architecture of flexible textile composites are investigated. Structural parameters within the fibrous matrix of several representative flexible textile composites are identified. Inclusions are considered because the physical properties of flexible textile composites are greatly affected by additives. Topical finishes and sizings are included, but not discussed in detail. Our discussion does not include dyes and chemical finishes because most of these are chemically bound to textile fibers at the molecular level. Similarly, co- and terpolymers also do not qualify as flexible textile composites whereas bicomponent fibers and yarns are flexible textile composites. Throughout this discussion, flexible textile composites are distinguished from rigid composites because their primary components are fibrous and they remain flexible during use.

2.2 Textile microscopy

2.2.1 Principles

Structural characteristics of flexible textile composites can be measured instrumentally, modeled mathematically or depicted graphically, but only with a microscope can the components of flexible textile composites be visualized with the human eye. Why analyze the components of flexible textile composites? Increasing competition within the textile composite industry and growth of global markets are incentives enough to prompt identification of the components of flexible textile composites and the fabrication methods used to produce them.

From the engineering and research points of view, optimizing the design of flexible textile composites requires full characterization of all components within a material and an understanding of how each component contributes to performance during different end-use scenarios. Such characterization is especially important when a composite must perform under adverse conditions or function as a protective barrier. Microscopy also can contribute to quality assurance and problem solving associated with routine production of flexible textile composites.

The microscope has an important role in the analysis of flexible textile composites because it provides both qualitative and quantitative data about the structure and performance of these complex materials. Fortunately for the microscopist, one of the most important characteristics of flexible textile composites is that each of the constituents retains its individual characteristics and these characteristics are distinguishable on a microscopic scale [7]. Thus, the microscope readily lends itself to analysis of flexible textile composites.

Microscopes may be classified as either optical or non-optical. In the optical category are stereo and compound microscopes illuminated with regular transmitted light, regular incident light or a combination of these. Plane-polarized light, crossed-polarized light and ultraviolet light also are sources of illumination used with compound microscopes. A stereo microscope has a lower range of magnification, greater depth of focus, larger field of view and longer working distance than a compound microscope. Higher magnification and resolution are obtained with a compound microscope than with a stereo microscope although these advantages are offset by a decrease in depth of field and field of view. The optics in some compound microscopes can be manipulated to introduce phase changes and interference phenomena that improve image contrast and appearance.

Non-optical microscopes use a source other than light to produce an image. Scanning electron microscopes (SEM), transmission electron microscopes (TEM) and confocal laser scanning microscopes (CLSM) are included in this category. These instruments provide images at higher magnification, contrast and resolution than optical microscopes. SEM and CLSM also provide greater depth of field than optical microscopes which is useful when viewing the three-dimensional characteristics of flexible textile composites. Maximum magnification with a TEM is higher than with a SEM, although specimen preparation is more complex because TEM analysis can only be performed on very thin specimens. Advantages of CLSM are that elaborate sample preparation is not required and incremental three-dimensional analyses are possible [14]. Elemental analysis is possible with both SEM and TEM. Other scanning, tunneling and probing microscopes can be used with components of flexible textile composites, but these are not frequently utilized during routine laboratory analysis [15].

In our experience, a stereo-zoom microscope, a compound polarizing microscope (PLM) and a scanning electron microscope (SEM) are the most useful instruments for microscope analysis of flexible textile composites [16–20,22]. Table 2.2 compares the characteristics of these microscopes [21] and Table 2.3 describes their practical laboratory utilization. Microscope selection

Table 2.2 Comparison of microscopes for analysis of flexible textile composites adapted from Sawyer and Grubb [21]

Ranges	Stereo	PLM	SEM
Useful magnification	5–100×	30–1500×	10–60000×
Typical resolution	3.0 μm	0.3 μm	10 nm
Depth of focus	large	small	very large
Field of view	very large	small	very large
Working distance	large	small	N/A
Elemental analysis	no	no	yes

Table 2.3 Laboratory utilization of microscopes for analysis of flexible textile composites

Consideration	Stereo	PLM	SEM
Appropriate for *	A,B,C	D	B,C,D
Can observe	surface and interior	surface and interior	surface
Relative cost	moderate	moderate	high
Sample preparation	minimal	moderate	moderate
Sample environment	ambient	ambient	vacuum
Destructive	no	no	varies
Photo-documentation	yes	yes	yes
Video capabilities	yes	yes	yes
Image processing & analysis	yes	yes	yes
Operator training	minimal	moderate	extensive
Maintenance	low	low	high

*A = multiple layer composite, B = fabric composite, C = yarn composite, D = fiber composite.

and sequence of use during analysis of textile composites depends on the size and visual characteristics of the material to be analyzed and the problem to be solved.

Generally, it is recommended that an analysis begin at low magnification and that other microscope techniques be added as the situation requires. A combination of the various microscope techniques generally provides the best insight into the structural characteristics of flexible textile composites.

Both optical and non-optical microscopes can be interfaced with a closed circuit video camera and digital imaging electronics. Microscope imaging techniques vary in their capabilities, complexity and cost. Video microscopy, in which the image of a flexible textile composite or its components is captured by a video camera and projected on a TV monitor, provides the most basic level of microscope imaging. With enhanced-video microscopy, the appearance of an image can be improved by manipulating contrast and color at the display end of the system. The basic components of this type of imaging system are shown in Fig. 2.1a. Most optical microscopes are easily fitted with video cameras. Laser, ink-jet or thermal printers interfaced with a video microscopy system can be used to document the appearance of flexible textile composites on the video monitor. Photographs also can be obtained directly from the video monitor or with conventional equipment used to obtain photomicrographs.

Enhanced video microscopy has been used extensively in our laboratory to identify and characterize the structure of flexible textile composites. This instrumentation also is useful for analyzing faults and impurities, examining the

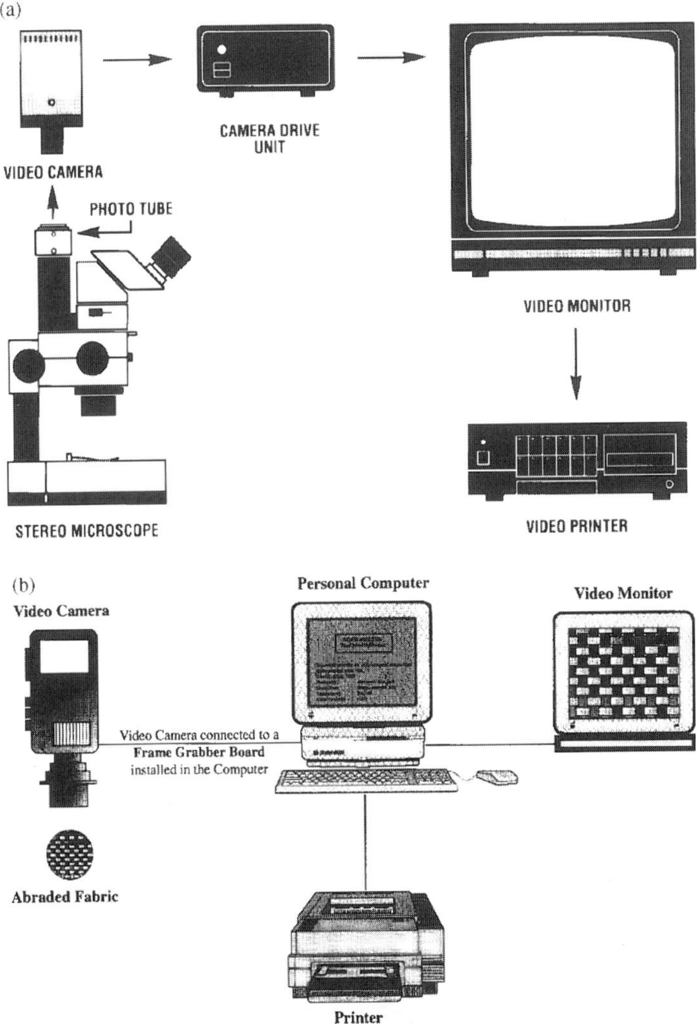

2.1 (a) Components of a video microscopy system. (b) Components of an image processing/image analysis system.

interface between layered composites and evaluating damage progression and fatigue phenomena. Applications of enhanced video microscopy to the analysis of flexible textile composites are discussed later in this chapter.

When a microscope is linked to a personal computer equipped with a digitizing board, more sophisticated manipulation and quantification of a microscope image is possible. Digital image processors modify video signals so as to enhance image contrast, sharpen edge features, improve signal-to-noise ratio and highlight features of interest on the surface or interior of a flexible textile composite. Image analyzers, on the other hand, extract statistical data, primarily on morphometric parameters such as the number, size, shape and length of fibers or inclusions, and the distributions of these features [23–30]. Both systems have the ability to edit images and to store, sort and retrieve images and related data. Interactive software is required to operate both image processing and image analysis systems. Important advantages of these systems for analyzing flexible textile composites are that they can quantify rapidly and non-destructively important structural characteristics.

The components of a typical image processing/analysis system are shown in Fig. 2.1b. Printers are interfaced with digital imaging systems to document image appearance and provide hard copies of data output. Internal and external high capacity zip drives are becoming useful hardware accessories during microscope analysis of flexible textile composites because of the large size of some image files. With these devices, images can be stored and retrieved for future reference. With the decrease in cost of personal computers, and increasing sophistication and usability of image analysis software, digital imaging is rapidly becoming an important tool in the analysis of flexible textile composites. Image processing and analysis techniques were applied to many of the materials depicted in the figures in this chapter and most figures were obtained with a high resolution jet printer.

2.2.2 Sample preparation and analysis

Procedures used to prepare flexible textile composites for microscope analysis depend on the form of the composite, goals of the analysis and the microscope to be used. Techniques such as polishing, casting, etching, drying and disintegration [21], which are routinely used to prepare rigid composites for microstructural analysis, generally are not required when preparing flexible textile composites for microscopy. Instead, traditional sample preparation techniques used with fibers, yarns and fabrics usually are appropriate for preparing flexible textile composites and their components for microscope analysis.

2.2.2.1 Flexible textile composites

Analysis of flexible textile composites usually begins with a stereo microscope. Examining a composite with a stereo microscope may be the only procedure required to complete an analysis or to solve a problem. The analysis of materials smaller than yarns, that is, fibers, whiskers, inclusions, adhesives and various types of contamination, may require use of a PLM whereas a SEM is required when viewing composite components at magnifications greater than about 1000×.

During initial examination of flexible textile composites, small samples can be placed directly on the stage of a stereo microscope or components can be extracted from the body of a composite and examined individually. Cutting samples with a blade or rotary cutter is preferable to cutting with shears because compression caused by the shears can deform textile components and pull them out of alignment. For ease of handling, intact composites can be mounted on a glass microscope slide and the slide secured to the stage of the microscope with stage clips. Double-sided adhesive tape or liquid adhesive can be used to secure samples to the glass slide if unimpeded transmission of light through the sample is not required. This procedure, however, eliminates the option to go back later and analyze the other side of the sample. If the surface being examined is uneven or creased, a second glass slide can be placed on top of the sample to level out the field of view.

Because of the thickness and density of many flexible textile composites, illumination of these materials is often most satisfactorily achieved using reflected light positioned at an oblique angle. To optimize reflected illumination, we use free standing fiber optic illuminators with spot lenses and/or a fiber optic ring illuminator attached to the objective of our stereo microscope. A mirror, placed under a flexible textile composite, can sometimes improve the illumination of the interior of a sample. Transmitted light in a brightfield or darkfield configuration also can be used to supplement reflected illumination and to improve the microscope image of flexible textile composites.

Examination of flexible textile composites at low magnification provides a wealth of information about fabrication method, that is, weaving, knitting, non-woven, braiding, molding, coating, laminating, etc., and about geometric structure of fibers, yarns and fabrics. Information about adhesives, interfacial bonding and delamination, and the presence and identification of contaminates also can be obtained. The overall response of a material to impact, loading and fatigue and response to thermal, chemical or abrasive damage also can be detected and documented at the levels of magnification required to visualize flexible textile composites, that is, 6× to about 100×.

Cross-sections of flexible textile composites can be mounted in an AATCC fabric cross-section holder [17]. This device, developed at the University of Georgia, enables the microscopist to examine each part of a composite without disturbing or distorting interfacial boundaries (Fig. 2.2).

2.2.2.2 Composite fabrics

The sample preparation techniques used to prepare composite fabrics are similar to the techniques described above for flexible textile composites with the

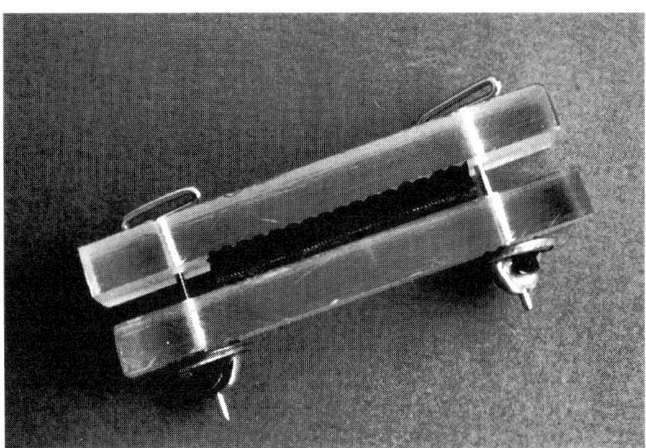

2.2 AATCC fabric cross-section holder, with flexible textile composite.

exception that individual fabrics frequently are not as cumbersome and dimensionally stable as intact flexible textile composites. Therefore, careful handling of a fabric composite during sample preparation is recommended to prevent raveling, yarn slippage and distortion. These problems are most likely to occur if a fabric is extracted from the body of a multilayer flexible textile composite. Scrim fabrics, with their low density or thread count, are particularly susceptible to distortion during sample preparation and microscope analysis.

An advantage of removing a fabric from the body of a flexible textile composite prior to microscope analysis is that a single layer fabric transmits more light than a multilayer composite. Sample manipulation, cross-sectioning and examination of surface and internal characteristics also are simplified when analyzing fabrics extracted from a flexible textile composite.

Microscope analysis can provide much useful information about the structural characteristics of fabric composites. At the macroscopic level, fabrication methods, interfacial bonding methods, thread count and interstices and stitch size can be determined. Mechanical problems associated with fabrication and wet processing also can be detected. Coating rubs, back coating migration, broken ends and non-uniform lamination are a few examples of production flaws in composite fabrics that have been identified with a stereo microscope. Contamination found in fabrics during microscope analysis includes motes, rust, oil, dirt, binder cord, mildew and even tobacco juice! The examination and identification of insect infestation also is possible with a stereo microscope.

PLMs are limited in their ability to analyze fabric composites because of their shallow depth of field and narrow working distance. However, the SEM, although not as simple or as inexpensive to operate as a stereo microscope, provides high magnification and depth of field for analyzing three-dimensional materials. Tilted side views and high resolution and contrast allow the microscopist to examine interfacial boundaries, structural anomalies and changes in fiber orientation and surface texture after processing and use [21]. X-ray analysis detects with certainty the presence of dyes, dyeing and finishing auxiliaries and contaminants at the nanometre level.

2.2.2.3 Composite yarns

Composite yarns or yarn components extracted from flexible textile composites can be evaluated with a stereo microscope, PLM or SEM. Stereo microscopes and PLMs equipped with both brightfield and darkfield illumination are useful when examining composite yarns. The increased contrast obtained with darkfield illumination often gives better results for surface analysis than brightfield illumination, especially with thick or tightly twisted yarns. However, brightfield illumination may be preferred to darkfield illumination for the better image it provides of the fibers in the cross-section of a yarn. Transmitted or reflected light, or both of these illumination sources, are useful when examining composite yarns.

Light microscopes offer advantages over SEM for analysis of composite yarns because these microscopes provide high enough magnification to produce a representative image without the effort and expense associated with preparing samples for SEM. In fact, the magnification required to visualize the surface of most yarns is similar to the magnification used to examine yarns with a SEM. Savings in cost and effort are, however, offset somewhat by the increased depth of field and better contrast provided by a SEM. Preparation of fibers for SEM analysis requires metallic coating techniques to improve image quality and to minimize a type of electronic static called charging.

The AATCC yarn holder, developed at the University of Georgia and designed to fit on the stage of both stereo and PLMs, simplifies longitudinal examination of yarns and helps reduce the frustration of positioning yarn samples on the stage of a microscope (Fig. 2.3) [17]. The hole in the center of the holder facilitates transmission of light through a yarn sample. Other methods for preparing composite yarns for longitudinal analysis include securing samples to a microscope slide using double-sided adhesive tape or liquid adhesive. In some cases, a yarn does not need to be removed from the body of a textile composite, but can be analyzed in its original position inside the composite.

Analysis of composite yarns provides much useful information about the composition, architecture, properties and performance of these flexible textile composites. Yarns can be examined with a stereo microscope to determine fiber length composition, spinning method, twist angle and uniformity. A stereo microscope also can be used to evaluate physical changes in yarns after exposure to tension, compression or torsion. Geometric data associated with linear density, packing fraction, twisting and texturing also can be obtained by examining the longitudinal image of a composite yarn.

2.3 AATCC yarn holder, with composite yarn.

Yarn cross-sectioning procedures vary in complexity and range from mounting yarns in a preformed structure to embedding yarns or untwisted yarn ply in thermosetting resins. However, in our experience, cross-sections obtained using perforated vinyl cross-sectioning plates and hand cutting techniques are adequate for microscope analysis of composite yarns (Fig. 2.4, [16,17]). An important consideration when preparing yarn cross-sections is that the sections are thin enough to transmit light partially. After sectioning, the perforated plate is attached to a glass microscope slide for viewing with a PLM. Mounting samples of flexible textile composites in a rigid thermosetting polymer and cross-sectioning with a microtome is another cross-sectioning option, but this is a labor intensive procedure that may not be worth the extra time and effort in a production setting.

2.4 Vinyl cross-sectioning plate used to obtain hand cut cross-sections of composite fibers and yarns.

Examination of yarn cross-sections can provide information on generic fiber content, fiber blend level and overall diameter of the yarn and its components. The penetration of dyes and encapsulation of sizings and coatings also can be determined. Biological stains specific for certain compounds or polymers assist with fiber identification and visualization of diffusants.

2.2.2.4 Composite fibers

Composite fibers are most effectively analyzed at high magnification using a PLM or SEM. A PLM generally is used for routine fiber analysis because of its ease of operation and relatively simple specimen preparation techniques.

Longitudinal mounts of composite fibers are easily prepared using a glass slide and coverslip. Although dry mounting techniques provide images of lower quality than images obtained using oil immersion techniques, dry mounting frequently is preferred, at least during initial examination, because sizings, particulate contaminants and finishes are not unintentionally altered. Casting techniques, which provide a permanent impression of the surface of a single fiber, also can be used to visualize and document the surface of composite fibers [31,32].

A variety of methods have been developed to obtain cross-sections [16,17] of composite fibers for microscope analysis. These procedures require anchoring fibers in some type of holder (cork or a perforated plate) or mounting medium (various resins are used) so that sections can be sliced thin enough to transmit light. Cutting methods for fiber cross-sections range from hand held razor blades and scalpels to glass knives mounted on an ultramicrotome. Embedding fibers in a thermosetting resin and cutting on an ultramicrotome provides the highest quality, most cleanly cut cross-sections. In our experience, however, extremely thin sections, although appropriate for biological specimens, are not appropriate for textile fibers because they transmit too much light (even at high magnifications) with a resultant decrease in contrast and loss of internal detail. Staining techniques do not usually remedy this problem. For routine analysis of composite fibers, we find that hand cut cross-sections usually suffice (Fig. 2.4).

The microscope image of composite fibers provides a wealth of information about generic fiber type, surface topography, cross-sectional shape, inclusions, pigmentation, imperfections and adhesion between bicomponent fibers. Measurement of length, width, cross-sectional area and three-dimensional configuration (crimp) is facilitated by inserting a micrometer into the optical path of the microscope. Image analysis provides a means to obtain quantitative data about fiber fineness, uniformity, shape and areal density.

Viewing fibers with regular polarized light improves the visual image of composite fibers and enhances their appearance. Polarized illumination is achieved by inserting a filter in front of the microscope's light sourcc so that thc direction of oscillation of the light varies only as a function of time [33]. Composite fibers viewed with polarized light have greater clarity and contrast than fibers viewed with regular light because visual aberrations associated with many different wavelengths are eliminated.

One of the most important applications of PLM to composite fiber analysis is the measurement of refractive index and birefringence. Internal morphological structure and information on crystallinity are indicated by the measurement of these optical properties. Birefringence is characteristic for a textile fiber and is, therefore, a quantitative and commonly used measure of generic fiber type. Birefringence may be measured with regular polarized light or by using a technique called cross-polar illumination which requires inserting a second polarizing filter into the optical path of a PLM.

Fiber damage and/or fatigue are readily detected by examining the longitudinal appearance of a fiber using crossed polarized light. In this application, the SEM also is a useful adjunct to PLM because of its ability to image fracture patterns of fiber composites and to provide elemental analysis of composite microstructures. Synthetic filaments also can be evaluated with PLM to determine the type and degree of mechanical–thermal deformation associated with processing and end use [33]. The appearance of composite fibers in crossed polarized light also is used to assess quality, to monitor production, and as a useful adjunct to other types of microanalytical data [34]. The PLM also can be fitted with a thermal hot-stage to determine fiber melting point and shrinkage characteristics.

2.2.3 Sample illumination

In our experience, illumination is the most critical variable associated with microscope analysis of flexible textile composites. In fact, illumination impacts upon almost every other variable associated with microscopy. Taking the time to adjust internal and external illumination sources properly is time well spent, and its importance to successful flexible textile composite microscopy cannot be over-emphasized.

2.3 Microstructural characterization of flexible textile composites

2.3.1 Composite fibers

Flexible textile composites are modified textile materials that have a textile structure as the major component or matrix. Traditionally, flexible textile composites are considered to be a class of materials composed of two or more planar materials, that is, fabrics. However, this definition is too limited because flexible textile composites also exist as fibres and yarns.

All of the natural fibers used for textile applications exist as individual flexible textile composites or are obtained from flexible biological composites. The natural protein fibers, that is, wool, mohair, cashmere, camel, and so on, are composed of two physically distinct parts, the cuticle and the cortex (Fig. 2.5a).

The cuticle is tricomponent in nature because it has three layers, the exocuticle, endocuticle and epicuticle, whereas the interior of the fiber is composed of two different types of cortical cells, the orthocortex and the paracortex. The cortex of some wool fibers also contains a third component called the mesocortex [35].

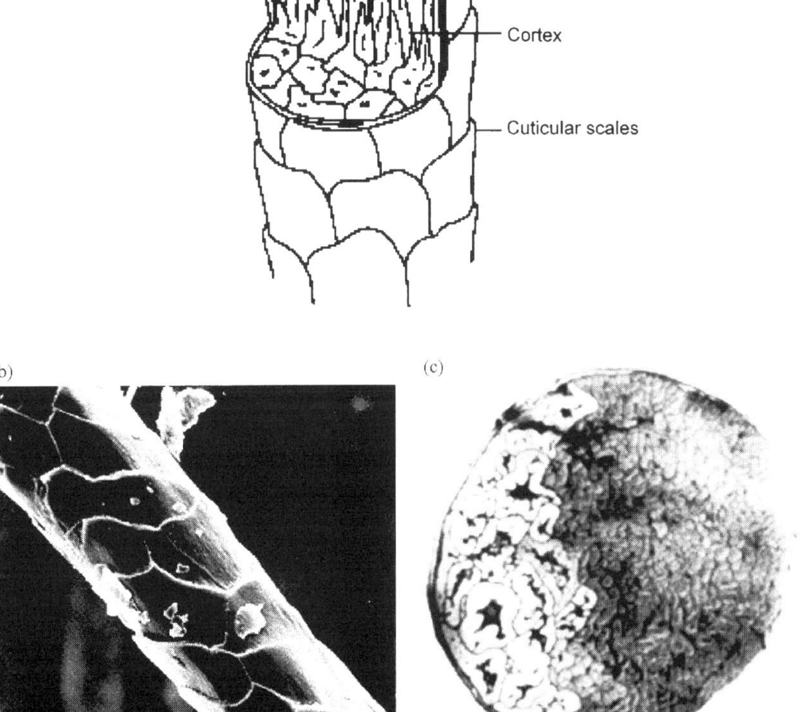

2.5 (a) Schematic of wool fiber (from Rippon [36]). (b) Scanning electron micrograph of wool fiber. (c) Transmission electron micrograph of wool fiber cross-section (from Pailthorpe [35]).

When the cortical cells in wool are broken down into their most basic elemental constituents, dibasic and diacidic amino acids are found. The composite structure of natural protein fibers, at both the micro and macro levels, is responsible for their amphoteric nature and directly influences physical properties such as crimp and cohesion, mechanical properties such as resiliency, elongation, thermal insulation and feltability, and chemical properties such as dyeability and alkali susceptibility. The longitudinal and cross-sectional appearance of representative wool fibers are shown in Fig. 2.5b and 2.5c, respectively [36].

Bast fibers, flax, jute, hemp and sisal, etc., are natural cellulosic fibers obtained from flexible biological composites. The SEM photomicrographs in Fig. 2.6a and 2.6b show the components of the flax stalk which is an aggregate of cuticle, epidermis, fiber bundles and woody core [37]. Natural pectins hold the components of the flax stem together during growth of the plant and processing

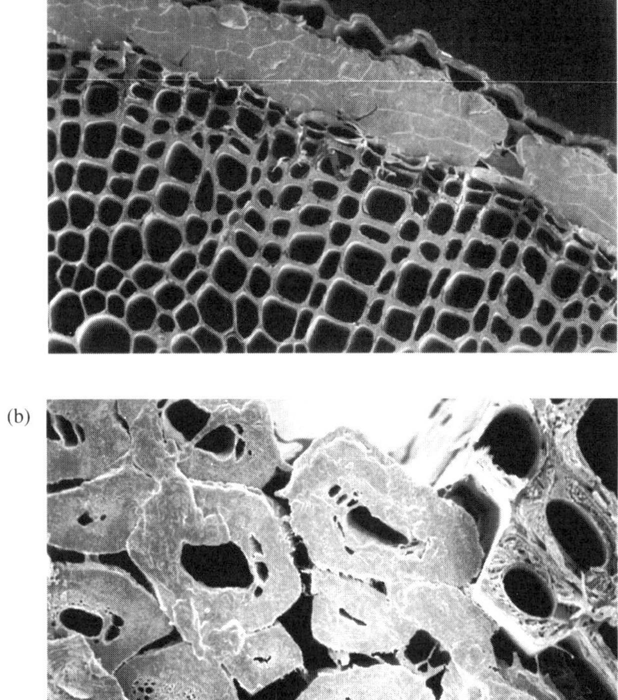

2.6 (a) Scanning electron micrograph of flax stalk (from Akin [37]). (b) Scanning electron micrograph of flax fiber bundles, after retting (from Akin [37]).

of the fiber. Although the fiber bundles (middle layer in Fig. 2.6a) are physically separated from the woody tissues of the stem during the retting process, bast fiber composites are used for textile applications without completely separating individual fibers from one another. Cotton, a seed hair fiber, is also a flexible textile composite because it is composed of multiple layers of cellulose encased by a multilayered cuticle.

Silk fibers, continuous filaments extruded by several species of caterpillars called silkworms, are produced as natural textile composites. During fiber formation, two silk filaments are produced by the worm simultaneously and coated with a sticky proteinaceous gum called sericin. In the longitudinal and cross-sectional views, the translucent sericin appears as an irregular and loosely attached film and as a distinct and separate layer around the triangular silk filaments (Fig. 2.7a and 2.7b, respectively) [31, 32]. Although raw silk is occasionally used for textile applications, degummed silk is preferred for most end uses because of its beautiful luster, smooth hand and luxurious drape.

The impetus for the development of the manmade fiber industry was the desire for easy care silk-like fibers with unlimited availability at a reasonable price. Polyester, nylon, olefin and other melt spun fibers are produced in a highly automated procedure that is remarkably similar to the laborious spinning performed by the silkworm and the fibers produced do resemble natural silk in many respects. By addition of inclusions and/or pigments prior to extrusion, the luster, color and electrostatic potential of manmade fibers can be modified and the resistance of these fibers to photodegradation, microbial agents, heat and flame improved. Manmade fibers containing such additives are flexible textile composites because fiber properties are enhanced without changing fundamental characteristics of the fibrous matrix. The photomicrographs in Fig. 2.8a and 2.8b show the longitudinal appearance of a delustered acrylic fiber and the cross-sectional appearance of delustered nylon, respectively [21].

During the last 50 years, manmade fiber technology has advanced to the degree that multicomponent fibers consisting of the same or different generic types can be spun simultaneously as one continuous filament. This process is called conjugate or composite spinning and fibers produced in this manner are flexible textile composites. Conjugate spinning produces 'individual filaments, each containing two or more distinct polymer phases' [38] that vary in configuration. Subsequently called bicomponent fibers, side-by-side, sheath core and sea-island configurations were initially developed to produce bulked continuous filaments after heat-induced differential shrinkage, thermal binding adhesives and artificial suede, respectively. Today, composite spinning can be so accurately controlled that multicomponent fibers can be obtained in a potentially unlimited number of product–feature configurations. Representative examples of bicomponent flexible textile composites are shown in Fig. 2.9a and 2.9b [39,40].

A variety of speciality fibers are produced by combining a traditional polymeric matrix with task specific components. For example, thermochromic fibers that

(a)

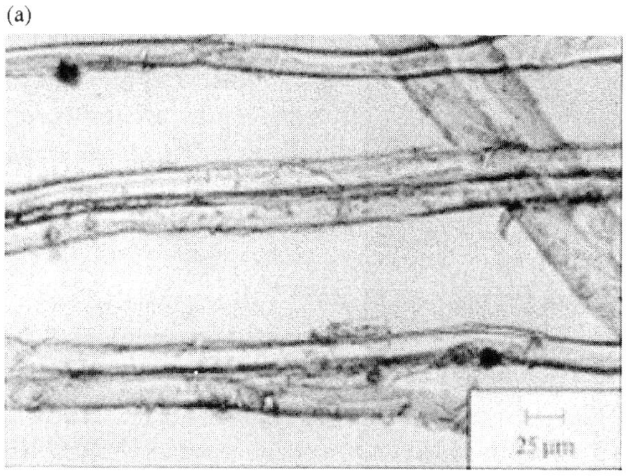

(b)

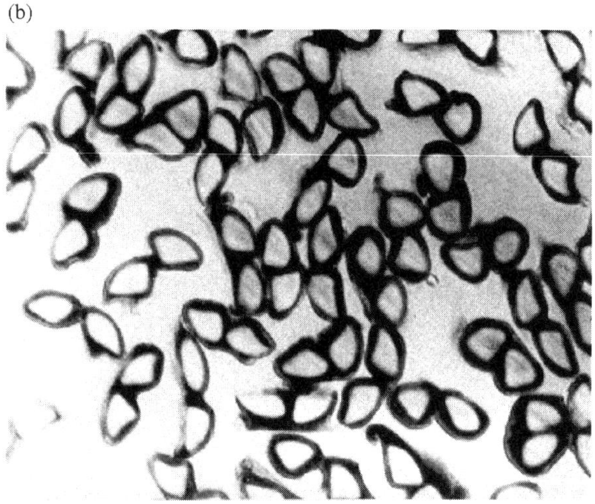

2.7 (a) Light micrograph of raw silk (from Mahall [31]). (b) Light micrograph of raw silk cross-section, after differential staining (from Reference [32]).

convert solar energy into infrared radiation can be obtained by encasing zirconium carbide particles inside a polymeric sheath. A cross-sectional view of a thermochromic fiber is shown in Fig. 2.10 [41]. Conversely, sun-shielding effects can be obtained using a variety of ceramic and other organic materials inside irregular or sheath–core polyesters [38]. Both of these bicomponent fibers are finding use in the outdoor, active wear and protective clothing markets.

Bicomponent fibres also can be modified to produce fragrant textiles that exude perfumes or insect repellents, bandages that dispense medicine over a wound, antimicrobial textiles which prevent growth of bacteria and mold, and yarns with improved electroconductivity. Fibers with ultralow density or extreme porosity

(a)

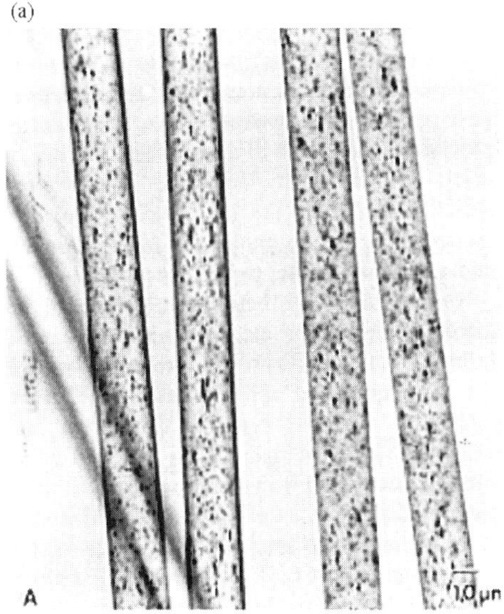

(b)

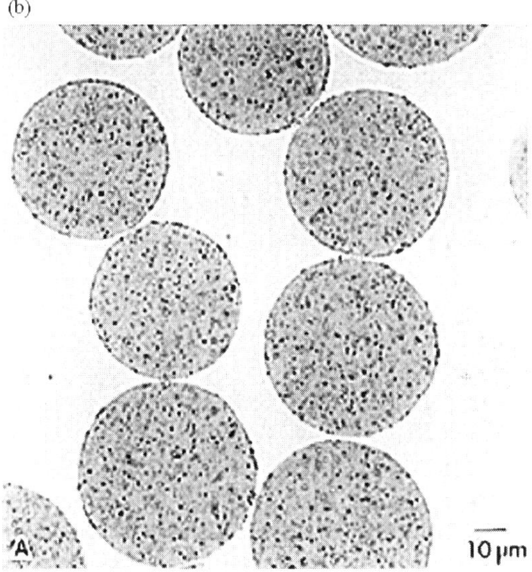

2.8 (a) Light micrograph of delustered acrylic (from Sawyer and Grubb [21]). (b) Light micrograph of delustered nylon cross-section (from Sawyer and Grubb [21]).

(a)

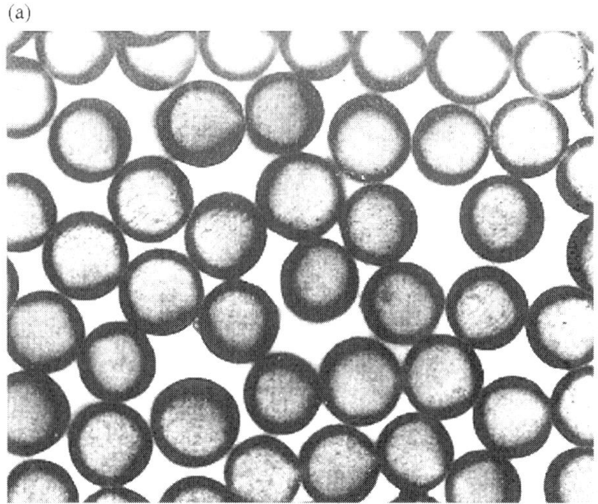

(b)

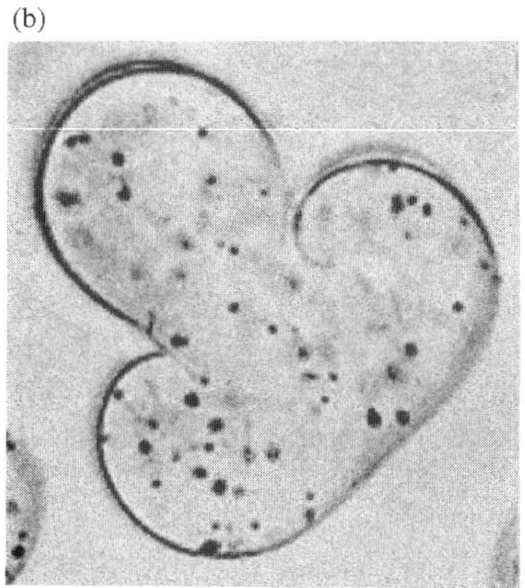

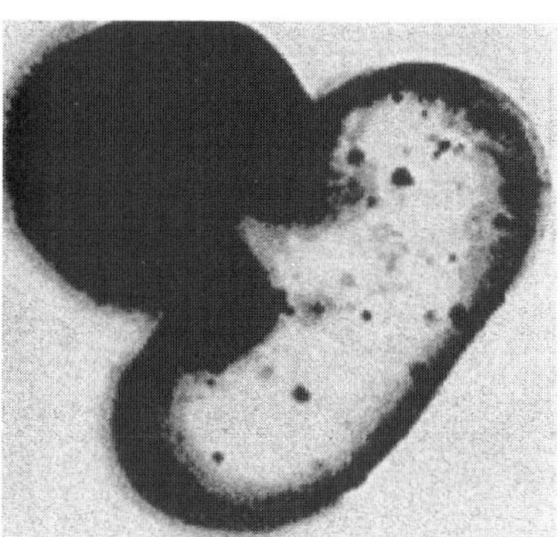

2.9 (a) Light micrograph polyethylene sheath/polyester core bicomponent fibers, after differential staining (from Reference [40]). (b) Light micrograph of side-by-side acrylic bicomponent fibers, before and after differential staining (from Moncrieff [39]).

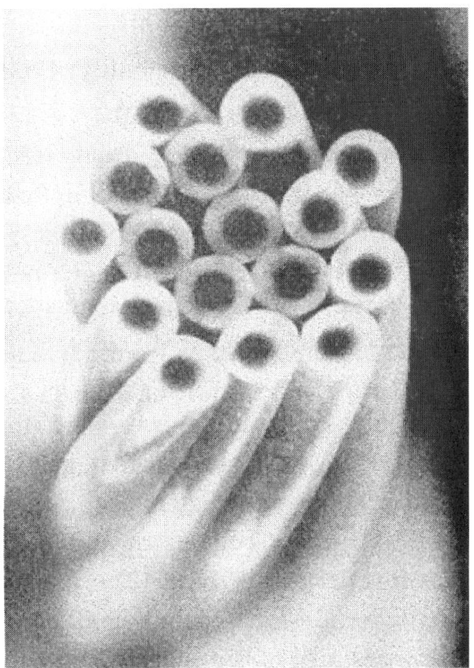

2.10 Light micrograph of sheath/core insulating bicomponent fibers (from Hongu and Phillips [41]).

are obtained from polyester bicomponent fibers with a soluble polymer core. Most speciality fibers in use today are sheath–core configurations with the active agent in either the sheath or the core depending upon end-use requirements [38].

A recent advance in conjugate spinning has been the development of microfiber technology. Microfibers, also called subdenier fibers, are extremely fine filaments (<1.0 dpf (denier per filament) or 30 μm) which are initially produced as flexible textile composites before they are split apart into individual fibers. The SEM photomicrographs in Fig. 2.11a, 2.11b and 2.11c illustrate sea-island and radial sheath–core bicomponents fibers, respectively, two of the many spinning configurations that are used to obtain microfilaments [38]. Applications for microfilaments are rapidly increasing because fabrics made with these fibers have superb filtration, barrier and insulation properties and when used in consumer products, a soft hand, supple drape and excellent cover.

2.3.2 Composite yarns

Because yarns are intermediate between fibers and fabrics, their importance in the textile industry in general and contribution to flexible textile composites in particular may go unrecognized by those not directly involved with this important segment of the textile industry. Yarns are flexible textile composites if they

(a)

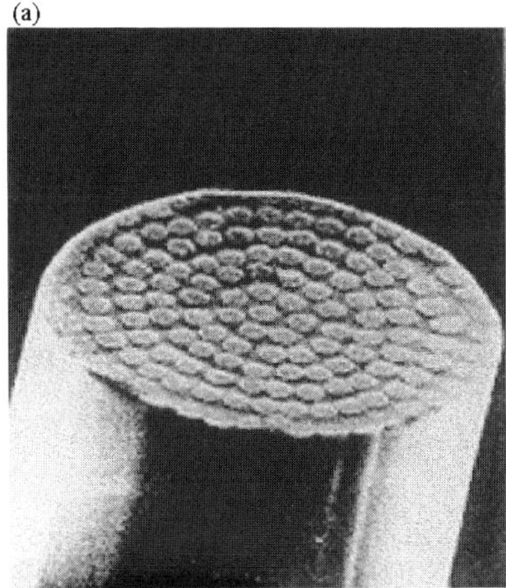

(b)

(c)

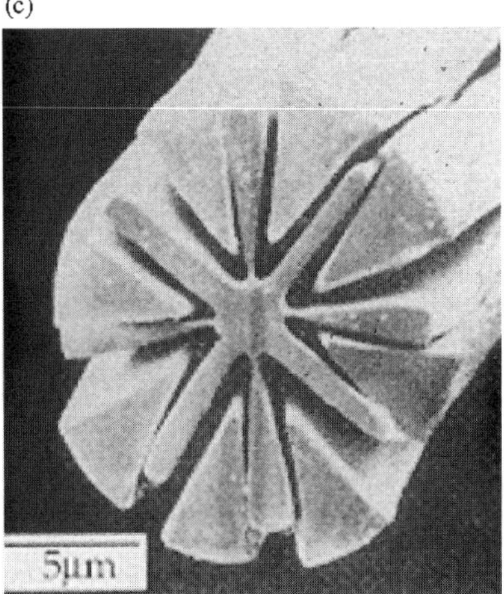

2.11 (a) Scanning electron micrograph of sea-island microfiber composite, before removal of matrix (from Berkowitch [38]). (b) Scanning electron micrograph of sea-island microfibers, after removal of matrix (from Berkowitch [38]). (c) Scanning electron micrograph of radial sheath/core microfiber composite, before sheath removal (from Berkowitch [38]).

contain two or more physically distinct components and if they are composed of fibrous and/or flexible materials. Many of the yarns used for industrial and consumer applications are flexible textile composites.

The most commonly used composite yarns are those composed of two or more generic fiber types in which the different fibers are physically blended so that they are uniformly distributed throughout the yarn bundle. Yarns constructed in this manner are called intimate blends. The composition of blended composites is virtually unlimited and ranges from the traditional cotton/polyester, wool/nylon and rayon/acetate, etc., to industrial yarns composed of stainless steel, copper or nickel and/or high performance fibers such as carbon, aramid, PBI (fibers produced from polybenzimidazole polymer) and FR (flame retardant) rayon. Reasons for using blended yarns include economical use of expensive raw materials, simplified care, enhanced durability and improved comfort, as well as development of special properties such as flame resistance and decreased static potential. An example of a nylon/rayon blend used for automobile upholstery is shown in Fig. 2.12a.

Other types of composite yarns are obtained by twisting two generically different strands together into a plied yarn or by twisting one yarn component around another to produce a sheath–core composite [42] (Fig. 2.12b). Intimate blends, plied yarns and sheath–core yarns are sometimes not suitable for high performance applications because they do not optimize the strength, elongation and wear properties of their various components [43,44]. These disadvantages, however, are not encountered with wrap spun composite yarns, also called core spun or parallel spun yarns [43–48].

Wrap spun yarns are composed of a central core of untwisted fibers that are combined with twisted or untwisted staple or filament fibers without using traditional twisting technology. The advantages of blending cited above can be achieved with wrap spun yarns and, compared with conventionally twisted yarn composites, wrap spun yarns also have better uniformity, higher bulk and yarn strength and can be more effectively engineered to meet desired end-use specifications, which include lower elongation, greater flexibility, fewer problems with torque, faster production speeds and a broader range of deniers [46]. The photomicrograph in Fig. 2.12c shows one type of wrap spun composite yarn [48]. The technology for producing wrap spun yarns is relatively new and the use of these yarns will increase in the future as their performance and economic advantages are recognized.

Laminated metallic yarns, although not fibrous in nature, qualify as yarn composites because they are multiphase textile materials. Composed of a thin film of aluminum laminated between two layers of polymeric film, metallic yarns are used for decoration in apparel and interior fabrics and for their insulating qualities in protective clothing and drapery linings [39].

Composite yarns for industrial uses differ considerably from yarns used for consumer application where appearance may be more important than durability. The opposite is true for industrial composite yarns whose purpose is to serve a special function, such as to improve wear resistance and/or performance, with less concern for appearance. Yarns coated with polymers and/or hot melt adhesives are primarily used for industrial applications [49] although they may have use in apparel because they resist raveling and provide good dimensional

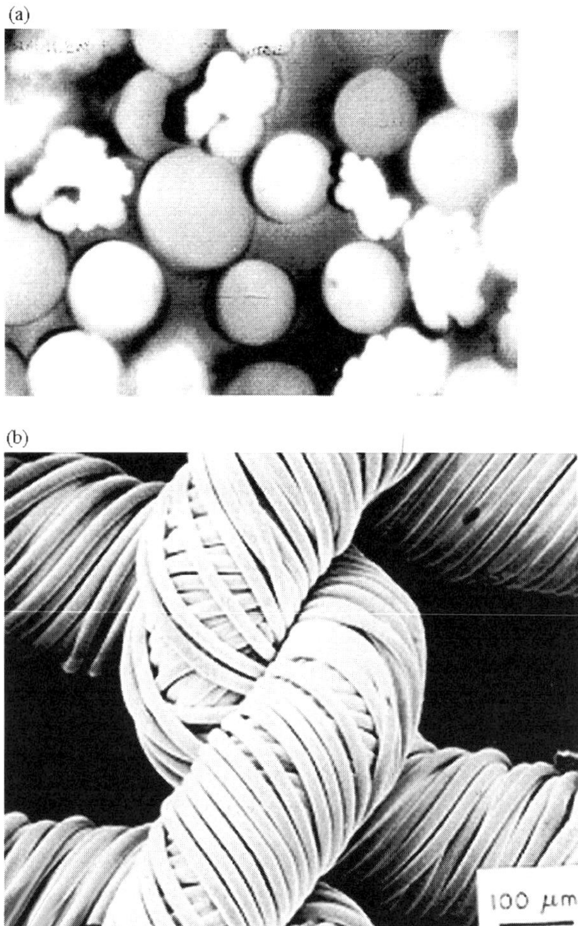

2.12 (a) Light micrograph of nylon/rayon blended yarn, nylon fibers are round. (b) Scanning electron micrograph of nylon/spandex core spun yarn, nylon on the outside (from Billica and van Veld [42]).

stability. The microscope images in Fig. 2.13a–d document the microstructural analysis of a coated composite yarn.

A low count woven fabric, composed of staple warp yarns and a heavy denier monofilament in the filling direction, was analyzed with the intent of documenting the structure of the composite yarn (Fig. 2.13a). Extracted from the fabric, the yarn had a shiny luster and an unusually high elongation. Stereo microscope examination of the yarn showed many short fibers protruding from its surface and regularly spaced indentations along its length (Fig. 2.13b). One of these indentations is shown at high magnification in Fig. 2.13c. Apparently the

(c)

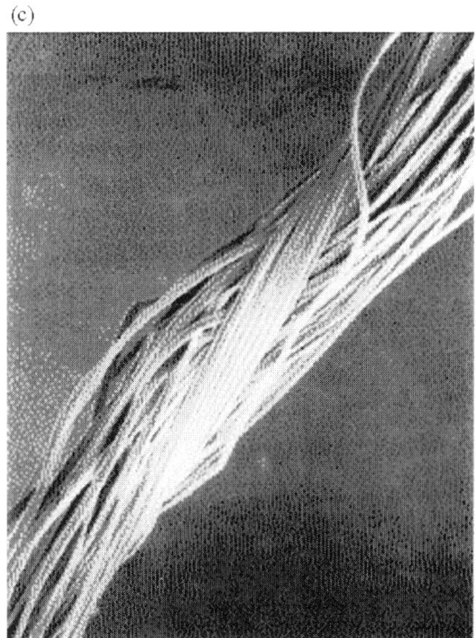

2.12 (c) Light micrograph of wrap spun yarn (from Reference [40]).

fabric was subjected to a thermal treatment that bonded warp and filling yarns together at each yarn crossover point. Raveling the fabric broke the spot welded points and left 'potholes' on the surface of the yarn.

A cross-section of a representative filament is shown in Fig. 2.13d. The filament is a composite yarn composed of a thick monofilament core coated with a thin layer of hot melt adhesive, probably a thermoplastic or crosslinked elastomer. The polymer sheath comprises only a small percentage of the bulk of the yarn. When the fabric was subjected to heat, the coating on the outside of the filament softened and, upon cooling, spot welded warp and filling yarns together at each contact point. Bonding warp and filling yarns together would increase strength and elongation of a fabric whose intended use is bandage springs in upholstered furniture.

2.3.3 Composite fabrics

Flexible textile composites engineered at the fabric level can be relatively simple planar materials, such as the bandage springs described above, which were constructed from two different types of yarn (Fig. 2.13a). More complex fabric composites are constructed using different fabrication methods or different fibers, yarns and/or fabrics layered or bonded together. Although the composition of fabric composites varies greatly depending upon end-use requirements, one consistent feature is that these materials appear and act as single layer units [8].

(a)

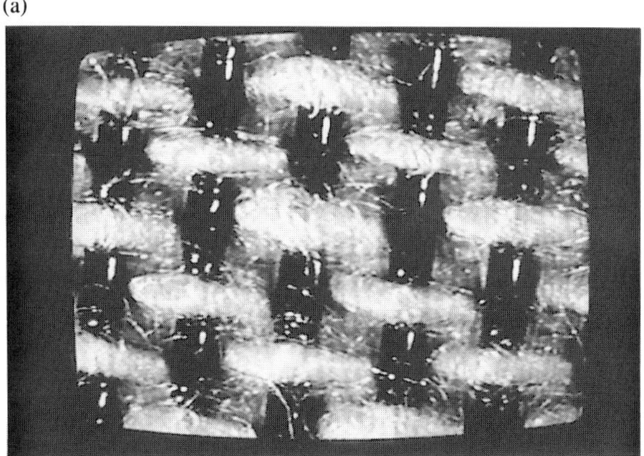

(b)

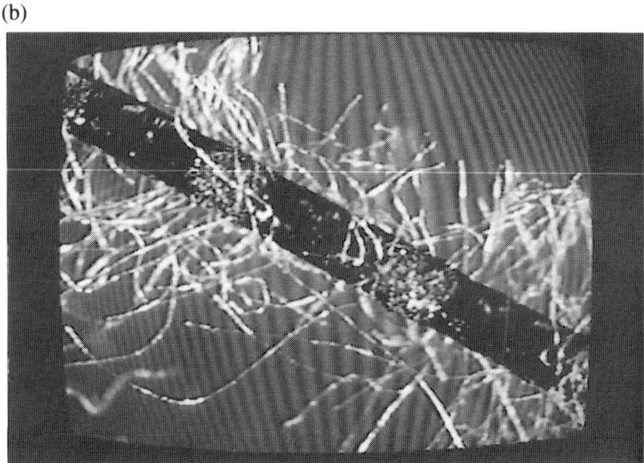

(c)

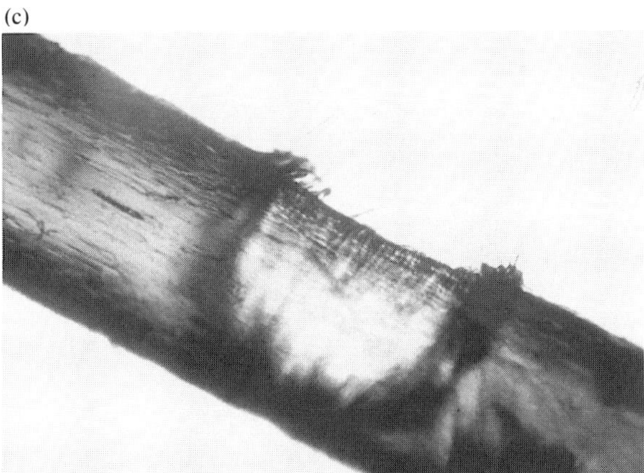

2.13 (a) Thermal print of woven fabric, with composite yarns. (b) Thermal print of composite yarn. (c) Thermal print of monofilament composite yarn, damaged at broken spot weld.

(d)

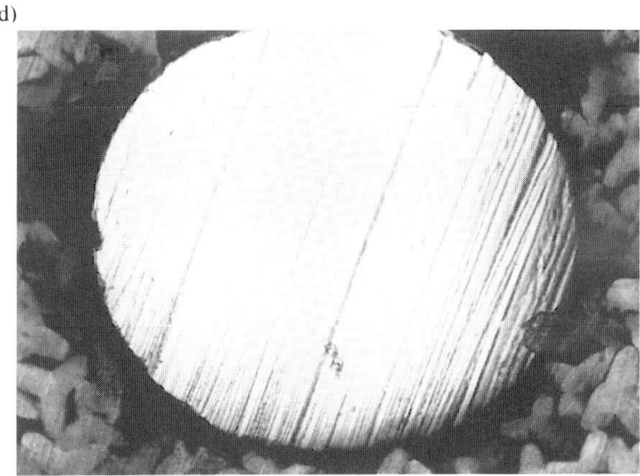

2.13 (d) Thermal print of monofilament composite yarn cross-section.

The fabric shown in Fig. 2.14 is a flexible textile composite produced using two different fabrication methods, non-woven and knitting. A non-woven material is composed of a web of entangled fibers whereas knitting utilizes interlocking loops of yarn. This particular composite was constructed from an air laid web that was stabilized with knit stitches and pigment printing for use as upholstery for outdoor furniture and umbrellas. Fabrics of this type are classified as stitched bonded. Roofing materials, medical slings, sleeping bags, vacuum cleaner bags and shoes also contain stitched bonded components.

An example of a multilayered fabric composite is wet suit fabric used by scuba divers, surfers and other water sport enthusiasts for thermal protection from cold water. Wet suit fabric generally consists of a lightweight fabric bonded to a continuous sheet of rubber or foam. Sometimes a second fabric is bonded to the outside of the foam which produces a three-layer flexible textile composite.

2.14 Thermal print of stitched bonded fabric.

The fabric side of a wet suit, which is worn next to the skin, must be non-irritating and lightweight yet sturdy enough to withstand flexing and compression. Cover must be excellent without excessive bulk or weight. Fabric components of wet suits must not ravel or run, lint or rub up or change in dimension. In addition, fiber components in wet suits must be compatible with bonding adhesives and resistant to salt water, photodegradation, heat, perspiration and other chemicals. The same requirements apply to the foam or rubber component of a wet suit.

A warp knit fabric bonded to a layer of foam is a flexible composite that satisfies very well the end-use requirements of a wet suit. Warp knit fabrics have thin crosswise ribs on the technical back and a stitch pattern formed in the vertical direction (Fig. 2.15a). They are thin, lightweight, strong and run resistant. When composed of continuous filament yarns, as in a wet suit, they resist surface wear and rubbing up. They also are low in cost and easily dyed. Closed cell foam is a good match for bonding with a warp knit fabric in a wet suit application because it contributes to durability, dimensional stability and flexibility as well as providing bulk and insulation.

A wet suit sample mounted in an AATCC fabric cross-section holder illustrates the composite nature of this material (Fig. 2.15b). The fabric layer is tightly and uniformly bonded to a layer of foam which makes up most of the bulk of the fabric. Microscope examination of the fabric cross-section shows the looped stitch formation characteristic of knit fabrics and reveals the presence of a third component at the fabric/foam interface: an elastomeric monofilament. Inclusion of an elastomeric yarn within the body of a fabric improves elongation and elastic recovery. Fabric density increases because the elastomeric filament contracts after stitch formation causing the fabric to decrease in dimension and increase in thickness thereby improving cover. This 'contraction effect' causes the yarns on the technical back of the warp knit fabric to buckle and form elongated loops that protrude from the surface of the fabric rather like the pile on the surface of a towel. However, the loops are so small and the fabric so tightly bonded to the foam, that snagging does not occur easily as in a towel. The monofilament yarn is securely held in place because it is interlooped with other knit stitches (Fig. 2.15c). The presence of the monofilament yarn within the knit qualifies this fabric as a composite. Thus, a composite fabric, the wet suit, can be constructed from other composite materials.

The soft, fuzzy texture of the knit component of this wet suit is enhanced by lightly napping with fine metal brushes. Napping entangles the fiber loops and densifies the surface of the fabric as well as softens its hand. These surface changes improve the functionality and aesthetics of a wet suit composite.

The structural components of the wet suit fabric described above are readily apparent using low power microscopy. However, the components of many flexible textile composites are not so easily discerned. Materials in this category may be laminated, coated or somehow concealed so that the layered material appears to

(a)

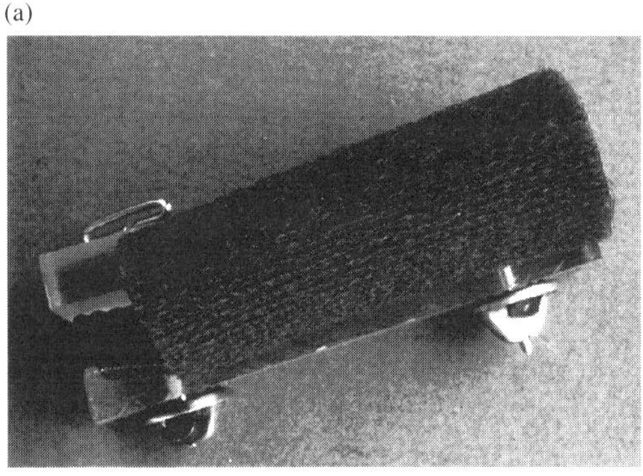

(b)

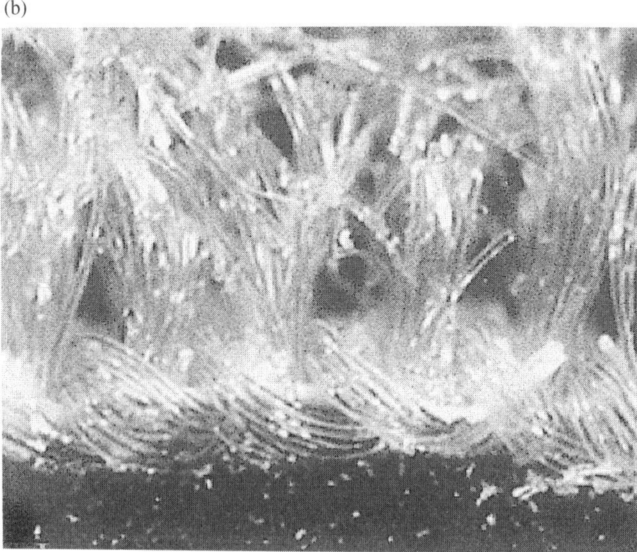

2.15 (a) Wet suit fabric in AATCC fabric cross-section holder. (b) Thermal print of wet suit fabric cross-section.

be homogeneous. An example of this type of composite, which finds widespread use in both consumer and industrial uses, is described below. The product is a four-layer laminated leather composite used as a watchband.

Viewed from the top, the outer surface of the band appears to be 'reptile-like' leather. However, when the band was received in our laboratory with the request to characterize its structural components, the surface of the band had split in a way which was not characteristic of natural leather. In fact, it was observed that a portion of the outer layer of the band had become detached from a fibrous substrate. Furthermore, the back side of the band was peeling away from its substrate.

2.15 (c) Thermal print of knit stitches and elastomeri[illegible]onofilament, at fabric/foam interface.

Dissection and microscope examination revealed that the outer layer of the band consisted of a polymeric film embossed and polished to resemble reptile skin (Fig. 2.16a). The embossed film was cast onto a fibrous non-woven layer called a scrim. Films used in flexible textile composites often are stabilized with scrims to improve their strength and durability. The scrim in this product consisted of white polyester fibers that were non-uniform in length and diameter. Fibers from the scrim are shown in Fig. 2.16a where a piece of the film became detached and torn near a size-adjustment hole. Holes and channels in the scrim suggested that it had been constructed using a needle punching technique. If the scrim had been thermally or chemically bonded, the fibers would have displayed a melted or glued-together appearance. The black background seen through the scrim is the adhesive that holds the laminated scrim to another layer of the composite.

The third and thickest layer of the band was a fine-finished leather shown in profile in Fig. 2.16b. Note how the cellular structure of the hide is finest next to the skin side of the hide and more coarse towards the center of the composite. The skin side of the leather, worn next to a person's wrist, was coated with a thin layer of finish or stain. The finish applied to the unbonded side of the leather was easily dislodged with a sharp tool to reveal a hide with variable size pores.

The function served by this flexible textile composite is economical rather than high performance. In fact, the collective cost of the components of this product and the cost of their assembly probably were less than the cost of a 'genuine' reptile product. The wear performance of this product evidently was sufficiently different from a 100% leather product for a consumer to observe that the components of the band were not homogeneous. There was a strong negative

(a)

2.16 (a) Light micrograph of watchband, polymeric film removed. (b) Light micrograph of watchband cross-section.

reaction when the owner of the watchband realized that a full price had been paid for a watchband that was not 100% leather, as apparently the vendor had claimed!

Considering the variety of materials that can be used to construct fabric composites, and the many different ways that these materials may be utilized, fabric composites can be quite complex in design and composition. The fabric cross-section in Fig. 2.17a illustrates the complexity of a fabric composite engineered to insulate and protect underground pipes and cables. This fabric composite has seven different layers and is composed of three different fabrications, two different types of yarns and at least three different generic fibers.

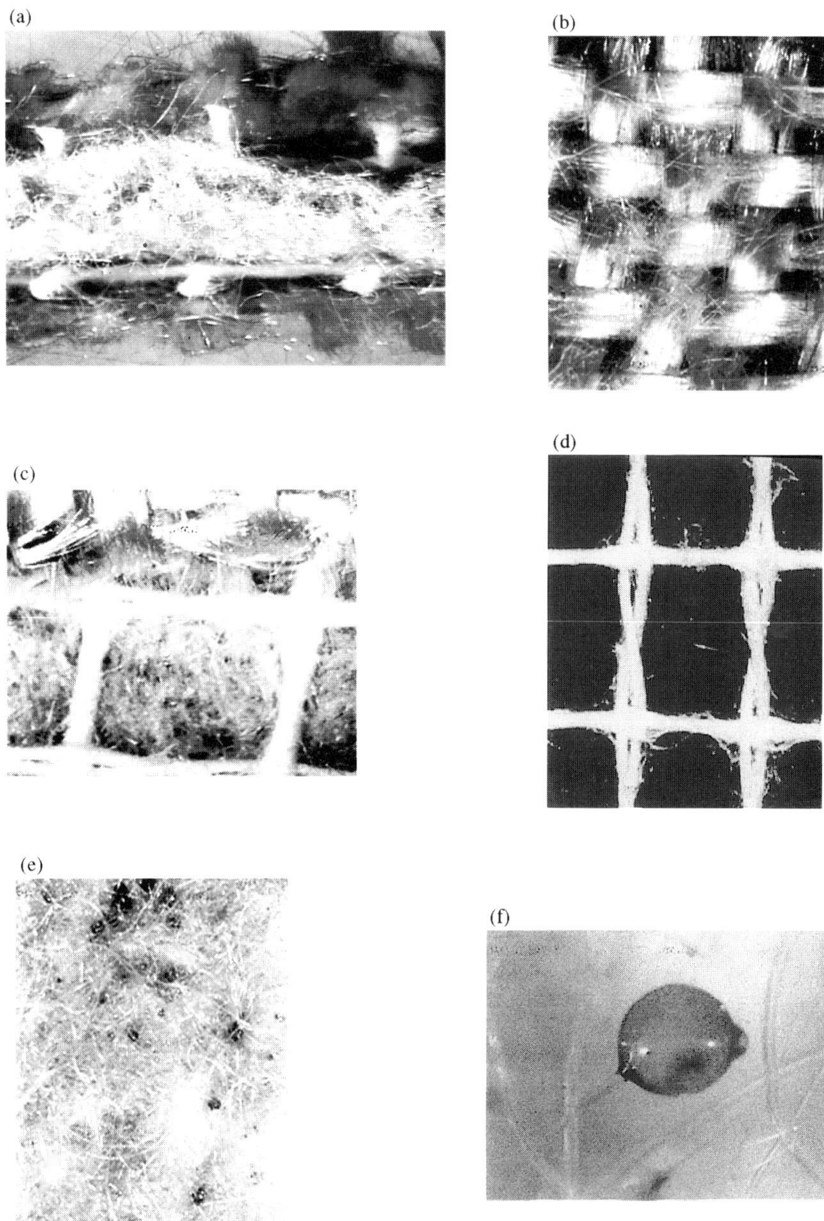

2.17 (a) Computer image of seven-layer composite cross-section. (b) Thermal print of pipe wrap, face fabric. (c) Thermal print of pipe wrap, after removal of face fabric. (d) Thermal print of pipe wrap, scrim fabric. (e) Thermal print of pipe wrap, non-woven web with spotty debris. (f) Thermal print of pipe wrap, glass fiber with bead formation in non-woven web.

The two outer layers of the pipe wrap (layers 1 and 7) are constructed of flat glass filament yarns (Fig. 2.17b). Although both face fabrics are woven using a 1×1 interlacing pattern, the yarns in these fabrics differ in number of ply and level of twist. When the face fabric is pulled back, layers 2 and 3 of the pipe wrap are exposed (Fig. 2.17c). Underneath both face fabrics are low density leno-woven scrims whose function is to stabilize and protect the inner layers of the composite without restricting overall flexibility (Fig. 2.17d). The scrim also contributes to the insulating capabilities of the composite by increasing dead air space between the composite layers. The inner layers of the pipe wrap, layers 3, 4 and 5, are non-woven webs. The location of the webs underneath the scrim is shown in Fig. 2.17c.

During the analysis of this composite fabric, layer 4, the innermost non-woven lap, was extracted from layers 3 and 5. Examination of this fabric revealed the presence of dark debris randomly distributed throughout the web (Fig. 2.17e). Under high magnification, layer 4 was found to contain tiny spheres of melted fibers in various stages of bead formation (Fig. 2.17f). 'Beading' indicated that this layer of the composite contained glass fibers that had been subjected to heat during some phase of manufacturing or during a previous end use.

Needle punching was used to stabilize the layers of this composite fabric as evidenced by numerous holes in the non-woven webs and the many fibers that are displaced from the web, pulled through the scrim and entangled with the fibers of the face fabrics (Fig. 2.17b). Staple fibers similar to those used in the non-woven webs were also found on the outer surface of the face fabrics. Horizontal splits in the yarns of the face fabrics and broken filament fibers also indicated that the composite had been subjected to needle punching after assembly.

2.4 Conclusion

In this chapter, flexible textile composites are defined as modified textile materials in which the textile structure is the major component or matrix. Flexible textile composites are found in every textile category including natural and manmade fibers, simple and complex yarns and woven, knitted and non-woven fabrics, or as a combination of any of these materials. Photomicrographs were used to illustrate the structure and composition of representative types of flexible textile composites. The use of various types of microscopes for analysis of flexible textile composites was discussed and the use of optical microscopes for routine laboratory analysis recommended. The advantages of utilizing video microscopy and image processing to analyze flexible textile composites were discussed and illustrated. Considering their many industrial and consumer applications, flexible textile composites probably are the most numerous category within a broad range of materials called composites.

References

Introduction to textiles and definition of flexible textile composites

1. *1997 Annual Book of ASTM Standards*, American Society for Testing and Materials, West Conshohocken, PA, 1997. ISBN 0-8031-2315-9.
2. M C Tubbs and P N Daniels (eds), *Textile Terms and Definitions*, The Textile Institute, Manchester, 1991. ISBN 0-900739-17-7.
3. F Scardino, An introduction to textile structures and their behavior, *Textile Structural Composites*, eds T W Chou and F K Ko, Elsevier Science, New York, 1989, Chap. 1, pp 1–24. ISBN 0-444-42992-1.
4. M Knight, Composite materials, *Encyclopedia of Physical Science and Technology*, Volume 4, Academic Press, San Diego, 1992, pp 17–27. ISBN 0-12-226930-6.
5. S Adanur, Textile structural composites, *Wellington Sears Handbook of Industrial Textiles*, ed. S Adanur, Technomic, Lancaster, PA, 1995, pp 231–270. ISBN 1-56676-340-1.
6. C G Namboodri and S Adanur, Coating and laminating, *Wellington Sears Handbook of Industrial Textiles*, ed. S Adanur, Technomic, Lancaster, PA, 1995, pp 181–197. ISBN 1-56676-340-1.
7. F K Ko, Three-dimensional fabrics for composites, *Textile Structural Composites*, eds T W Chou and F K Ko, Elsevier Science, New York, 1989, Chap. 1, pp 1–24. ISBN 0-444-42992-1.
8. B D Agarwal and L J Broutman, *Analysis and Performance of Fiber Composites*, John Wiley and Sons, New York, 1980, Chap. 1, pp 1–14. ISBN 0-471-05928-5.
9. P Lennox-Kerr (ed), *Flexible Textile Composites*, The Textile Trade Press, Manchester, 1973. ISBN 0-903772-02-7.
10. T L Vigo and A F Turbak (eds), *High-Tech Fibrous Materials – Composites, Biomedical Materials, Protective Clothing, and Geotextiles*, American Chemical Society, Washington, DC, 1991. ISBN 0-8412-1985-0.
11. T-W Chou, *Microstructural Design of Fiber Composites*, Cambridge University Press, Cambridge, 1992. ISBN 0-521-35482-X.
12. J Wypych, *Polymer Modified Textile Materials*, John Wiley & Sons, New York, 1988. ISBN 0-471-83959-0.
13. C C Evans, *Whiskers*, general ed, J G Cook, M & B Monograph ME/8, Mills & Boon, London, 1972.

Microscopy and imaging

14. H Shuman, J M Murray and C DiLullo, Confocal microscopy: an overview, *BioTechniques* 1989 **7**(2) 154–163.
15. W I Miller, III and B Foster, Fluorescence and confocal laser scanning microscopy: applications in biotechnology, *Amer. Lab.* 1991 **23**(6) 73–78.
16. P A Annis, T W Quigley and K E Kyllo, Hand techniques for cross-sectioning fibers and yarns, *Textile Chem. Colorist* 1992 **24**(9) 78–82.
17. P A Annis, T W Quigley and K E Kyllo, Useful techniques in textile microscopy, *Textile Chem. Colorist* 1992 **24**(8) 19–22.

18. P A Annis, Applications of microscopy to textile analysis: Part 1 – Fabrics, *Amer. Dyestuff Reporter* 1991 **80**(9) 46–48,94.
19. P A Annis, Applications of microscopy to textile analysis: Part 2 – Yarns, *Amer. Dyestuff Reporter* 1992, **81**(9) 38–43,82,87.
20. P A Annis, Textile microscopy, *Colourage* 1993 **26**(1) 16–19.
21. L C Sawyer and D T Grubb, *Polymer Microscopy*, Chapman and Hall, New York, 1987. ISBN 0-412-25710-6.
22. P A Annis, Textile analysis using video microscopy, *Amer. Dyestuff Reporter* 1990 **79**(9) 79,82.
23. J C Russ, *The Image Processing Handbook*, 2nd edn, CRC Press, New York, 1994. ISBN 0-8493-2516-1.
24. G Baxes, *Digital Image Processing: Principles and Applications*, John Wiley & Sons, New York, 1994. ISBN 0-471-00949-0.
25. R Stevenson, Bioapplications and instrumentation for light microscopy in the 1990s, *Amer. Lab.* 1996 **28**(6) 28–51.
26. S Inoue, *Video Microscopy*, Plenum Press, New York, 1986. ISBN 0-306-42120-8.
27. S Inoue, Video enhancement and image processing in light microscopy: Part 1 – Video microscopy, *Amer. Lab.* 1989 **21**(4) 52–57.
28. S Inoue and T Inoue, Video enhancement and image processing in light microscopy: Part 2 – Digital image processing, *Amer. Lab.* 1989 **21**(4) 62–70.
29. B Gunning, Video enhancement and image processing in light microscopy: Part 3 – Display processing, *Amer. Lab.* 1989 **21**(4) 72–81.
30. D Lansing Taylor, M Nederlof, F Lanni and A S Waggoner, The new vision of light microscopy, *Amer. Sci.* 1992 **80**(4) 322–335.
31. K Mahall, *Quality Assessment of Textiles – Damage Detection by Microscopy*, Springer-Verlag, Berlin, 1993. ISBN 0-540-57390-9.
32. Sub-committee of the Textbook Committee of the Textile Institute (eds) *Identification of Textile Materials*, 7th edn, The Textile Institute, Manchester, 1985. ISBN 0-900739-18-5.
33. R Hesse, Microscope methods for textile testing, *Melliand Textilberichte* 1986 **67**(7) E212–E216.
34. W C McCrone, The case for polarized light microscopy, *Amer. Lab.* 1996 **28**(9) 12–21.

Flexible textile composites

35. M T Pailthorpe, The theoretical basis for wool dyeing, In *Wool Dyeing*, ed D M Lewis, Society of Dyers & Colourists, W Yorkshire, UK, 1992, Chap. 2, pp 52–87. ISBN 0-901956-53-8.
36. J A Rippon, The structure of wool, In *Wool Dyeing*, ed D M Lewis, Society of Dyers & Colourists, W Yorkshire, UK, 1992, Chap. 1, pp 1–51. ISBN 0-901956-53-8.
37. D E Akin, Photomicrographs, US Department of Agriculture, Agricultural Research Service, Richard B Russell Agricultural Research Center, Athens, Georgia.
38. J C Berkowitch, *Trends in Japanese Textile Technology*, U.S. Department of Commerce, Office of Technology Policy, Asia-Pacific Technology Program, Washington DC, December 1996.
39. R W Moncrieff, *Man-Made Fibers*, John Wiley & Sons, New York, 1975. ISBN 0-470-61318-1.

40. Celbond Type 255, Trevira, 2300 Archdale Drive, Charlotte, NC, 1993.
41. T Hongu and G O Phillips, *New Fibers*, Ellis Horwood, New York, 1990. ISBN 0-13-613266-9.
42. H R Billica and R D van Veld, Scanning electron microscopy of synthetic fibers, In *Surface Characteristics of Fibers and Textiles*, ed M J Schick, Marcel Dekker, New York, 1975, Chap. 7, pp 295–328. ISBN 0-8247-6316-5.
43. K Wilson, Composite yarns, *Textile Horizons* 1986 **6**(7) 28–29.
44. R J Harper, Jr and G F Ruppernicker, Jr. Woven fabrics prepared from high tenacity cotton/polyester core yarn, *Textile Res. J.* 1987 **57**(3) 147–154.
45. A Parker, Yarns without true twist, *Textile Horizons* 1988 **8**(12) 30–35.
46. D W Palmer, Wrap spinning: A new technology for knitting yarn, *Knitting Times* 1986 **55**(6) 48–52.
47. W C Smith, HP puts high tech, profit in yarn makers' future, *Textile World* 1989 **139**(2) 44–48.
48. *Parallel Yarn End Products*, American Sussen Corporation, P.O. Box 7147, Charlotte, NC, 1987.
49. Textile trade finds new uses for Raufil, *Textile Month* 1986 (**10**) 47.

3

3D confocal microscopy of glass fibre-reinforced composites

ASHLEY R CLARKE, GEOFF ARCHENHOLD AND NIC C DAVIDSON

3.1 Introduction

The aim of this chapter is to introduce the reader to the novel technique of confocal laser scanning microscopy (CLSM):

- To discuss those issues that affect the quality of the three-dimensional (3D) structural measurements of composites with particular attention to glass fibre-reinforced polymer composites.
- To review recent developments in the determination of 3D spatial distributions of fibres in composites.
- To speculate on future application areas for the CLSM technique in composites research.

Any prediction of the mechanical and thermal performance of composite materials by finite element models assumes that the micromechanics of such heterogeneous, highly interacting fibre systems is fully understood. Considerable progress has been made, but there has been a dearth of good quality, 3D structural data to validate the micromechanical models. Also, the main aim of the composites manufacturer is to improve the quality of the part by tuning the processing conditions to achieve the optimum part performance. The research work to be described in this chapter has concentrated solely on the analysis of processed parts (rather than attempting on-line, rheological measurements) by optical microscopy.

Composites usually contain long (continuous) fibres or short fibres. Long or continuous fibres cannot be considered as rigid straight rods – they exhibit local curvature and often a wave-like nature at the millimetre scale in the matrix. However, each smooth short fibre within a composite could be specified uniquely by a set of fibre parameters $f_i\{\mathbf{r}_i, l_i, d_i, \theta_i, \Phi_i\}$ where $\mathbf{r}_i$ is a vector from the origin of an arbitrary, but convenient, orthogonal coordinate system to one end of the fibre, l_i is the length of the fibre, d_i is the diameter of the fibre and the angles (θ_i, Φ_i) define the orientation of the fibre with respect to the coordinate system, as shown in Fig. 3.1a. Until recently, high accuracy measurements of these individual fibre parameters were impossible to achieve. Therefore, experimentalists have made certain simplifying assumptions about the fibres and they have used a number of measurement techniques to describe the ensemble of fibres within a small sample subvolume, δv. These techniques usually yield specific frequency distributions, for example, fibre orientation distributions, FOD$\{\theta_i\}$, FOD$\{\Phi_i\}$ and a fibre length distribution, FLD$\{l_i\}$. As frequency distributions are somewhat unwieldy, single parameters have been proposed which are derivable from the frequency distributions, for example geometrical coefficients or orientation tensor components. Researchers have then sought to correlate these characterisation parameters with either processing conditions or material properties.

The advent of the confocal laser scanning microscope technique (described fully in Section 3.2) has placed researchers in an interesting position. For the first time, all of the five fibre parameters – $\mathbf{r}_i$, l_i, d_i, θ_i and Φ_i – may be measured non-destructively within a subvolume, δv of the fibre-reinforced composite sample. The size of this subvolume depends upon the opacity of the sample at the laser wavelength used (which determines, to a large extent, the depth of penetration Z into the sample), the objective lens magnification, numerical aperture and working distance and the degree of automation for scanning over large regions in X and Y (i.e., over the sample surface).

Hence the major issues which need to be addressed in the next few years are:

- To what extent can the analysis of 3D structure be achieved in a non-destructive way – what maximum depth is achievable with the CLSM technique?
- What novel parameters could be devised to characterise different aspects of composite material structure?
- What sample subvolume constitutes a reasonable characterisation of the whole sample?
- How fast and efficiently can 3D data be obtained?

and this chapter presents at least some initial answers to these questions.

3.1.1 Definition of the measurement problem

Typical diameters of reinforcing fibres in composite materials are in the range 5–15 μm and their lengths could vary from 10 μm through to tens of millimeters. Thermal and mechanical properties are governed by voids in the range of a few micrometres to many hundreds of micrometres diameter, mean fibre orientation states, fibre bundles or clusters over 50–100 μm regions and fibre waviness with waviness amplitudes of 10–40 μm over fibre 'wavelengths' of a few millimetres. Hence there is considerable interest in composite structure in the region of a few micrometres to many millimetres and Haritos *et al.* [1] have coined the phrase 'mesostructure' to describe these artefacts between the micro- and macro-structural scale sizes. Ideally, such 3D mesostructural measurements must be made with submicron spatial resolutions in $\delta x, \delta y$ and δz and it would be desirable for suitable measurement techniques to scan large volumes (millimetres by millimetres by 100 µm) within reasonable experimental timescales.

As 3D data are to be captured, reduced and reconstructed, a suitable orthogonal coordinate system must be specified. Coordinate systems in some research papers assume that one axis is parallel to the mean fibre direction or to some external reference direction, for example the direction of 'draw' for extruded samples. However, in this chapter, a right-handed orthogonal coordinate system has been adopted where the *XY* plane always corresponds to the sample surface section and *Z* represents depth into the sample, irrespective of the orientation of the fibres or any other entity within the composite sample, see Fig. 3.1.

The traditional method used to infer fibre orientations $\{\theta_i, \Phi_i\}$ is to scan a single two-dimensional (2D) section plane and to measure the ellipticities of each fibre image. The out of plane angle is given by

$$\theta_i = \cos^{-1}(b/a) \tag{1}$$

where b is the semi-minor axis length of each elliptical fibre image and a is the semi-major axis length of that image. The angle Φ_i is given by the orientation of the major axis of the elliptical image in the section plane. Note that there is an inherent ambiguity in the angle Φ_i because the same fibre image would be produced by a fibre with angle $(180^\circ + \Phi_i)$. It is possible to remove this ambiguity with extra effort, for example some researchers etch the sample surface to produce a shadowing effect from which the ambiguity can be removed (by manual analysis rather than by fully automated instrumentation) but this technique only works for fibre images with high θ.

An alternative method of deducing each fibre's $\{\theta, \Phi\}$ is to take the shift of fibre image centre coordinates (x_1, y_1) and (x_2, y_2) on two parallel registered section planes

$$\theta = \tan^{-1}\left(\frac{\sqrt{(x_1 - x_2)^2 + (y_1 - y_2)^2}}{\Delta z}\right) \tag{2}$$

$$\Phi = \tan^{-1}\left(\frac{x_1 - x_2}{y_1 - y_2}\right) \tag{3}$$

where Δz is the separation between the two *XY* section planes. Note that this method yields unambiguous angular information and is the basis of current research at Leeds to use both the 2D high-resolution large-area image analyser and the confocal laser scanning microscope to remove the Φ ambiguity by pattern-matching fibre images between section planes.

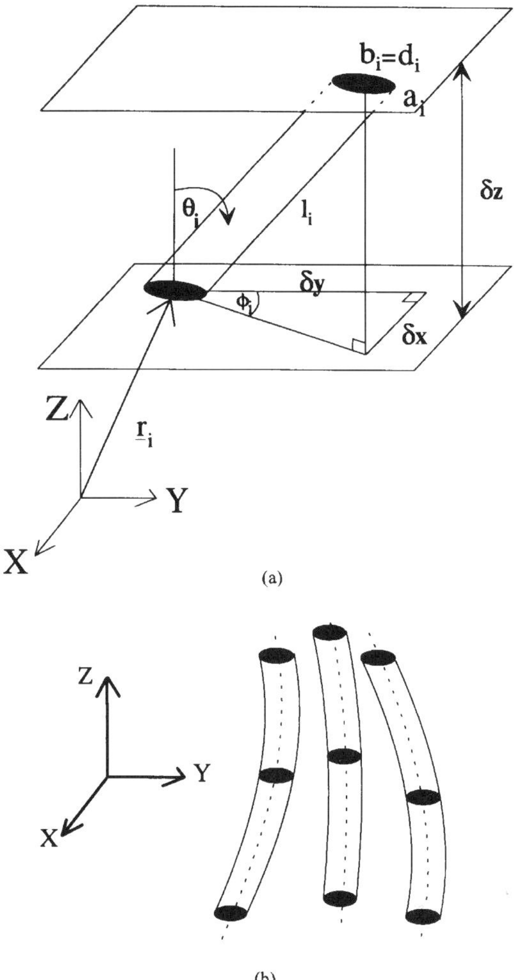

3.1 (a) Complete characterisation of a short straight fibre in 3D space. (b) Real fibres or fibre segments (length $\geq 100\,\mu m$) are curved, making characterisation more difficult.

The only other possible alternative would be some kind of tomographic technique where a series of scans are taken at different orientation planes and the 3D structure is inferred by finding the best fits to the orientation data. These techniques are currently very expensive because of the computing power required and the associated precision engineering of the system. Apart from their obvious attraction for non-destructive testing of a sample, it is difficult to see how the tomographic technique would be advantageous to research groups when compared with cheaper 2D and 3D microscopical techniques.

3.1.2 Comparison of measurement techniques

The spatial resolution achievable by different experimental techniques is shown in Fig. 3.2. Nowadays, the highest spatial resolutions (at atomic scale sizes) are being achieved by the new scanning probe techniques (see for example the review by Grim and Hadziioannou [2]). However, these techniques are too sensitive and the inherent scale size is too small to have an impact on the mesostructural domain.

One of the earliest papers on fibre orientation distributions, Darlington *et al.* [3], used X-ray contact microradiography which was capable of imaging all fibres within a 100 μm thick slice of the sample. Typical regions studied were 600 μm by 600 μm by 100 μm in *X*, *Y* and *Z*, respectively. Preferred fibre directions in the *XY* plane were then obtained by manual analysis of the overlapping glass fibres in both polypropylene and polyamide 66. In high packing fraction fibre-reinforced composites it is difficult to separate the fibres and impossible to obtain true 3D characterisation by contact microradiography.

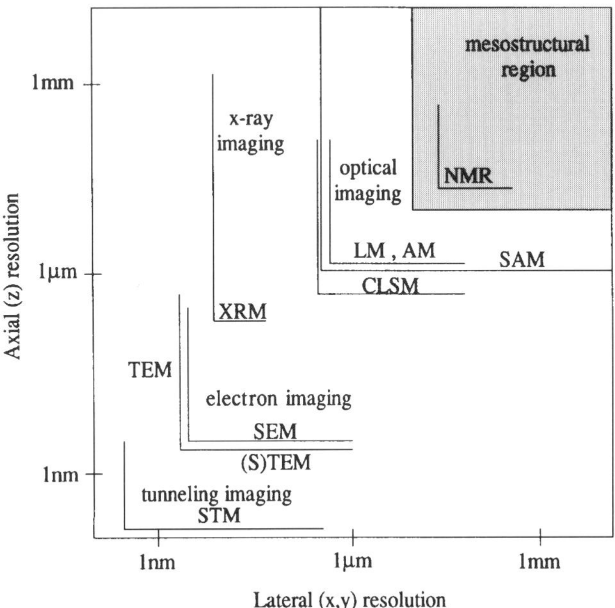

3.2 Overview of the different measurement techniques showing the ultimate resolution achievable, both laterally (in *XY*) and axially (in *Z*). Key for the techniques: AM, acoustic microscopy, LM, light microscopy; NMR, nuclear magnetic resonance; SAM; scanning acoustic microscopy; SEM, scanning electron microscopy; STM, scanning tunnelling microscopy; TEM, transmission electron microscopy; XRM, X-ray microscopy.

However, more recent investigations have been conducted into 3D, X-ray scanning tomography (see for example Geier [4]). Medical scanners operate with spatial resolutions of the order of 200 μm, but enhancements can be made to reduce the voxel resolutions down to 10–20 μm, as discussed by Maisl *et al.* [5].

Nuclear magnetic resonance (NMR) has been used to see localised defects, to image local stress distributions associated with filler agglomerates and to image diffusion and non-Newtonian flow in polymer melts, as described in Blumich and Blumler [6]. NMR microscopy achieves NMR imaging at spatial resolutions δx, δy and $\delta z \geq 10\,\mu m$, which is a much poorer spatial resolution than for optical microscopy. However, the use of larger field gradients and multipulse sequences might improve the NMR spatial resolution (see Segre *et al.* [7]).

C-SCAN is a common technique for non-destructively testing very large sample volumes for evidence of voidage, but once again, spatial resolutions at the 20–50 μm level are common with these systems. Scanning acoustic microscopy (SAM) has made significant strides over the past few years. Transducers operating at frequencies $v \leq 200\,MHz$ and capable of focusing the ultrasonic waves may be used to scan samples in 3D with δx, δy and $\delta z \geq 20\,\mu m$ spatial resolutions (see for example Lisy *et al.* [8]). Higher spatial resolutions (a few μm) can be obtained by operating at 1 GHz, but these signals suffer high attenuation and the attainable sample penetration decreases significantly. Image sections are reconstructed from variations in the intensity of the reflected ultrasonic signals. Good contrast can be achieved, but other image artefacts seem to appear which would make fully automatic 3D operation extremely difficult (e.g. the presence of interference fringes around fibre images below the surface, due to surface acoustic wave reflections from the matrix–fibre interface).

Urabe and Yomoda [9] describe a non-destructive, 4 GHz microwave techique for detecting fibre orientation and fibre content in both carbon and glass fibre-reinforced composites. The conductive carbon fibres give better sensitivity than glass fibres, but the technique has poor spatial resolution owing to the long wavelengths ($1\,cm < \lambda < 10\,cm$). Their technique is reminiscent of the optical diffraction techniques (see for example McGee and McCullough [10] and Polato *et al.* [11]) for deducing fibre orientation distributions directly (rather than from the averaging of a large ensemble of fibre data sets).

Young [12] has published extensively over the past few years on the applications of the Raman spectroscopic techniques to aramid and carbon fibre-reinforced composites. By mapping the stress-induced Raman band shifts in the fibres, the relationship between structure and deformation processes can be investigated (see Chapter 8 in this book by Galiotis [13]).

However, the low cost and versatility of microcomputers and image processing systems has enabled fibre orientation distributions to be derived from high spatial resolution measurements of hundreds (or, in some cases, thousands) of fibre images on a polished, 2D section viewed with a conventional optical reflection microscope. Recently, the current state of research into the use of image processing for the characterisation of polymer composites was reviewed by Guild and Summerscales [14].

Initially, digitiser pads were used to take manual measurements (for example Fakirov and Fakirova [15], Yurgartis [16] and Toll and Andersson [17]). As the technology improved, image frame grabbers attached to economical micro-computers running commercial image processing software have allowed more rapid measurements to be performed, for example Fischer and Eyerer [18] for fibre orientations and Yurgartis and Morey [19] for yarn shape and crimp angle in woven composites. At Leeds, Clarke *et al.* [20–22] have developed a special purpose, high spatial resolution, large area, fully automated 2D image analyser for fibre length and fibre orientation research (see Hine *et al.* [23]), which is shown schematically in Fig. 3.3. The small network of INMOS T805 transputer microprocessors ensure optimum speed of data acquisition and decoding of fibre images – over 20 000 fibre images can be analysed within half an hour (>10 images/second).

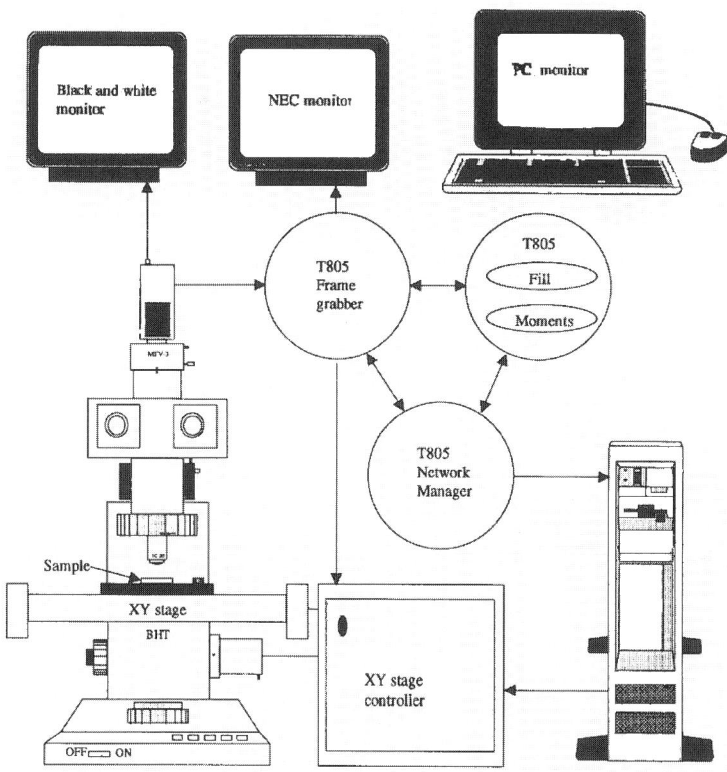

3.3 Overview of the Leeds, 2D large area image analyser for fibre orientation studies.

Unique features of the Leeds fibre orientation image analyser are:

- automatic fibre image splitting
- automatic merging of fibre images between neighbouring *XY* image frames
- automatic calibration in *X* and *Y* (by following images between consecutive *XY* frames)
- automatic focusing in *Z*
- a robust 'first and second moments' algorithm to derive elliptical parameters in a single pass through the image data.

A typical complex region of an automotive part covering an area of 2 mm × 2 mm is shown in Fig. 3.4a. Over 200 overlapping *XY* frames of data have been used to reconstruct the best-fit elliptical images. The fitted elliptical parameters yield the out-of-plane angle, θ, and in-plane angle, Φ, for each fibre and from these data the frequency distributions in both of these angles are created, as shown in Fig. 3.4b and c. A detailed discussion of the measurement errors associated with the 2D image analyser and its capabilities can be found in Clarke *et al.* [24].

3.1.3 Characterisation parameters for fibre orientation states

Clearly, statistically significant fibre orientations within a composite sample can be represented by frequency distributions in θ and Φ, as shown in Fig. 3.4b and c, provided that the 2D image analyser has the ability to scan over large areas. However, it is easier to handle the data, and to use it for modelling, if only a few parameters can be devised to represent all of this information. For example, if it were necessary to highlight orientational misalignments from the flow direction over a large area, the weighted mean of the cosine function of θ_x, $\langle \cos^2 \theta_x \rangle$, in each subregion could be computed and displayed graphically using a 'grey scale' to represent those mean values, as shown in Fig. 3.5.

Each of the chapter authors in this book will no doubt have their own favourite parameter(s) to characterise the fibre orientation states. The tensor representation first described by Advani and Tucker [25] is one popular compact method of representing the orientation state of an ensemble of fibres within a processed part. However it has only been possible to deduce fibre orientations from a single 2D section (until recently) and the ambiguous fibre orientations do not allow complete tensor formulations. Also, the tensor coefficients do not give any information on fibre lengths or correlations between lengths and orientations. The general form of the second order orientation tensor, a_{ij}, is shown in equation [4], where $\Psi(\mathbf{p})$ is the fibre orientation distribution function and $\mathbf{p}$ is a vector parallel to the fibre

$$a_{ij} = \int_p p_i p_j \Psi(\mathbf{p}) \mathrm{d}\mathbf{p} \tag{4}$$

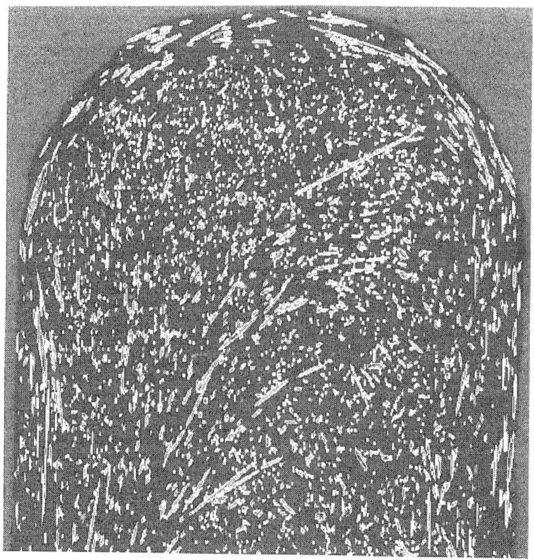

(a)

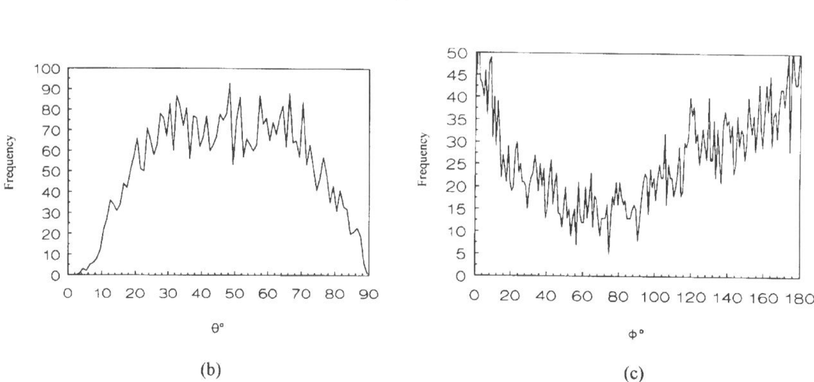

(b) (c)

3.4 (a) Example of a reconstructed large area complex region of a glass fibre-reinforced epoxy (area 2 mm × 2 mm) with all elliptical fibre images identified. (b) A typical frequency distribution of out-of-plane angles, θ. Note the cut-off at 15° due to pixellation noise and small solid angle for detection of fibres at 0°. (c) A typical frequency distribution of in-plane angles, Φ. Note that the range is 0° to 180° because of the ambiguity in Φ.

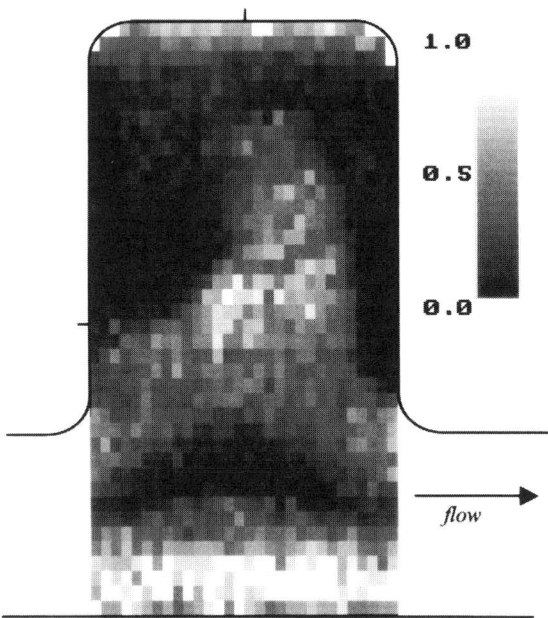

3.5 Mosaic of subregions is shown with the mean value of $\cos^2 \theta_x$ plotted as an intensity. The lighter the region, the greater the proportion of fibres oriented parallel to the flow direction shown, i.e. $\langle \cos^2 \theta_x \rangle$ tends to unity. The total area is 4 mm × 7 mm and has been reconstructed from over 1300 *XY* frames.

Geometrical coefficients have been preferred in the Polymer Group at Leeds because of predictions (see Ward [26]) relating fibre orientation states via these coefficients to stiffness constants and 3D elastic moduli. For example,

$$\langle P_4(\cos \theta) \rangle = (35\langle \cos^4 \theta \rangle - 30\langle \cos^2 \theta \rangle + 3)/8 \tag{5}$$

is one such orientation function where the orientation averages are denoted by $\langle \ \rangle$.

Taya *et al.* [27] have explored the use of fractal morphological quantities to represent, not only the orientation of fibres in the part, but also the descriptors of fibre or particulate filler distributions within a 2D section. A number of groups have shown interest in relating the clustering of fillers to modified sample properties, including the effects of interparticle distances, distribution of Voronoi cells around each filler and distribution of nearest neighbours (e.g. Pyrz [28] and Clarke *et al.* [29]).

3.1.4 Measurement bias and experimental error with the 2D optical technique

The importance of the 2D optical imaging technique cannot be overstated because there are many other important parameters (besides fibre orientation state) that can be obtained from the 2D images, especially when mesoscale regions are scanned, that is areas of many millimetres times many millimetres. For example, the derivation of void volume fractions from area fractions is described by Shi and Winslow [30] and the determination of fibre length distributions after pyrolysis (i.e. after the burn-off of the matrix) by Davidson *et al.* [31].

The 2D optical technique is the preferred technique for such mesostructural measurements, but apart from tomographical techniques, 3D structure is inferred from one or more 2D sections. There is therefore the problem of measurement bias and experimental error associated with 2D measurements and renewed interest in stereology (i.e. those mathematical techniques which can turn 2D section measurements into unbiased estimates of 3D quantities).

As discussed in a previous section, the 2D section plane may be scanned and the fibre data analysed to give the number of fibres within a certain angular range, θ to $\theta + \Delta\theta$ and within Φ to $\Phi + \Delta\Phi$. However, this is a 'height weighted' set of fibre orientations because those fibres which are oriented perpendicular to the section plane have a higher probability of being intersected by the plane than those fibres which are oriented in the plane of the section, see Fig. 3.6a. Also, as shown in Fig. 3.6b, the available solid angle for detection of fibres is smaller for near perpendicular fibres than for fibres which lie close to the section plane. These issues have been discussed by a number of authors recently, see for example Moginger and Eyerer [32] and Bay and Tucker [33] who propose the application of model-based correction factors to the raw 2D fibre orientation distributions in order to derive more representative 3D fibre orientation distributions. (An alternative, design-based approach is discussed in Section 3.4.3.)

Many researchers must exercise caution when using commercial image analysers to measure fibre orientations and interpret fibre orientation distributions. The most natural approach is to section the composite in a plane perpendicular to the most probable fibre orientation, but this introduces a significant measurement error. Consider the fibre images in Fig. 3.7a. If a spurious noise pixel is added to the circumference of the circular fibre image (even at high magnifications), the image will be indistinguishable from the elliptical image of a fibre oriented at $\theta = 10^\circ$ to 15°. Therefore, the orientation distribution of an ensemble of fibres will have a 'spurious' peak in the distribution at approximately 15°. In Fig. 3.7b, the error in angle θ (as θ tends to zero) is plotted as a function of different fibre image sizes in pixels. Also from equation [1] note that the error, $\delta\theta$, decreases as a function of increasing angle, θ (because an extra noise pixel at the edge of an elliptical image has less effect). Hence, one should section at 45° or 60° to a recognisable, preferred fibre orientation axis in order to minimise this potential source of error. If the fibre orientations relative to this section plane are given by $\{\theta', \Phi'\}$, a mathematical transformation can be performed back to an orthogonal coordinate system which

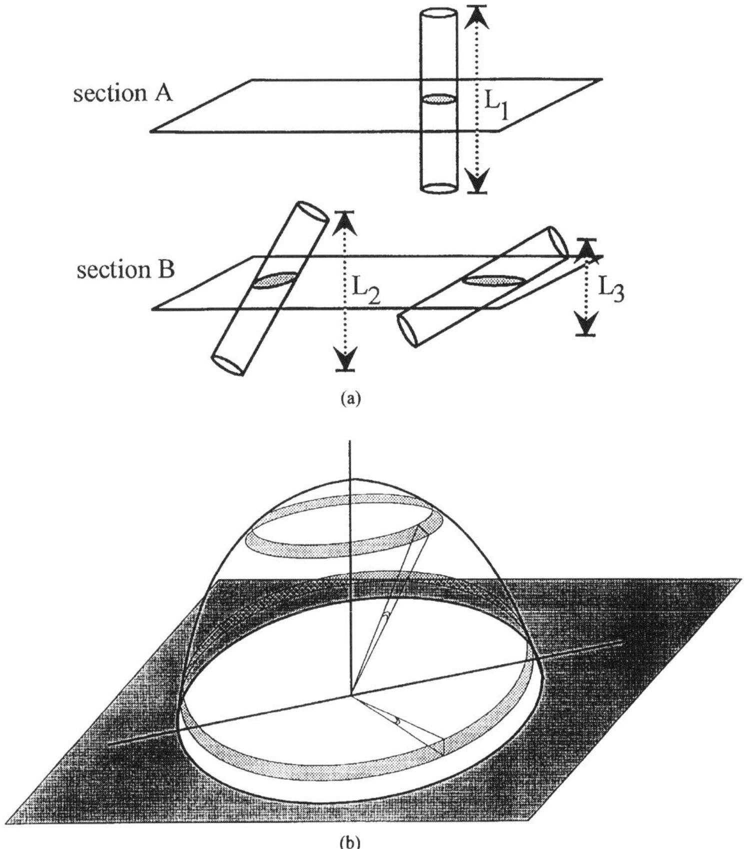

3.6 (a) Height bias in single 2D sectioning, i.e. the probability of detection will depend upon the effective length of the fibre perpendicular to the section plane. (b) Solid angle bias effect, i.e. the number of fibres detectable will depend upon the solid angle within a range of angles, $\Delta\theta$. Near $\theta = 0^\circ$, the solid angle is small whilst near $\theta = 90^\circ$ the solid angle is large.

has the fibres oriented close to the Z axis and with orientation angles $\{\theta, \Phi\}$, using the following equations:

$$\theta = \cos^{-1}(\sin\alpha \sin\theta' \sin\Phi' + \cos\alpha \cos\theta') \tag{6}$$

$$\Phi = \tan^{-1}\left(\frac{\cos\alpha \sin\theta' \sin\Phi' - \sin\alpha \cos\theta'}{\sin\theta' \cdot \cos\Phi'}\right) \tag{7}$$

where $\alpha =$ section angle (e.g. 45° or 60°).

Yurgartis [16] has attempted to use this sectioning idea to achieve the smallest orientation errors for investigations of fibre misalignments in unidirectional fibre-

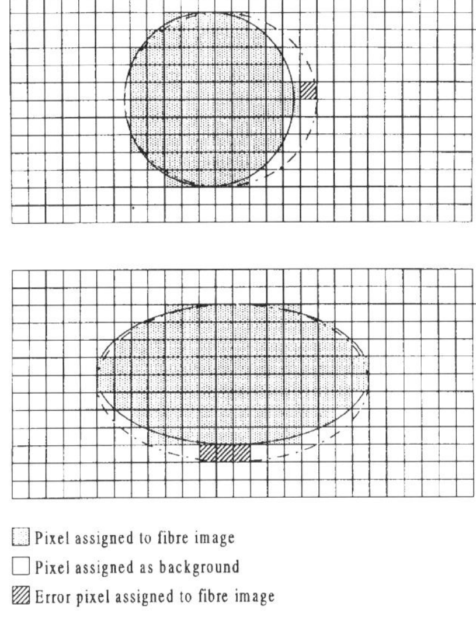

(a)

(b)

3.7 (a) Effect of a single noise pixel at the circumference of a fibre image is to make it look more elliptical. (b) The error in the out-of-plane angle, $\delta\theta$ (as θ tends to zero degrees) as a function of fibre image radii in pixels.

reinforced composites, as shown in Fig. 3.8. Subdegree orientation errors have been claimed when the sample is sectioned at an angle of 5° to the mean fibre orientation, but there are a number of assumptions which are not strictly valid (see Clarke *et al.* [34]). The Yurgartis assumptions of fibre cross-sectional circularity, straight fibre segments (over 150 μm length of fibre) and lack of correlation

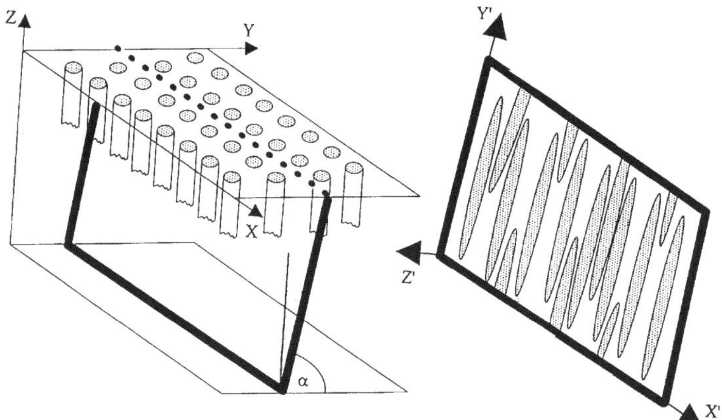

3.8 Yurgartis technique for estimating small angular displacements of well aligned fibres from highly elliptical images on section planes at a small angle (90 – α)° to the mean fibre direction.

between orthogonal misalignment planes are all suspect and these issues are discussed in more detail in Section 3.4.4.

There have recently been many biological science research papers (see for example reviews by Gundersen *et al.* [35], Cruz-Orive and Weibel [36] and Howard [37]) on stereological tools which help biologists or pathologists who are concerned with manual counting of objects under a microscope to obtain more reliable 'unbiased' estimates of 3D structure from several 2D sections with less effort! The stereological methods discussed in these references include 'unbiased counting frames' and unbiased estimates of anisotropic surface areas, numbers and individual mean volumes of cells and nuclei without any shape assumption. Figure 3.9a illustrates the idea behind the unbiased counting frame. If one wants to count objects in a 2D image plane, an area, A, within the image plane should be defined and a counting rule used which states that objects are only counted if part of the object appears within area A but does not cross two sides of the region A. (This procedure prevents the double counting of objects in space, see Gundersen *et al.* [35]). In Fig. 3.9b the counting rule for the 'disector' is shown, which is the union of a reference plane of area A and a perfectly registered look-up plane separated by a distance h from the reference plane. Fibres denoted by ' + ' would be counted, but those denoted by ' – ' would not be counted because either they fail the unbiased counting frame rule on the reference plane or they are intersected by the look-up frame.

If the total number of valid fibres is represented by Q, an unbiased estimate of the number of fibres per unit volume, N_v/V may be inferred from the expression

$$\frac{N_v}{V} = \frac{\sum_i Q}{\sum_i A \cdot h} \tag{8}$$

whereas the fibres seen in a single 2D section are a biased non-representative sample of all fibres because they are height weighted. Note that these arguments also apply to any 'objects' within the sample, for example voids or particulate fillers.

The determination of the distribution of mean volumes of objects within the image frames could be best accomplished, as shown in Fig. 3.10a and b, by a combination of both the traditional Cavalieri method and the disector technique (see Howard [38]). There are two ways to achieve this goal; either a systematic

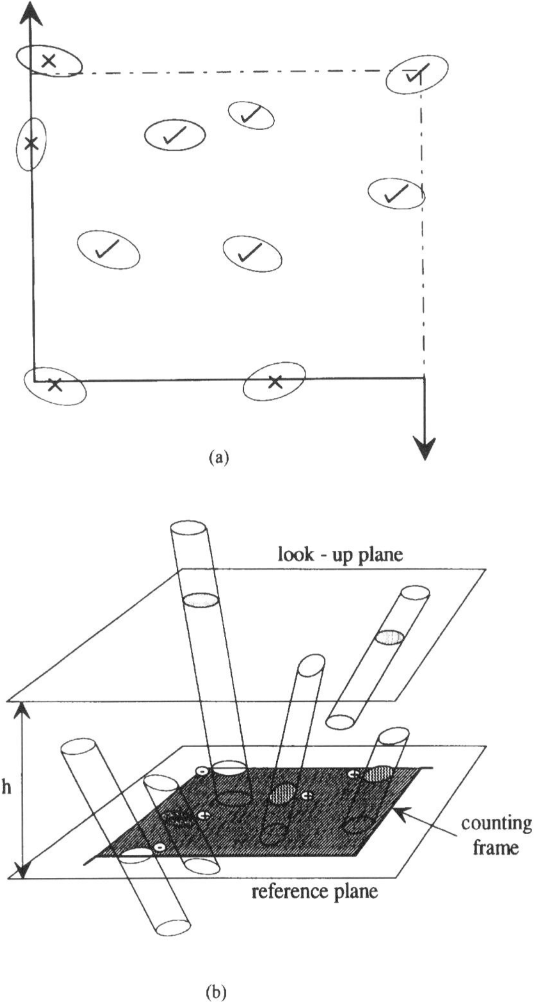

3.9 (a) The unbiased counting frame concept – only objects which exist within the region of interest and which do not intersect the solid lines around that region should be counted. (b) The 'disector' concept for unbiased volume estimation.

sectioning, polishing and pattern matching using a 2D system or optical sectioning in 3D using a confocal laser scanning system.

The challenge is to automate the image data collection and data reduction for both the 2D large area analyser and also the 3D confocal laser scanning

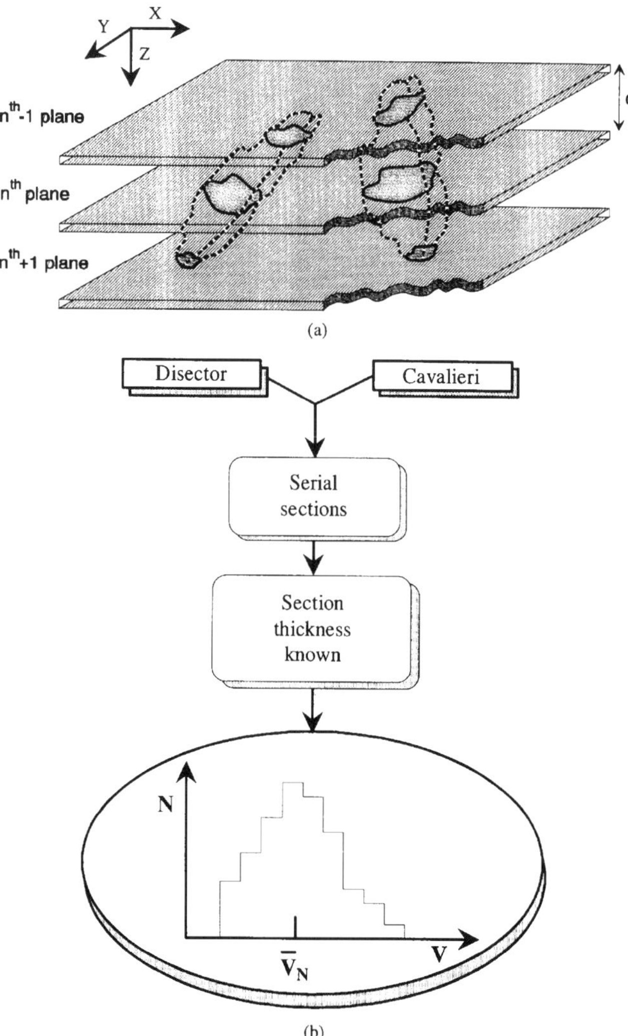

3.10 (a) Objects in the sample can be reconstructed by taking optical sections each separated by a distance *d* and recording both the centre coordinates and the cross-sectional area A_n on each plane – this technique mimics Cavalieri's method of volume determination. (b) By combining the disector concept and the Cavalieri approach, an unbiased estimate of the mean volume of objects in 3D may be deduced.

microscope so that results can be obtained within reasonable experimental timescales.

3.1.5 2D optical microscopy using many sections and pattern matching

The question is whether or not some of these inherent measurement errors with a single 2D section could be removed by developing more complex procedures. It is a small logical step from analysing a single polished 2D section to attempt to take a number of images on a set of polished sections and by pattern matching between sections, to reconstruct fibres or voids in 3D. However, the practicalities are awesome, which probably explains why literature searches have only uncovered a few published papers on this topic and only one paper within composites research!

Paluch [39] describes a pattern matching technique for deriving the waviness of unidirectional carbon fibres. In this technique, a set of 40 sections was systematically microtomed and analysed using an optical reflection microscope. As shown in Fig. 3.11a, the section planes were cut perpendicular to the mean fibre orientations. Between the scanning of each section 20 μm of material was removed. Using high magnification, the same 50 fibres were imaged on a single *XY* frame on each section. A pattern matching technique, involving the fibre centre coordinates, was used to find the most probable registration of each new section with respect to the previous section. In this way, each fibre image was followed as the material was removed. However, the technique is repetitive and time consuming.

The difficulties with Paluch's technique have been discussed by Clarke *et al.* [34]. Basically, because the sample must be removed from the *XY* scanning stage for microtoming and polishing and then returned to the *XY* stage for rescanning. In order to reconstruct in 3D with any confidence, the following problems must be resolved:

- accurate registration (in dx, dy and rotation) of one polished section relative to the next section,
- accurate, absolute fibre centre coordinates over a large area of sample, and
- accurate knowledge of the separation, dz, between neighbouring sections.

Paluch's technique sought to find the best registration of two neighbouring planes by minimising the errors in the position of fibre centres between sections. The problem is that this technique fails when neighbouring fibres have correlated movement, as discussed in Section 4.4 and shown in Fig. 3.11b. Many research papers involved in pattern matching between different image planes refer to special 'control points' to aid the matching process (e.g. Merickel [40]). If certain

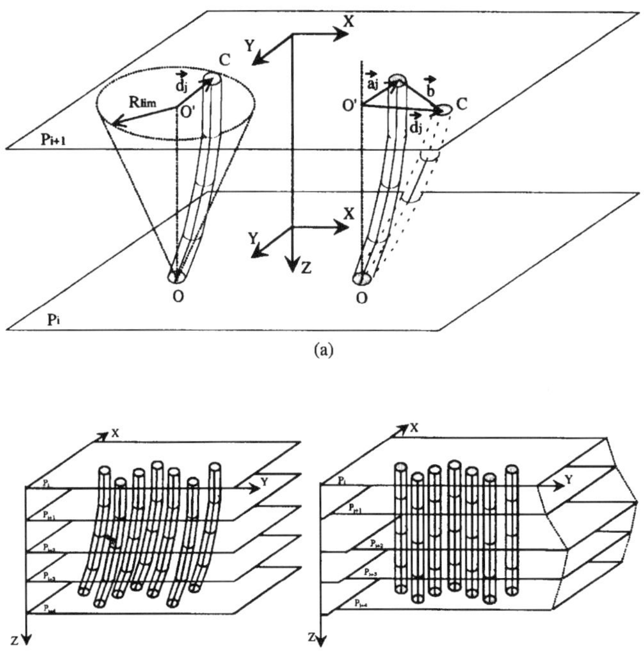

3.11 (a) Paluch's technique assumes that two sectioned planes can be registered by determining the centre coordinates of each fibre within the sections, search for matching fibre centres within a limit R_{lim} of where the centre should be and assume that a fixed misregistration vector **a** acts on all fibre images. If the true movement of a fibre between sections is given by the vector **d** Paluch assumes that a randomly oriented vector **b** applies to the fibre data set as a whole. (b) illustrates the problem, that is that localised fibres are often moving in synchronism with each other and hence Paluch's technique will tend to take out the real waviness because of the non-random vector **b**.

fibres were, in effect, 'control fibres' with orientations quite distinct from other fibres, the ellipticity of the image and hence angles $\{\theta, \Phi\}$ could be correlated to a prediction of where the fibre image should be on the next section plane. However, by definition, if 'unidirectional' fibres are being investigated there are no such high angle 'control fibres' seen in the image data set.

We are aware that a number of composite research groups have looked into these problems but no one has published anything except internal reports and all groups seem to have abandoned the technique! However, work is underway at Leeds, using the fully automated 2D image analyser to attempt to resolve these difficulties. Initial findings are reported by Clarke *et al.* [24] and Davidson *et al.* [41].

3.1.6 3D optical microscopy

It would be better if the 3D mesostructure of composite materials could be obtained from one single polished section, rather than many microtomed and polished sections. At the moment there would appear to be only two active groups who are tackling the issue of 3D fibre orientations using novel optical techniques.

The first of these techniques, Wille and McGrath's optical scanning method (OSM) uses 0.1 w/o opaque carbon 'tracer' fibres within a 30 w/o glass fibres in thermoplastic composites whose matrix refractive index is a close match to the refractive index of the reinforcing glass fibres. The OSM technique can focus on the tracer fibres which are at different depths within this specially prepared composite sample. By controlling the *Z* movement of the objective lens, the focal plane moves within the sample. Unlike the Paluch technique, discussed in Section 3.1.5, these focal planes are guaranteed parallel and perfectly registered, enabling unambiguous correlation of fibre locations in 3D.

Spatial resolutions of a few μm in δx and δy and approximately 10 μm in δz have been achieved. The matching refractive index requirement limits the range of matrix materials that can be investigated, but the technique can obtain 3D fibre orientation distributions from significant depths (1.3 mm has been quoted) and at high processing rates (between 1 and 10 $\text{mm}^3\,\text{min}^{-1}$). An overview of their technique is shown in Fig. 3.12. A transparent window in the side of a mould has enabled them to explore the movement of fibres during processing and the measurement errors associated with their technique are discussed by McGrath and Wille [42].

An alternative technique has been explored by Clarke and co-workers [24,34,43] who have shown that the confocal laser scanning microscope (CLSM) technique can reduce experimental errors and bias, especially for glass fibre-reinforced polymer composites at high packing fractions. Stereologically sound measurements are now possible with the CLSM because of the accurately registered optical sectioning ability, as described in the next section. Subdegree fibre orientations may be obtained and also true 3D fibre waviness in significant sample volumes (i.e. 200 μm × few millimetres × 50 μm in *X*, *Y* and *Z*, respectively) within reasonable timescales (i.e. two working days). Although a wide range of glass-reinforced polymer composites can be investigated with this technique, the maximum effective depth that can be achieved ultimately depends upon the 'working distance' of the high numerical aperture objective lens. In the case of an air objective, the maximum effective depth will be a few millimetres, but for oil immersion objectives, the maximum effective depth reduces to around 200 μm. The CLSM technique is described in more detail in the following section.

3.2 Confocal laser scanning microscopy

One of the best reference texts available on all aspects of confocal laser scanning microscopy is edited by Pawley [44]. This textbook contains contributions from

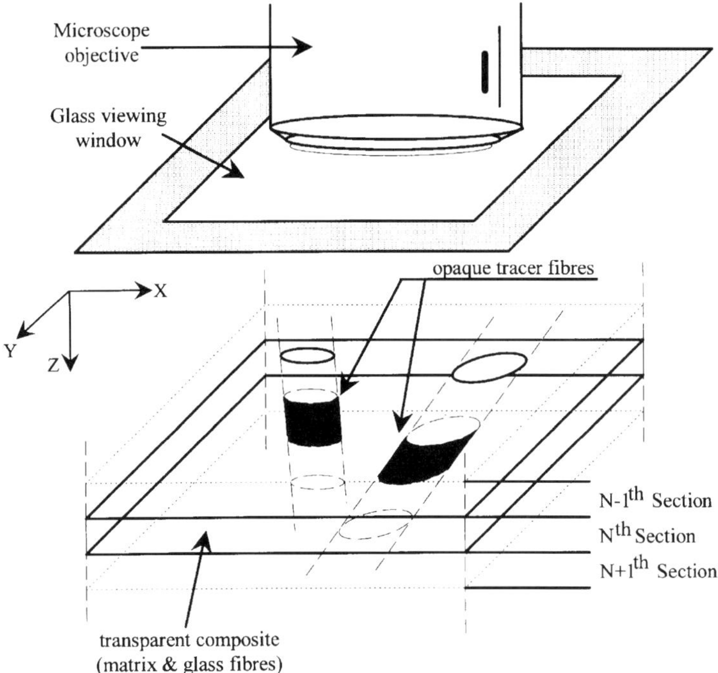

3.12 McGrath and Wille determine the 3D location of a small percentage of tracer fibres in glass reinforced polymer composites with their optical sectioning method. The thickness of the optical sections is tens of micrometres and therefore their volume resolution is inferior to the resolution achievable with the confocal laser scanning microscopy technique.

international experts on confocal designs, 3D computer visualisation and applications of the CLSM to biological and physiological research.

3.2.1 Basic principles

A highly schematic view of the basic principle behind the confocal attachment to a standard microscope is shown in Fig. 3.13. Laser light, from either an argon-ion laser (or more expensive types of laser) at a wavelength λ is directed through a beamsplitter. In a practical system, a selectable neutral density filter would be inserted in the optical train before the beamsplitter in order to control the laser power onto the sample. After passing through a scanning mirror arrangement, the laser light enters a conventional microscope body and passes through a high numerical aperture objective lens before reaching the sample. The mirrors scan the laser beam in both X and Y directions so that an optical XY plane is defined.

The reflected laser light and any fluorescence emission from the sample passes back through the microscope body, meets the beamsplitter and is directed towards a sensitive photodetector via an emission filter and a 'confocal' aperture. The emission filter selects either the reflected light or the longer wavelength fluorescence light and therefore determines the mode of operation. Until recently, a photomultiplier tube (PMT) has usually been the first choice for a highly sensitive detector but charge-coupled-devices (CCDs) are also making an impact in the latest designs. As the ray tracing in Fig. 3.13 implies, the effect of the confocal aperture is to remove unwanted stray light from above and below the focal plane of interest. By reducing the aperture size, a thinner region about the focal plane is imaged, and the intensity of the signal at the detector is reduced. The detected signal is enhanced electronically and the best optical sectioning is also achieved by using a high numerical aperture (NA) lens for the microscope objective.

Hence, by moving the *Z*-drive on the microscope, the focal plane may be placed at the sample surface or within the sample. Unlike conventional

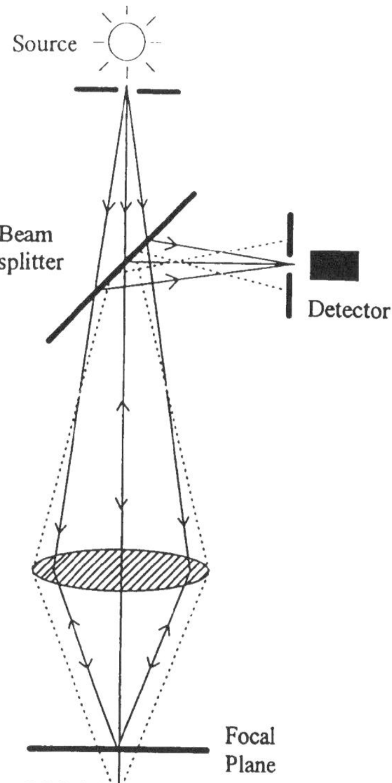

3.13 Highly schematic view of the confocal microscopy technique. The solid ray shows the path for light reflected from the chosen focal plane. The beamsplitter directs the light through the small confocal aperture and onto the detector. However, very little light reflected from below the focal plane (shown dotted) reaches the detector and also very little light from above the focal plane. Hence the axial resolution of the confocal technique is better than conventional microscopy.

microscope systems, the best surface focus does not have to be found because every *Z* position will focus a different plane at the detector. By systematically scanning in *Z*, a set of *XY* planes can be stored and hence the 3D structure of the sample can be reconstructed, provided that, at the laser wavelength, the sample is sufficiently transparent.

There are many commercial CLSM systems on the market but the research work to be described in the next two sections has been undertaken at Leeds with a Biorad Microscience instrument (an upgraded MRC500) which is an early model in their MRC series. The typical folded optics arrangement is shown schematically in Fig. 3.14. Many laser launch subsystems are available for the CLSM machines. Longer wavelength laser lines are attractive for materials science because they should penetrate deeper into a sample than shorter wavelengths (i.e. Rayleigh scattering is proportional to $1/\lambda^4$). However, near infrared laser systems are more costly and produce lower power than the more

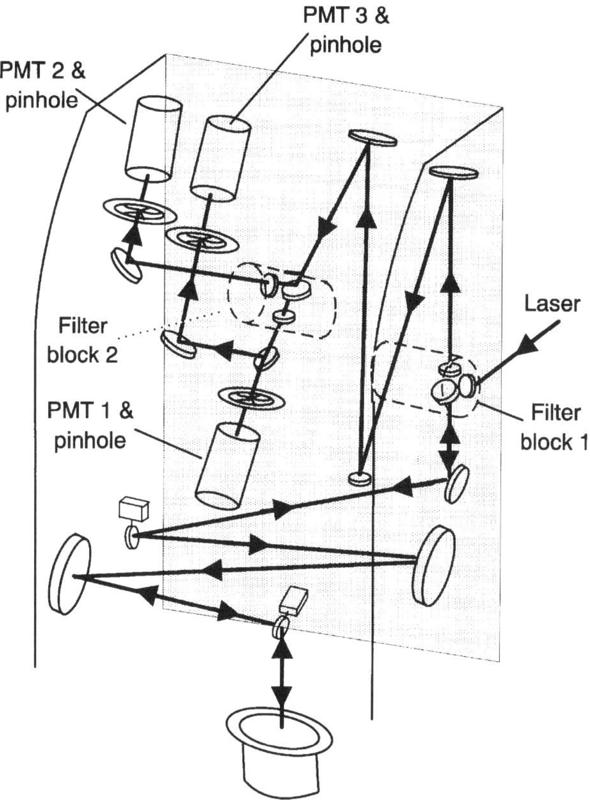

3.14 Schematic of the folded optics within a Biorad MRC1000 confocal laser scanning attachment to a conventional microscope. PMT = photomultiplier tube.

common argon-ion lasers. For materials science, higher power lasers also achieve greater depths and the resulting photobleaching effect can be most useful as an indication of the position of a previous scan line (see Section 3.2.4). Regardless of the potential depth of penetration, as measured by the number of photons detected from a certain depth, the major issue for research groups is 'what image processing is required/available to enhance the raw images and improve the automatic identification of 3D features down to a depth, Δz_{max}, where Δz_{max} is the 'maximum useable depth' for composite studies?'

Ultimately, the depth of penetration will be limited by the 'working distance' of the objective lens. As can be seen in Table 3.1 below, the oil immersion lenses have shorter working distances around 200 μm whereas the air lenses have working distances up to a few millimetres. However, this penetration will not be achieved in practice because of both refractive index mismatching, see Section 3.3.2, and the light scattering/attenuation due to number density and clustering of the reinforcing fibres in real composites.

3.2.2 Biorad MRC series confocal laser scanning microscopes

With any commercial CLSM system, there are a number of control functions which must be mastered in order to optimise the image quality. For example, the manual controls would include:

- confocal aperture size (influences the thickness of optical sections)
- electronic gain and dark level (influences the contrast and brightness achievable)
- neutral density filter selection (controls the laser power incident upon the sample)
- user selectable filter blocks (so that one can view in one operational mode and easily switch to another mode before scanning the same sample region).

Whereas earlier confocal systems had manual controls, the latest confocal systems generally have the capability for computer control of all of these system functions.

Table 3.1 Working distances of typical objectives used in CLSM

Manufacturer	Type of objective	Numerical aperture	Magnification	Working distance
Nikon	planapo, oil	1.4	× 60	220 μm
Zeiss	planapo, oil	1.4	× 100	100 μm
Zeiss	achroplan, water	0.9	× 63	1.45 mm
Zeiss	eriplan, air	0.75	× 100	3 mm

Usually, CLSM systems allow the user to display a set of captured image frames on the computer screen and to rotate the frames so that the reconstructed sample subvolume can be seen from all directions as in a slow motion movie. Some CLSM systems software allows the user to visualise objects within this volume at any 2D section plane orientation.

The light collection efficiency, ε, as a function of objective lens numerical aperture, NA, and the refractive index, n_i, of the immersion medium is given by

$$\varepsilon = 0.5\left(1 - \sqrt{1 - \left(\frac{NA}{n_i}\right)^2}\right) \quad (9)$$

Therefore, the higher the *NA* the greater the solid angle for collection of the reflected light and the higher the collection efficiency. The objective should be flat fielded and, as the composite samples have refractive indices close to 1.5, it is best to use an oil immersion objective lens (to minimise the refractive index mismatch as described in Section 3.3.2). Hence, most of the images shown in this chapter have been taken with a Nikon, oil immersion objective, ×60 magnification, *NA* 1.4 planapochromat. (When studying porous materials like foams, a more suitable lens with good optical sectioning capability is a Nikon planapochromatic, ×40 magnification, *NA* 0.95 air objective.)

3.2.3 Calibration issues

In most commercial CLSM systems, a point detector is used and therefore mirrors are scanned in *X* and *Y* in order to generate a 2D image at a particular depth within the sample. The *XY* image area scanned will depend upon the combined magnification of the microscope lenses. When the Biorad CLSM system is attached to a Vickers microscope and a ×60 oil immersion lens is used as the objective, the area of scan is 186 μm by 124 μm and the pixel size in both *X* and *Y* is 0.242 μm. The linearity of the scan in both *X* and *Y* may be assessed by imaging a 2D grating, as shown in Fig. 3.15.

A ×10 objective will give an *XY* image field of 1.1 mm by 0.76 mm with a pixel spatial resolution of 1.5 μm in both *X* and *Y*. Therefore, if in a particular application the user wants to cover a sample area of many mm^2 (or cm^2) and achieve higher spatial resolution, customised software can be written to scan systematically over a large sample area by controlling a motorised *XY* stage and using, say, a ×60 oil (or air) objective lens.

The issue of the *Z* calibration of a CLSM system is more problematical and a number of researchers have explored the manufacture of a suitable standard with which to assess the *Z* calibration (see for example Pawley *et al.* [45]). Backlash and other effects, as discussed in Section 3.3.1, mean it cannot be assumed that, by sending a fixed number of pulses to a stepper motor driven *Z* movement, the same linear movement is obtained every time! (However, expensive piezo-

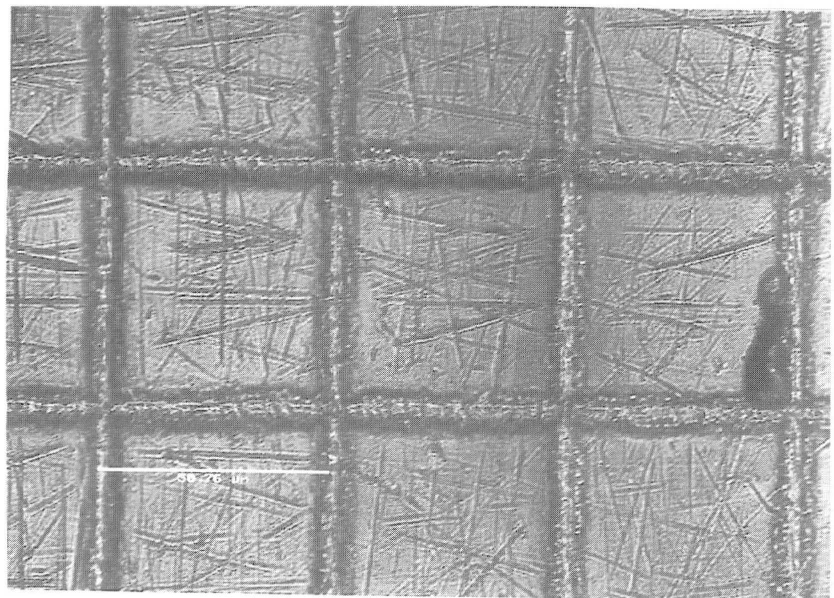

3.15 *XY* scan of a square grating used for calibration of the field of view. This image shows the high spatial resolution achievable. A Biorad MRC500 with ×60, *NA* 1.4 oil objective lens was used to obtain the image. The scale bar represents 50.76 μm (overall image area is 186 μm × 124 μm).

electric transducers are now being used to give high precision movement and better linearity of movement.) Usually, for materials science applications, the issue of the *Z* calibration is of secondary importance when compared with the potential problem of the 'apparent depth' of artefacts below the sample surface. This effect is due to refractive index mismatching and is discussed in more detail in Section 3.3.2.

3.2.4 Optical sectioning capability of the CLSM

The optical sectioning capability is determined primarily by the size of the confocal aperture. The relationship between the aperture size and the off-axis rejection of light is shown in Fig. 3.16a. A useful measure of the optical sectioning capability is the width of the plot at which the intensity falls to a half of its maximum value, $\Delta z_{1/2}$ (used to characterise the thickness of the optical section).

The ultimate limit to the optical sectioning ability of the CLSM, characterised by the parameter $\Delta z_{1/2}$, depends upon the objective lens *NA*, the laser wavelength

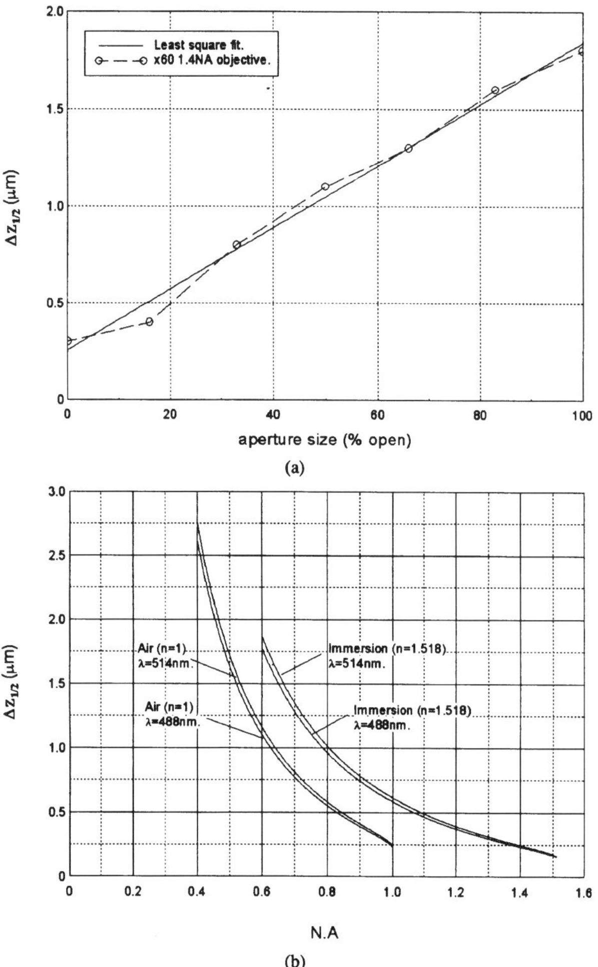

3.16 (a) Axial resolution is characterised by $\Delta z_{1/2}$ which is plotted for two objectives as a function of confocal aperture size. (b) The axial resolution is plotted in terms of the numerical aperture of the two types of objective lens.

λ and the immersion medium refractive index n_i as described by Archenhold [46]

$$\Delta z_{1/2} = \frac{0.43\lambda}{\left\{n_i - n_i \cos\left[\sin^{-1}\left(\frac{NA}{n_i}\right)\right]\right\}} \tag{10}$$

The function described by equation [10] is shown in Fig. 3.16b for both air and oil immersion objectives. The excellent surface imaging properties of the CLSM systems are apparent from both the *XY* image of the calibration grating (shown in Fig. 3.15) and also a typical surface level image of a complex polymer composite microstructure containing fibres and voids which is shown in Fig. 3.17.

The user must select the required separation, Δz, between *XY* image fields before 3D data acquisition can be undertaken. When a ×60, *NA* 1.4 oil immersion objective is used on a CLSM system, an appropriate Δz to choose is 0.3 μm which gives almost cubic 'voxels', that is volume elements (because of the *X* and *Y* pixel resolution of 0.242 μm at this magnification).

CLSM systems also have an electronic zoom facility which allows the user to display on the computer screen only part of the normal image area magnified by a factor between ×1 and ×10. However, when the ×60 magnification objective lens is used, no more useful information can be obtained from these zoomed image areas, because of the Rayleigh optical resolution limitations due to the wavelength λ of the laser light. For conventional microscopes, the optical resolution is shown in equation (11a) and, for comparison, the confocal optical resolution (in the *XY* plane) is given in equation (11b)

$$\text{Resolution}_{\text{conv}} = 1.22\lambda/2(NA) \tag{11a}$$

$$\text{Resolution}_{\text{clsm}} = 0.8\lambda/2(NA) \tag{11b}$$

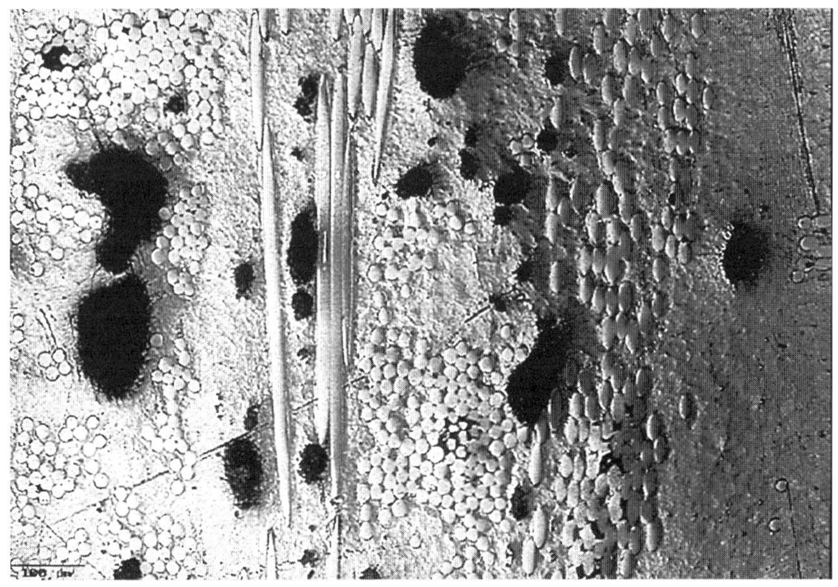

3.17 Air objective surface scan (×10, 0.3 *NA*) of a typical carbon fibre-reinforced composite showing dark voids, circular and elliptical fibre cross-sections and hence a complex fibre orientation state. This is a reflection mode image, area 1.1 mm × 0.74 mm.

Clearly, the refraction of light in sample volumes will cause aberrations that influence the imaging properties of the CLSM and, especially in heterogeneous two-phase structures like composites, it is difficult to devise analytical models to take these aberrations fully into account. Image degradation is quantified in terms of the 'point spread function', which is discussed in detail by Carlsson [47]. However, research is in progress to give higher optical resolutions through the use of special optical masks, see for example Pike [48], and also with special confocal designs (Hell and Stelzer [49]). Another source of image degradation is the scattering of light through spatially complex structures and also the attenuation of the laser light by both fibres and matrix.

A typical *XY* image and *XZ* image obtained in the fluorescence mode of operation is shown in Fig. 3.18a and b. Note that these images are noisier than the reflection mode image in Fig. 3.17. Also note the dark line at constant *Y* value in Fig. 3.18a which was caused by the photobleaching effect (discussed further in Section 3.3.4). This photochemical damage was produced by the laser scanning over the same *Y* line in order to create the *XZ* optical section shown in Fig. 3.18b.

3.2.5 Reflection/fluorescence modes of operation

The laser launch subsystem that has been used in the composite studies research described in the following sections is the popular argon-ion laser system, providing two main wavelengths at 488 nm and 514 nm giving laser powers of tens of milliwatts.

An *XY* image produced by the reflection mode of operation, shown in Fig. 3.19a gives a better spatial resolution at the surface of the sample than the fluorescence mode, as can be seen by comparing Fig. 3.19a with Fig. 3.19b. If 3D topological surface details are to be explored, the reflection mode is obviously preferred. Surface image detail may be acquired with low laser power and low system gain. However, if reflection mode signals are to be acquired from below the surface of the sample, the electronic gain of the system must be increased and at the highest gains, the Biorad CLSM MRC series of machines suffer from an instrumental artefact, shown in Fig. 3.26a, and discussed more fully in Section 3.3.1.

Despite the fact that the clearest images are reflection mode images, it has been found that the fluorescence mode is preferred for investigations into 3D fibre orientations. As the intensity of the fluorescence is lower than the reflected light, the signal-to-noise ratio is poorer. However, most CLSM systems have the facility for 'Kalman filtering,' that is by integrating over N consecutive *XY* frames of the same sample region, the signal-to-noise is improved by $N^{1/2}$, see Fig. 3.20 and 3.21, if the electronic noise is Gaussian in form. In Fig. 3.20, the same region has been scanned in *XY* but with different levels of integration and it is clear that the

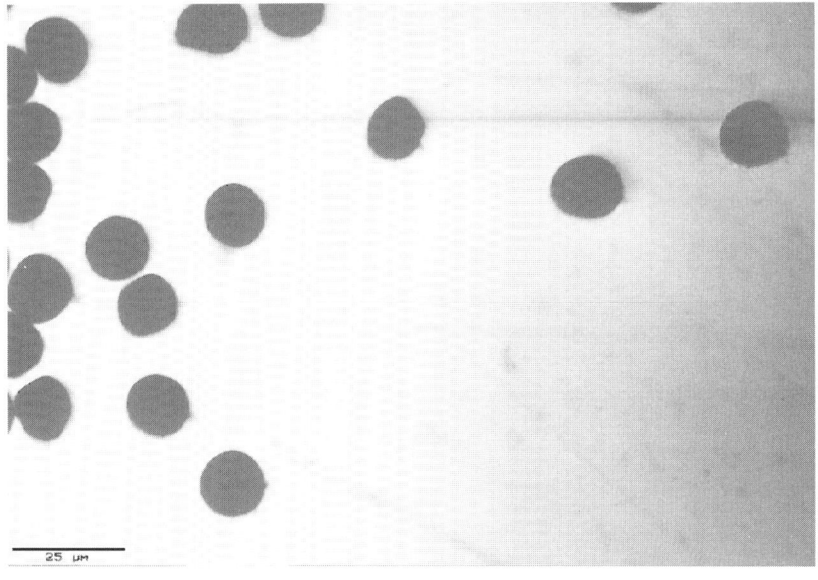

(a)

(b)

3.18 (a) An *XY* scan in fluorescence mode of glass fibres in an epoxy matrix. Note the dark horizontal line due to an *XZ* scan photobleaching that region and (b) the *XZ* scan (186 μm × 30 μm) that was taken. Although the fibre cross-section looks circular, the fibre is not perpendicular to the *XY* plane but is at an angle of 20° as computed from the *XZ* section.

sharpest definition *XY* image is obtained by choosing the largest value of *N*. (However, at some stage, any further improvement in image quality will be marginal and not warrant the extra time required to integrate over more frames.) Figure 3.21 shows how the frequency distribution of the pixel intensities changes. As *N* increases, the 'low intensity peak' in the frequency distribution which represents those pixels within the fibres and the 'high intensity peak' which represents those pixels within the matrix become increasingly well defined.

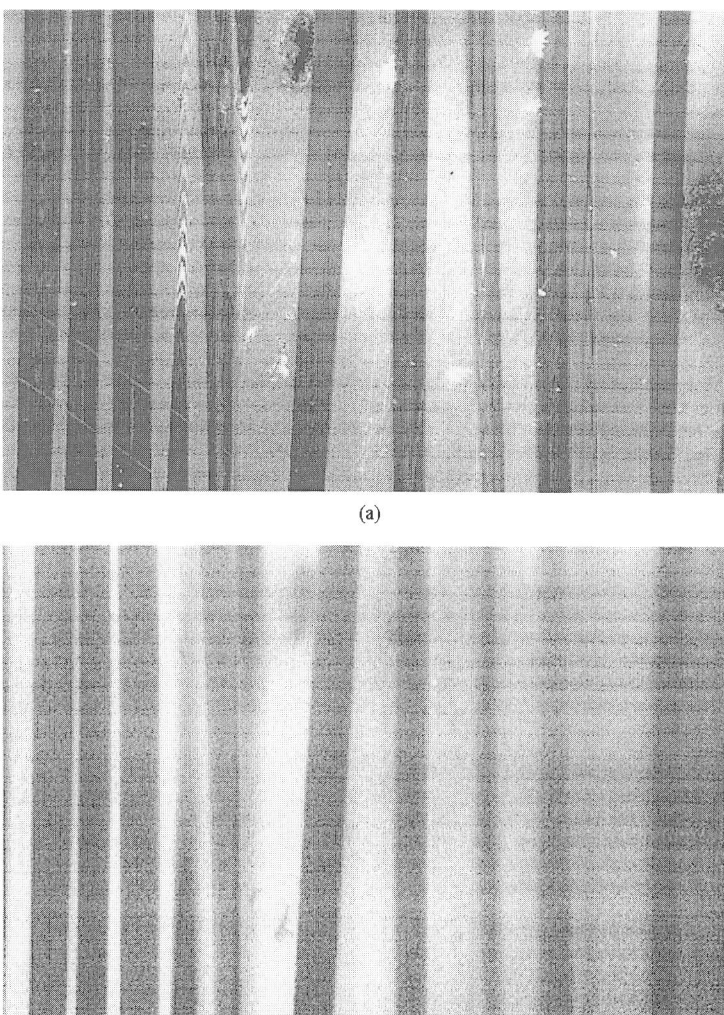

(a)

(b)

3.19 (a) High resolution *XY* image taken at the surface of a glass fibre-reinforced epoxy in reflection mode. Note the interference rings and the two voids. (b) The same region taken in fluorescence mode illustrating the poorer image quality. Both images were taken with a ×60, *NA* 1.4 Nikon oil objective giving an area of 186 μm × 124 μm.

Another useful facility on CLSM systems is the 'extended focus' capability. The CLSM can be instructed systematically to scan an *XY* area, move by Δz, scan in *XY*, add this new *XY* image to the previous *XY* image and move in *Z* again. In this way, without operator intervention, a sample region which may be tens or a

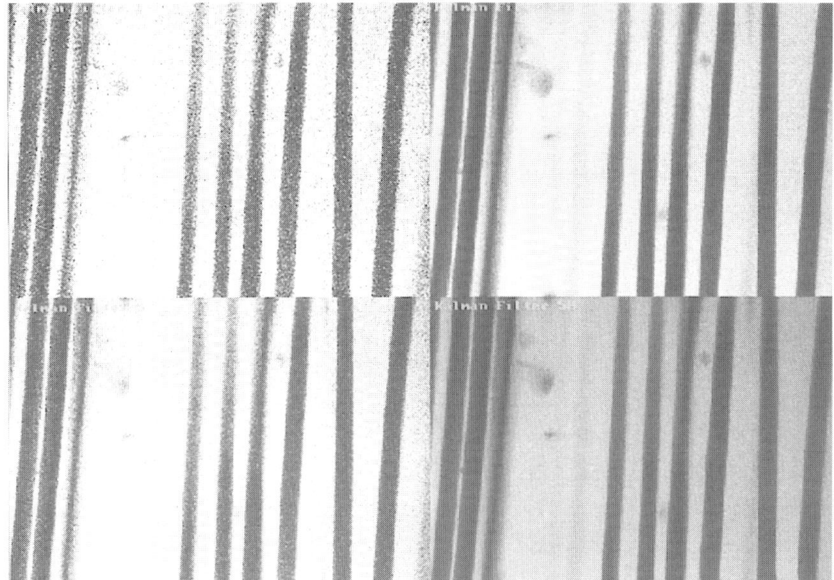

3.20 Four *XY* images of the same scene are shown in fluorescence mode. The technique of Kalman filtering is illustrated by the improvements in image quality when multiple images are combined. The fuzziest image is for a single image frame (top left), then 5 frames are shown (lower left), then 20 frames (top right) and finally 50 frames (lower right). The improvement is obtained at the expense of the time needed to accumulate the images.

hundred micrometres thick can be scanned and a single composite 2D image of the whole region reconstructed, as shown schematically in Fig. 3.22.

Some CLSM machines also offer a third mode of operation, the 'photon counting' mode, which can be used to detect very weak signals, as discussed by Pawley [50]. Although a very sensitive technique, images are built up over long time periods (compared with the time needed to create images using other modes of operation).

3.2.6 Methods for 3D reconstruction

The 'best' technique for the 3D analysis of composite materials depends to a large extent on the actual measurement to be made and on the type of composite. High resolution surface topological studies must be undertaken in reflection mode of operation but, in our experience, fibre orientations are best deduced using the fluorescence mode of operation. There are two fundamental methods to reconstruct a 3D sample volume: either from a set of *XY* frames taken at

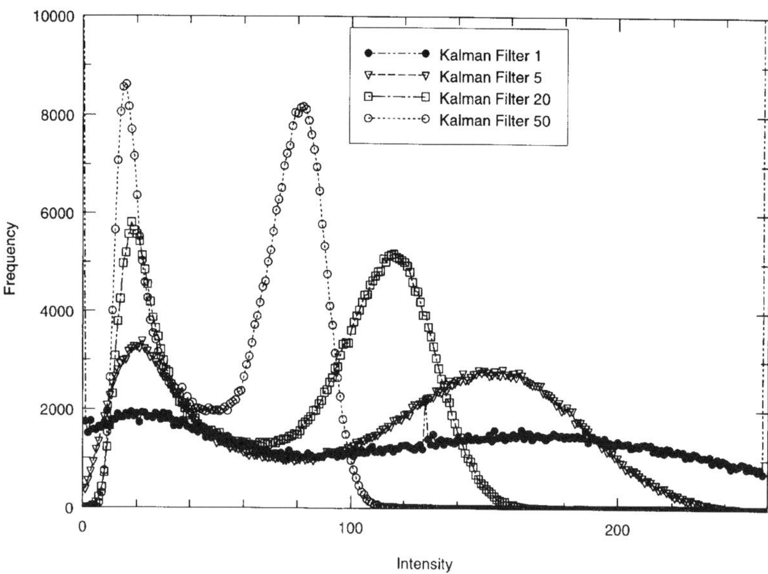

3.21 Frequency distribution of pixel intensities is plotted for each of the four *XY* image frames shown in Fig. 3.20. As the number of image frames being combined is increased, the two peaks (the lower one representing fibres and the higher value peak representing matrix elements) become sharper and there is more certainty which pixels are within fibre or matrix.

different *Z* or from a set of *XZ* frames taken at different *Y*. Confocal systems create *XY* image frames by using mirrors which scan the laser spot in both *X* and *Y* for each *Z* position, see Fig. 3.23a and b. There is another mode where the *Y* scanning mirror can be stopped at a preselected position along its scan – enabling the user to speed up the data acquisition and minimise memory requirements by scanning in *X* only for different depths, *Z*. This procedure generates an *XZ* image frame as shown in Fig. 3.24a and b.

Fibre orientations may be derived from either ellipticity of surface images (as in 2D microscopy) or movement of fibre centres between *XY* frames or *XZ* frames. Although it is believed that glass fibres exhibit cross-sectional circularity, other types of reinforcing fibres often have non-circular cross-sections, which implies that movement of centres is inherently a more sound method of deducing fibre orientations, if it can be achieved.

Therefore, by manually or automatically locating the fibre centre coordinates on these optical sections and pattern matching (which is made easy by the perfect optical registration), the loci of fibre centres may be reconstructed and hence fibre orientation states, *in situ* fibre lengths, fibre curvature and 3D fibre waviness may be investigated, as described more fully in Section 3.4. Recent CLSM data from

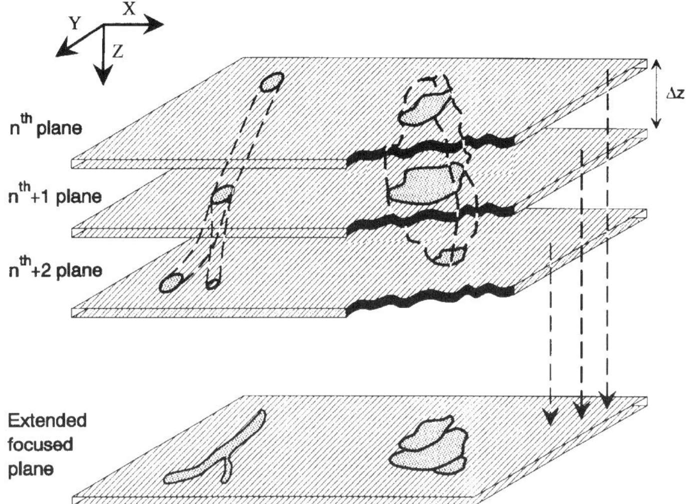

3.22 Schematic view of the idea of an 'extended focus' image. All confocal systems have the facility to merge a number of image planes. If a number of planes at depths separated by Δz are scanned and thresholded to pick out the objects of interest, the objects above threshold can be represented by a certain intensity value and plotted on a 'composite' image frame. Hence an intensity coded extended focus image may be produced which gives a 3D impression of the scene (see also Fig. 3.44).

a number of glass fibre-reinforced polymer composites are described in the next two sections.

3.3 CLSM measurements of polymer composites

The multitude of references to Pawley's book [44] indicates the extent to which the CLSM technique has made an impact in the biological sciences. Relatively little work has been published in the research field of materials science, especially in composites research. However, a number of Dutch groups have been active (see for example Thomason and Knoester [51] and Brakenhoff *et al.* [52]). For a review of the CLSM in materials science, see King and Delaney [53] and for applications of the CLSM, atomic force microscopy and scanning acoustic microscopy (SAM) in organic coatings research, see Lange *et al.* [54].

All commercial systems have basic image processing algorithms built into the user menu options in order to improve raw images by thresholding and image filtering, to determine point-to-point distances on *XY* or *XZ* image frames manually, and to determine the areas of regions of interest, simple counting statistics, etc. These facilities may be of some interest to the materials scientist

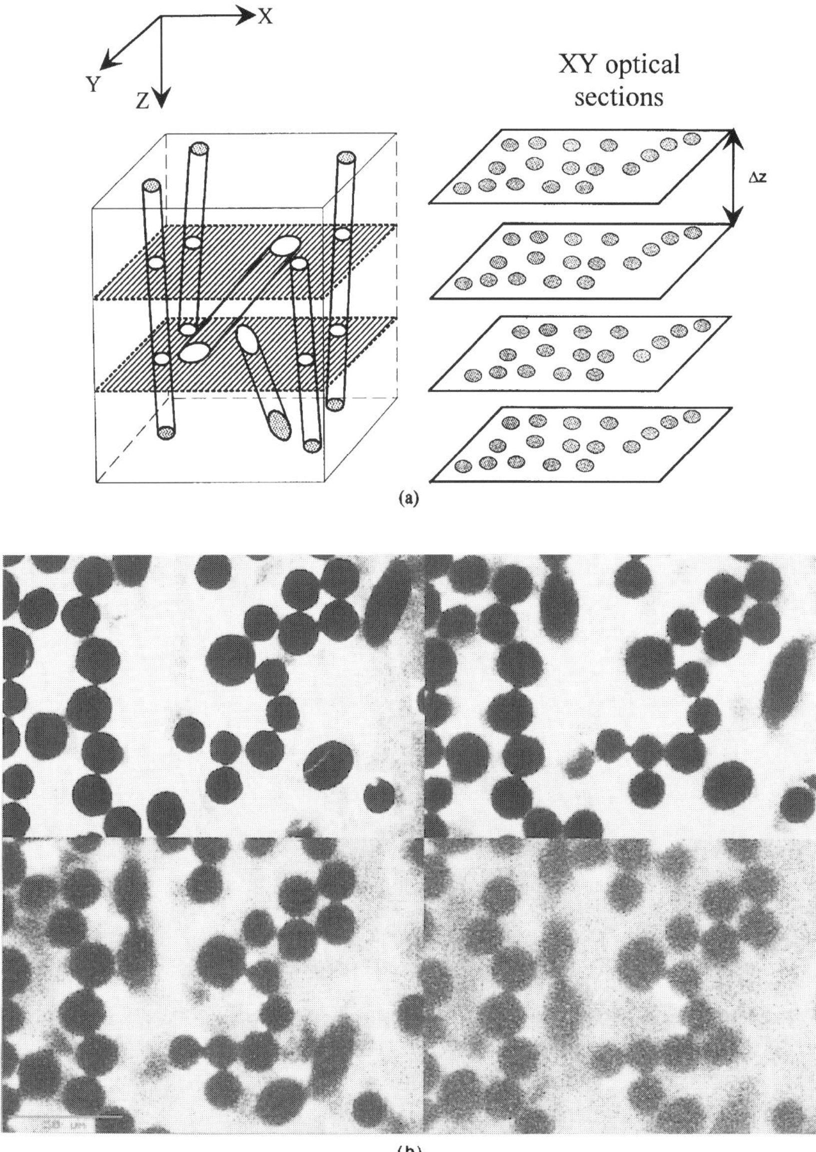

3.23 (a) Schematic of one technique for 3D reconstruction using a CLSM system. A series of *XY* optical sections are taken, each separated by Δz. By pattern matching object centre coordinates on one plane with those centres on the next plane, the orientation of fibres or the size distributions of objects may be deduced. (b) A typical set of fluorescence mode *XY* images for a short glass fibre-reinforced liquid crystalline polymer at the surface (top left), 15 μm (top right), and at depths of 30 μm (bottom left) and 45 μm (bottom right).

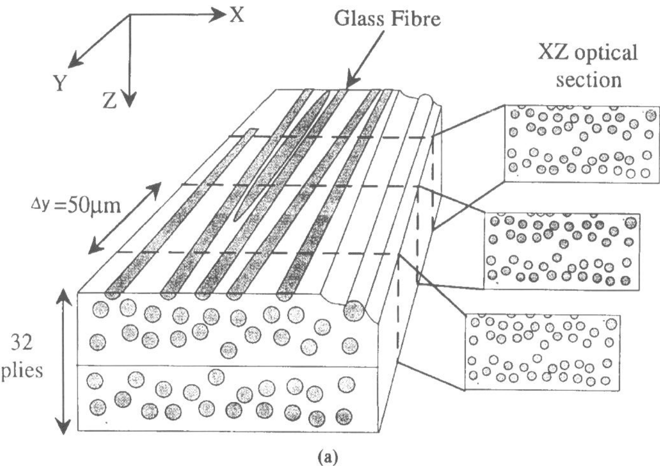

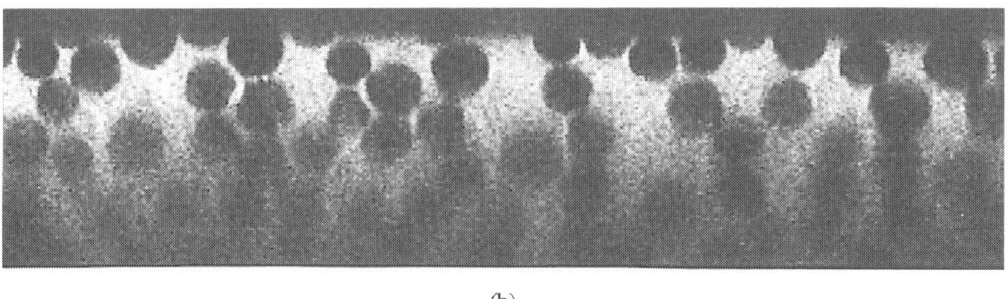

(b)

3.24 (a) Another technique for 3D reconstruction using the CLSM which involves taking a series of *XZ* sections, each separated by Δy. If, as shown in this schematic diagram, the sample is cut with the fibres lying almost parallel to the surface, the fibre cross-sections will appear as circular images in the *XZ* planes. By pattern matching between *XZ* sections, the 3D waviness of fibres may be determined. (b) A typical fluorescence *XZ* image of well-aligned glass fibres in an epoxy matrix. Note the worsening of the image quality with depth into the sample.

but invariably it is the automated scanning over large regions as quickly as possible that would be of greater interest. Commercial CLSM systems also allow the user to scan automatically over isolated subvolumes, once the user has selected the required separation, Δz μm, between consecutive *XY* planes, the number of *XY* planes, the zoom factor, the Kalman filter setting (i.e. the number of integrations at a particular *XY*), etc. However, for scanning isolated subvolumes over large regions (millimetres × millimetres), unless the sample is perfectly horizontal with respect to the vertical motion of the z-drive, a macro must be written to pick up the surface level at each new *XY* location (ensuring that image acquisition from the same depth is achieved over the whole *XY* area).

The situation is even more complicated if the user wants to acquire image data over contiguous *XYZ* subvolumes because of the vast amounts of image pixel data to be handled and the matching of features across subvolume boundaries!

3.3.1 Automation of image acquisition with CLSMs

This section is really a cautionary tale for users who are thinking of implementing large area scanning into the confocal software environment. Recently piezo-electric transducers have become available that provide fine movement, but most commercial *XY* microscope stages which have the capability of movement in tens of millimetres edge length use a stepper motor for each axis. CLSMs have operating systems within which the user can write a macro to move the *XY* stage systematically over a large sample area.

Stepper motors give an accurate angular rotation for each pulse sent to the motor, but when these rotations are converted to linear movements, the distance travelled for a given number of pulses is governed by the eccentricity of the lead screw. The true distance travelled by the stage can be determined by the movement of images within the *XY* image frame and, as shown in Fig. 3.25, a cyclic response is produced. Presumably this is why manufacturers prefer to quote the repeatability of an *XY* stage (typically $\pm 0.25\,\mu m$) between any two points, rather than the linearity of movement (which can exhibit variations from linearity of the order of $\pm 3\,\mu m$ depending upon the lead screw eccentricity).

Therefore, the successful scanning over large sample areas, with matching of partial fibre images at the edges of the *XY* frames requires an active monitoring of image movement to verify the true stage movement.

An interactive software package MPL (MacroProgrammingLanguage) is available on the Biorad MRC machines which allows the user to tailor the basic in-built system control routines to perform a set of sequential repetitive operations. Macros must be written if non-standard operations are required. As an example of the use of this facility, our solution to the unwanted Biorad instrumental artefact in reflection mode is discussed below.

When the Biorad machines are operated in reflection mode at the highest electronic gains, an annoying instrumental artefact occurs in the form of a bright diffuse region in the centre of the *XY* image frame. Crosspolarisers used in the optical train can reduce this effect to some extent, but all versions of the MRC series seem to suffer from this artefact to some degree. Using a macro specially written in Leeds, it is possible to remove this artefact completely by moving in the *X* direction and using control points such as objects in unaffected regions of the *XY* image frame. The artefact before removal is shown in Fig. 3.26a and Fig. 3.26b shows the cleaned up image by this scanning technique for a thin film sample containing particulate fillers.

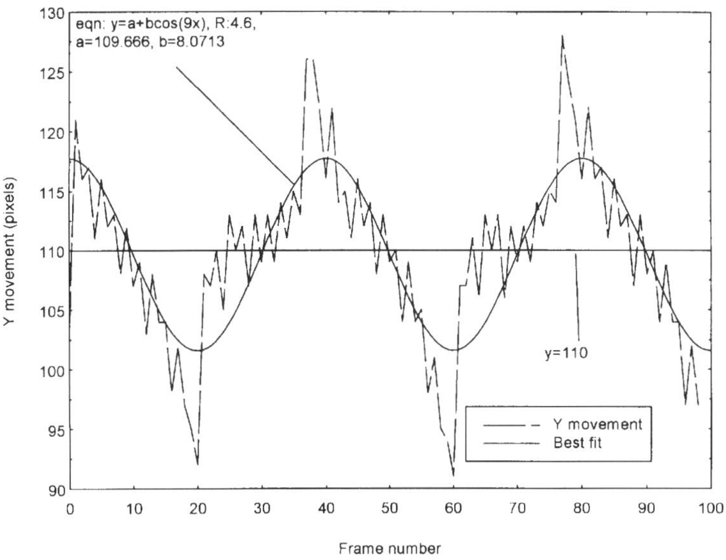

(a)

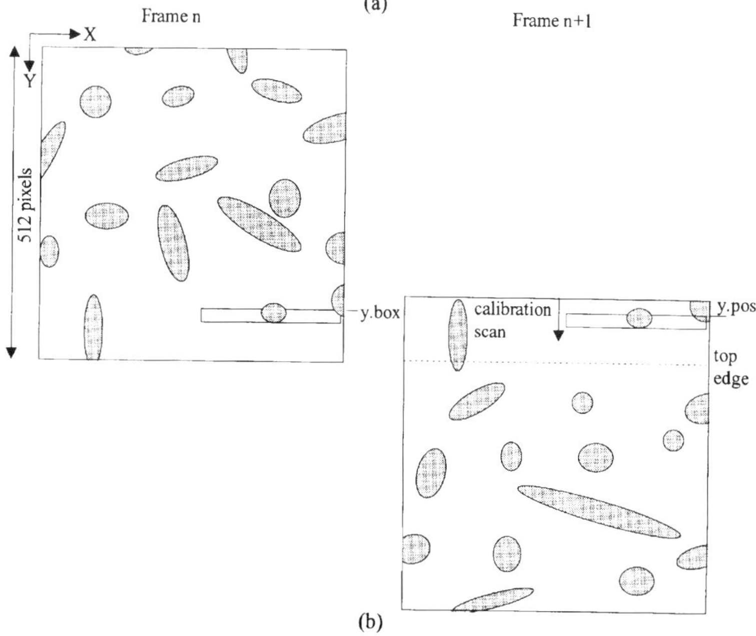

(b)

3.25 (a) Stepper motor controlled *XY* stage may be used to scan a sample automatically but if the same number of pulses are sent to the stepper motor, the actual linear movement in *X* or *Y* will vary in a cyclic manner due to the eccentricity of the lead screw converting rotary motion to linear! The observed peak-to-peak variation corresponds to 4.5 μm. (b) Illustrations of how (a) was produced. Recognisable images in one *XY* frame were followed in an overlapping *XY* frame and knowing the *XY* calibration, the true linear movement for a set number of pulses could be determined. This following of fibres from frame to frame is used at Leeds to reconstruct absolute fibre centre coordinates in the 2D large area image analyser.

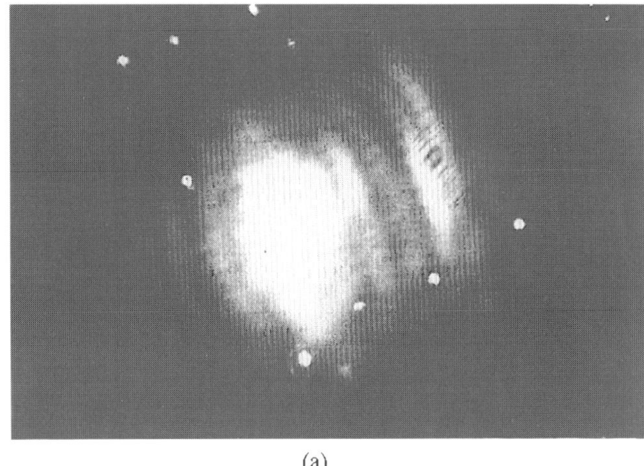

(a)

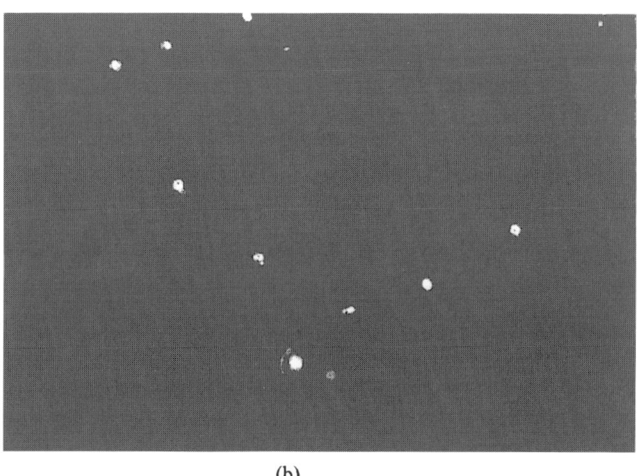

(b)

3.26 (a) At high electronic gain settings, the Biorad MRC series of confocal laser scanning microscope systems show an instrumental artefact – a central diffuse 'blob' – when the reflection mode of operation is selected. (b) The reflection mode *XY* image shown is the same region as shown in (a), but the Biorad artefact has been removed by systematically moving the sample in *X* and pattern matching the objects not obscured by the artefact between overlapping *XY* frames.

3.3.2 Refractive index and the apparent depth problem

Apart from the opacity of the sample, the refractive index is the next most important sample parameter. To maximise the light collection efficiency and increase the optical resolution in *X*, *Y* and *Z*, a large numerical aperture (*NA*) objective lens is required. The largest *NA* lenses are the oil immersion lenses with *NA* 1.4. Following Visser *et al.* [55], and as shown in Fig. 3.27, a simple ray tracing through the objective lens matching oil and into the sample illustrates that there is an apparent depth problem. Although the microscope stage may move vertically by an amount Δz, the movement of the focal plane is actually Δf. The equation linking the two quantities is

$$\Delta f = \frac{\tan[\sin^{-1}(NA/n_i)]}{\tan[\sin^{-1}(NA/n_s)]} \cdot \Delta z \qquad (12)$$

where the refractive index of the immersion medium is n_i and the refractive index of the sample is n_s. Visser's correction is based upon three important assumptions:

- Only the marginal rays in the objective acceptance cone are used to determine the focal point.
- Errors including spherical aberrations in the microscope's optical system or the interface between immersion medium and sample are ignored.

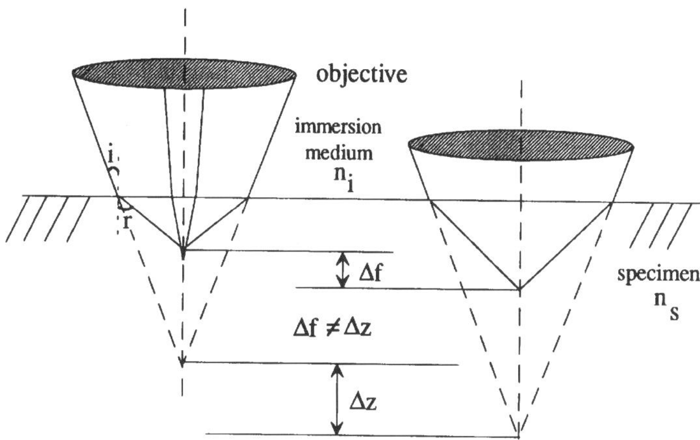

3.27 Schematic view of the 'apparent depth' problem for a CLSM where the objective lens immersion medium has a different refractive index, n_i, from the sample refractive index, n_s. Although the *Z* drive moves by an amount Δz the actual movement of the focal plane is given by Δf. Similarly, apparent object cross-sections in *XZ* frames will differ from the true object cross-sections if a refractive index mismatch occurs.

- Both the immersion medium and the sample are considered to be homogeneous.

A more schematically rigorous description of the effect of lens aberrations in confocal fluorescence microscopy can be found in Hell and Stelzer [56].

If there is a mismatch between the immersion medium and the sample refractive indices, it would be expected that, on *XZ* sections, circular fibres aligned parallel to the surface (along the *Y* direction) would appear to be elliptical in cross-section! Using equation (12), it is possible to compute the expected ratio of axial (*Z* direction) to lateral (*X* direction) 'diameters' of the circular fibre and this ratio is plotted as a function of sample refractive index in Fig. 3.28. Note that when the sample has the same refractive index as the oil (assuming oil immersion objectives), circular cross-section fibres in the *XZ* plane will appear circular, but if the sample refractive index is smaller than that of the oil, the fibres will appear stretched in the *Z* direction and when the sample has a higher refractive index than the oil, the fibres will appear stretched in the *X* direction. Therefore this is one potential method of determining the refractive index of the matrix.

Fortunately, most polymer composites studied to date have refractive indices which are well matched to the immersion oil refractive index of 1.518 at the laser wavelengths of interest. The approximate refractive indices for a few common matrices are shown in Table 3.2. As the fibre orientations are estimated from the shift in fibre centre coordinates, a spurious ellipticity value is not a problem. Hence it is only the error in 'apparent depth' which is important for the spatial resolution and hence for the fibre orientation resolution.

As highlighted by Hall [57], the refractive index varies over the range of optical wavelengths but the variation is typically only 0.75% between 450 nm and 650 nm. Indeed, a second method to determine the refractive index of the matrix (and also the glass) of glass fibre-reinforced polymers is to measure the interference fringes seen around the fibre image (and within the glass fibre image) when a polished surface section is viewed in high spatial resolution, reflection mode.

The interference pattern is formed when the sample is sectioned with the glass fibres at glancing angles with respect to the *XY* plane, that is for $80^\circ < \theta < 90^\circ$. If the fibre surface is just below the sectioned surface, a thin 'film' of matrix is formed and light will be reflected from both the surface of the sample and also the surface of the fibre, as shown in Fig. 3.29a, giving interference fringes outside the fibre image. As the light is coherent, the two beams will interfere and the fringe spacing (by analogy with the classical physics 'air wedge' experiment) is related to the refractive index of the matrix, n_m, and laser wavelength, λ, by the expression

$$n_m = \frac{\lambda}{2(\text{fringe spacing in matrix}) \cdot \tan\theta_0} \tag{13}$$

where $\theta_0 = (90 - \theta)^\circ$.

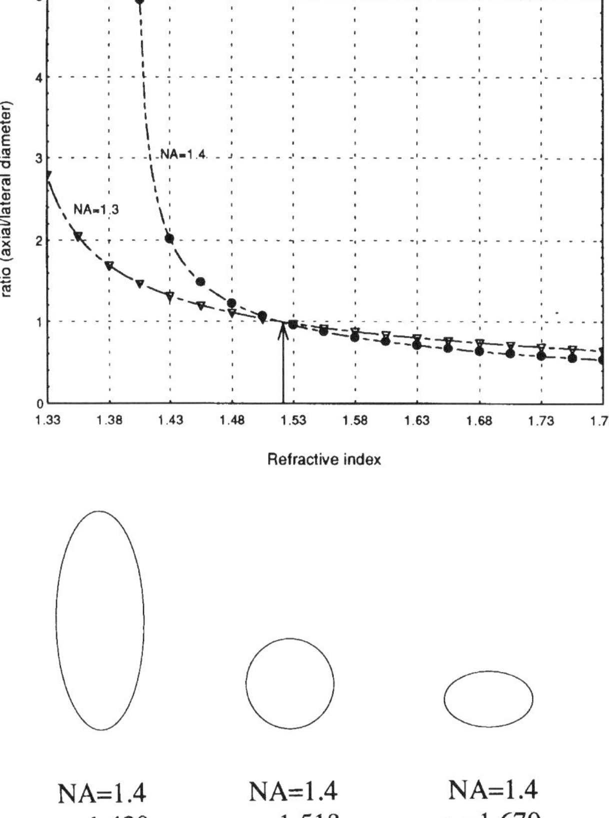

3.28 Distortion of object cross-sections due to the apparent depth problem may be characterised by the ratio of axial to lateral diameters for spherical objects within the sample. A ratio of unity implies that spherical objects will appear in *XZ* sections as circular in cross-section. Ratios above unity imply that a spherical object will appear elliptical in an *XZ* frame and will be elongated in the *Z* direction. Ratios less than unity imply that a spherical object will appear elliptical in an *XZ* frame and will be elongated in the *X* direction. The plots are for two different *NA*, oil immersion objectives.

Similarly, fringes will also be seen inside the glass fibre image where the matrix is just below the surface and the fringe spacing gives an estimate of the glass refractive index, n_g, by an identical relationship.

$$n_g = \frac{\lambda}{2(\text{fringe spacing in glass}) \cdot \tan\theta_0} \tag{14}$$

Table 3.2 Refractive indices of polymer matrices

Polymer matrix	Refractive index
PTFE	1.35
PVDF	1.42
POM	1.48
PMMA	1.49
PP	1.49
PVC	1.54–1.55
PVDC	1.60–1.63
PEEK	1.67

Figure 3.29b shows both of these effects and also shows two regions where there are fringes but no fibre image. These fringe images are due to glass fibres which happen to be curved just below the sample surface!

Similar fringes are seen in scanning acoustic microscope images and it might be argued that the fringes could be used to remove the Φ ambiguity of fibre directions when fibre images from a single 2D section plane are obtained. However, devising robust algorithms to detect the fringes, especially in high packing fraction samples, is not a trivial exercise. The Leeds approach is to use the fluorescence mode of operation and to correlate the relative movement of fibre image centre coordinates between registered image planes in order to infer the orientation state of each fibre.

It is instructive however to compare the fibre orientations, $\{\theta_{ab}, \Phi_{ab}\}$, deduced from the ellipticity of the fibre image at the sample surface to the fibre orientations, $\{\theta_{xy}, \Phi_{xy}\}$, deduced from the shift of centres between two, parallel optical planes. In Fig. 3.30, the difference between the in-plane angle deduced from the ellipticity of the fibre image and the in-plane angle deduced from the shift of fibre centres is shown as a function of the out-of-plane angle of each fibre. Note the two regions of the plot which illustrate the Φ ambiguity problem with 2D analysis and the trend towards zero error between the two estimates of the in-plane angle when the out-of-plane angle is large. The increasing differences for fibres nearly perpendicular to the section plane are a result of the pixellation errors associated with the 2D analysis (and to a lesser extent of the CLSM pixellation errors).

3.3.3 Optimisation of image processing

A number of groups are using significant computer power to improve image clarity well below the surface by deconvolving the CLSM images or allowing for ray tracing through upper layers of the sample material (e.g. Visser *et al.* [58] and Kriete [59]).

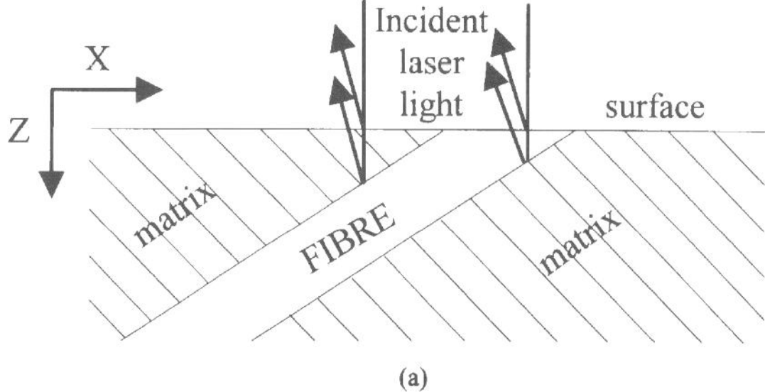

(a)

(b)

3.29 (a) If the axis of a glass fibre is oriented at a small angle to the section plane, CLSM *XY* images may show the presence of black and white patterns due to interference between the monochromatic laser light reflected off the surface matrix and off the matrix/fibre interface. Also, black and white fringes will be produced by interference between reflections off the surface glass and the glass/matrix interface. (b) A typical oil immersion objective, reflection mode *XY* image for a glass fibre-reinforced epoxy showing the interference fringes and also two pure fringe 'objects' which are due to two fibres curving just below the surface! (Area of scan is 186 μm × 124 μm.)

Our approach is to simplify the image processing algorithms as much as possible, both for increased speed of execution as well as minimising cost. The human brain is adept at locating fibre centres way below the surface in a rapidly deteriorating *XZ* image and deducing the best-fit radii of the fibre images. When

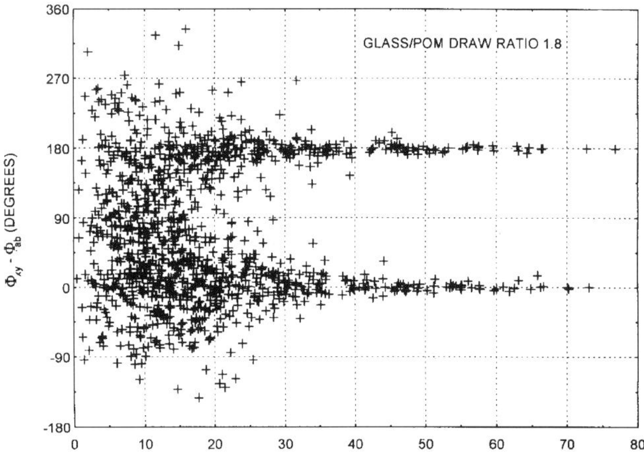

3.30 Short glass fibre-reinforced polyoxymethylene sample has been investigated to deduce fibre orientations from a series of *XY* frames taken at different depths. From the surface *XY* image frames, the ellipticity of each fibre has been computed and this yields an in-plane orientation angle, $\Phi_{a,b}$. By determining the shift of fibre centres between two *XY* frames at different depths, the in-plane angle, Φ_{xy}, has been computed. The plot shows the difference between these two estimates as a function of the out-of-plane angle, θ. The 'splitting' into two distinct regions is due to the Φ ambiguity of the 2D moments calculation and both techniques give excellent agreement at large θ angles, as expected.

the *XZ* images are being taken, improved signal-to-noise is achieved by using the Kalman filter facility on the CLSM, but at greater depths within the sample, the light attenuation is severe and the fibre images tend to merge, cluster and become indistinguishable from noise.

When the signal-to-noise is high, commercial systems have the standard basic facilities to improve image quality, for example contrast stretch, spatial filtering of the images (see for example Russ [60]), thresholding of image planes to binarise the image into, say, fibre and matrix regions. However, the quality of the CLSM image decreases rapidly with depth and ways must be found of enhancing the objects of interest within the image, for example either improving the contrast so that the human operator can identify the objects more reliably, or following these pre-processing strategies by routines which automatically find the fibre-centres or voids, etc. The following text highlights a set of procedures applied to a typical *XZ* image in order to achieve fibre centre and diameter data automatically from the greatest sample depths.

First, the change in fluorescence intensity with depth should be compensated for, as shown in Fig. 3.31 and 3.32.

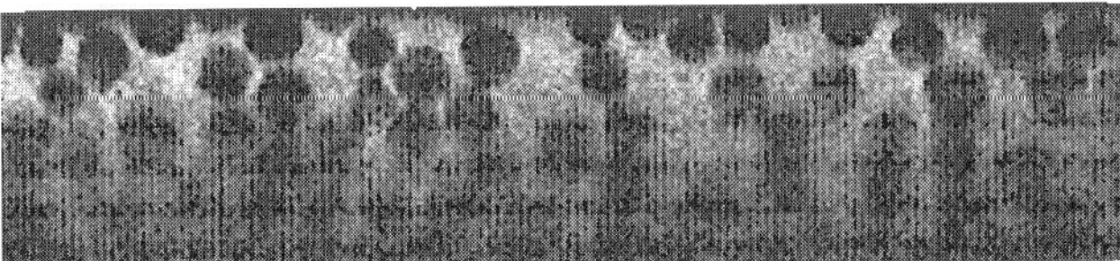

3.31 Typical *XZ* optical section through the unidirectional, glass fibre-reinforced T800 epoxy sample using the Biorad MRC600 in fluorescence mode. (The *XZ* dimensions are 186 μm × 40 μm and all fibres are lying parallel to the surface along the *Y* direction.) Note the degradation of the image quality with depth, but the eye is still capable of distinguishing fibres at the lowest levels.

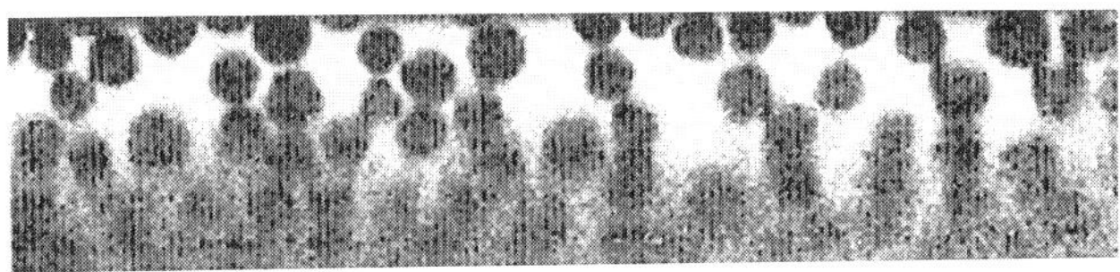

3.32 The first stage in retrieving the fibre centre locations automatically is to correct for the attenuation of the fluorescence intensity with depth. The enhanced contrast achieved in the normalised *XZ* image frame is shown.

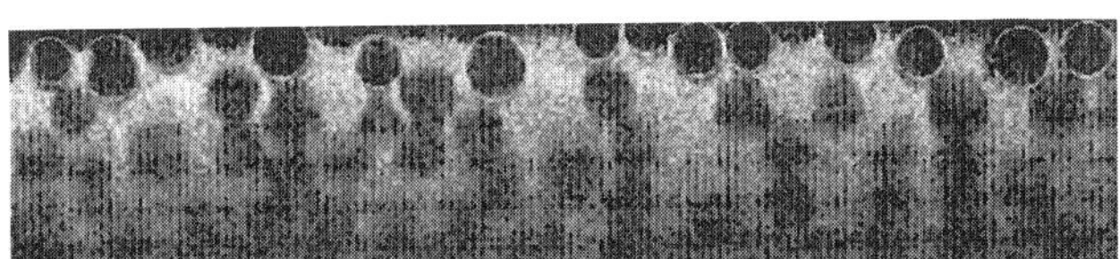

3.33 The second stage is to fit the fibres at (and close to) the surface where there is no doubt about the fibre/matrix boundary. The result of using the 'incomplete fibre least squares algorithm' (see Section 3.5.1) on the *XZ* frame and the best fit circular fibre cross-sections is shown overlaid on the original *XZ* of Fig. 3.31 (for clarity).

Next, the surface level should be located from the XZ image and a least squares fitting procedure used to identify the partial fibre images at the surface and also those clear fibre images in the mid-section of the *XZ* image where the matrix pixel intensity is far greater than the fibre pixel intensity. In Fig. 3.33, the fitted circular fibre cross-sections are shown over the original *XZ* image frame.

As it is known that the fibre images are essentially circular, a circular spatial filter whose radius is slightly less than the known minimum radius of a fibre in the sample can be used. The procedure is to sum all of the pixel intensities within the circular filter and assign that summation value to the pixel position of the centre

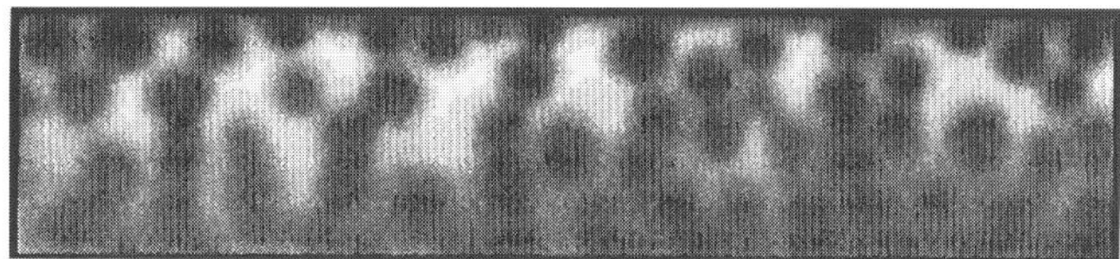

3.34 The third stage is to use a circular filter (whose diameter is slightly smaller than the expected smallest fibre cross-section) to improve the signal-to-noise ratio at the lower levels. In this way, another 'layer' of fibre images deep within the sample may be identified with a higher confidence level.

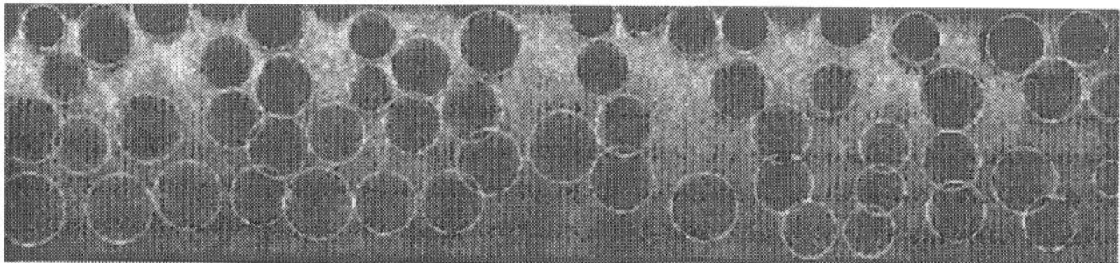

3.35 Work is in progress to extend the identification of fibre images to greater depths. The final version of the *XZ* frame which has been analysed by a fully automatic routine at Leeds is shown. Each fitted circle represents a fibre and it is clear that the automatic routine is as good as a human operator's judgement on the location of all fibre centres in the field of view.

of the filter. Hence a map is created of those regions which are most probably fibres, even deep within the sample, where the eye is just aware of the presence of a potential fibre, see Fig. 3.34. All that remains is to define a threshold below which a fibre region may be indicated.

The ability of these procedures to fit fibre cross-sections on the *XZ* image is shown in Fig. 3.35. Developments are on-going to incorporate this fibre following algorithm into an interactive program and hence speed up the analysis of fibre orientations and fibre waviness studies (as described in Sections 3.4.3 and 3.4.4).

3.3.4 Maximum depth achievable

The most important parameter for any experimentalist who is considering using the CLSM technique for investigations of composite materials is the 'maximum effective depth' which could be achieved for a particular fibre/matrix combination. Unfortunately, an exact value for such a parameter is impossible to fix, even for one type of composite. All that can be done is to define a range of values for that particular fibre/matrix combination. It is clear that there is a good correlation between maximum effective depth and the fibre packing fraction, but the clustering of the fibres, the strength of the fluorescence from the matrix and ultimately the quality of the image processing facilities available must all be taken into account. However, in this section, estimates of this clusivc parameter are presented for a range of composites.

As the fluorescence mode has been used at Leeds to determine fibre orientation states, the maximum effective depth attainable refers to this mode of operation. Clearly, if fibres are to be distinguished from matrix, either the polymer matrix must autofluoresce when excited by the laser wavelength or the matrix must be doped in some way to enhance the fluorescence response. Standard epoxies used in composite research fluoresce strongly but potential matrices, for example polyethylene terephthalate (PET) and polyoxymethylene (POM), fluoresce weakly. Fortunately it has been found that the 'sizing agents' coating glass fibres to improve their adhesion to the matrix also fluoresce strongly. Therefore, as shown in Fig. 3.36, the sizing agents produce 'haloes' around glass fibres reinforcing POM in the CLSM optical sections, making it easier to locate the fibre centre coordinates. Also note that the images could be analysed to determine the dispersion of the sizing agents into the matrix during processing.

The fluorescence signal intensity received at the photodetector is a function of the laser power incident upon the composite sample and a function of the attenuation processes affecting the transmission of light within the sample. A typical graph of the exponential decrease of fluorescence intensity, $I(z)$, versus depth is shown in Fig. 3.37 and the best fit equation is given by

$$I(z) = 174.6\, e^{-0.0055z} \tag{15}$$

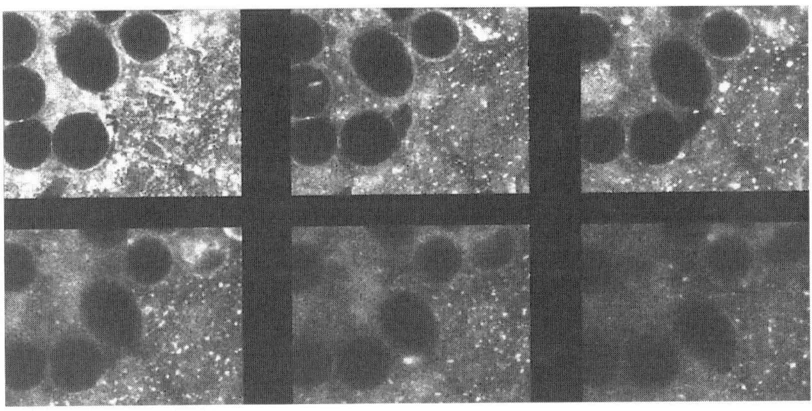

3.36 Six typical *XY* scans (55 μm × 40 μm), each separated by 5 μm in Δz, of a glass fibre-reinforced POM sample in fluorescence mode showing the 'haloes' around each fibre due to the concentration of sizing agent at the fibre/matrix interface. Hence, even though the POM does not fluoresce, the CLSM technique can still be used to identify and locate the glass fibre centres.

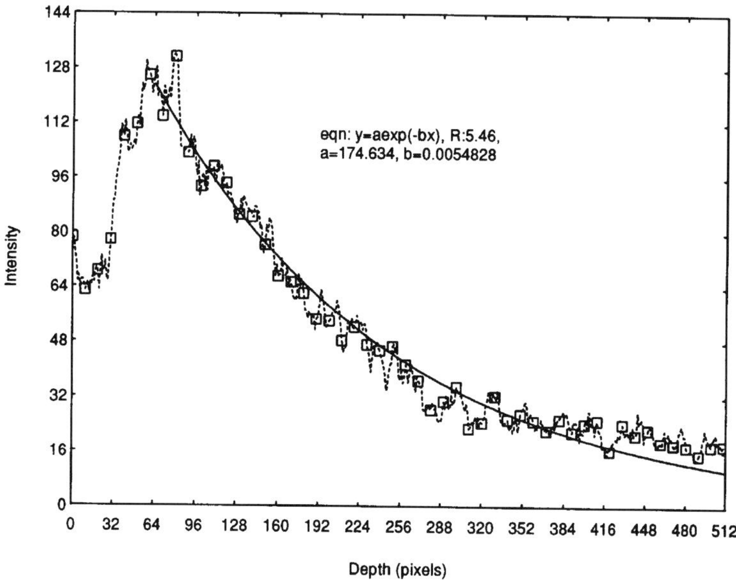

3.37 Plot of the pixel fluorescence intensity as a function of *Z* position within an epoxy matrix. The curve shows the peak intensity at the surface level and the exponential nature of the attenuation of the fluorescence signal with depth of penetration into the sample.

Therefore, in order to maximise the depth penetration, it might be thought that all one had to do was to increase the laser power. However, if the laser power is increased, there will be irreversible chemical changes to fluorescing molecules which inhibit the fluorescence. This photochemical damage, called photobleaching, is one of the most important and yet least understood aspects of the use of fluorescence microscopy (see Tsien and Waggoner [61]). The practical effect of photobleaching is gradually to decrease the intensity of the fluorescence signal from a sample region as it is continuously scanned.

Note that all of those sample regions within the light cone during confocal scanning are exposed to large laser powers even though they are not being imaged and therefore photobleaching limits the number of frames which can be integrated to improve the signal-to-noise, S/N, ratio (i.e. by Kalman filtering) and the clarity of sets of *XY* frames at different depths. For example, if 16 optical sections are to be recorded at 16 different depths within the sample, the sixteenth section will probably have been photobleached by 10–40% by all of the previous 15 scans! Also, if an *XZ* image frame is to be built up, the *Y* scanning mirror is stopped and the laser dwells on the chosen *X* scan as the *Z* drive is moved. Hence, as shown in Fig. 3.18a and b, if an *XY* frame is taken after the *XZ* section, the photobleached line of pixels confirms the location of the *XZ* scan. Although the photobleaching limits the ultimate S/N ratio, it can be most useful for large area scanning algorithms, providing proof of the location of previously scanned regions.

Within most composites, especially at higher packing fractions, fibres tend to cluster and this clustering has the effect of reducing the intensity of the fluorescence signal from the matrix immediately beneath such fibres, that is 'shadows' are cast which reduce the maximum achievable depth of penetration. The shadowing effect is most important when opaque carbon fibres are used as the reinforcements, but there is no evidence of shadowing beneath the separate glass fibre images for the low fibre packing fraction composite shown in the *XZ* image frame of Fig. 3.38.

It is impossible to obtain an analytical solution for the relationship between maximum achievable depth and fibre packing fraction. However, by analysing a number of different composite samples, it is possible to obtain a reasonable estimate for the maximum achievable depth (to within a certain range). The form of the correlation between fibre packing fraction and depth of penetration is shown in Fig. 3.39. An oil immersion objective lens, *NA* 1.4, × 60 magnification was used to make the measurements. The two curves represent, on the one hand, glass fibres embedded in strongly fluorescing matrices and, on the other, glass fibres embedded in non- or weakly fluorescing matrices like POM (see Sections 3.4.3 and 3.4.4).

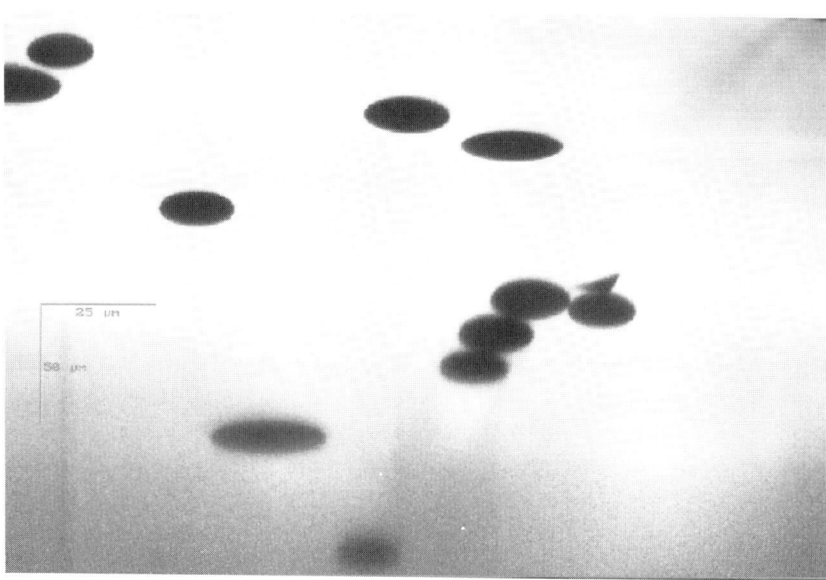

3.38 An *XZ* image plane through a low packing fraction, glass fibre-reinforced epoxy composite, illustrating that information over the full working range of the oil immersion objectives can be obtained at these lower packing fractions.

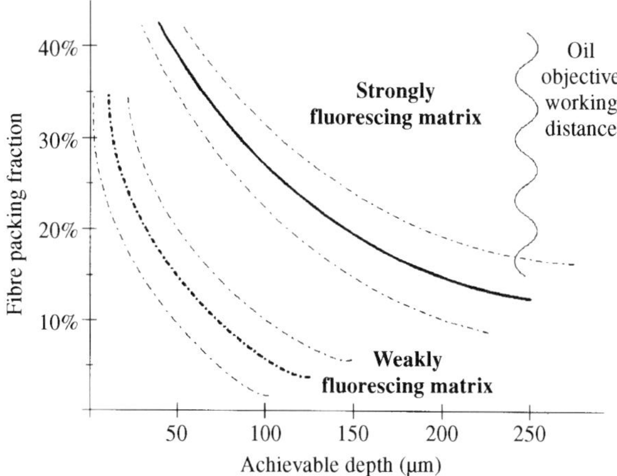

3.39 Based upon a number of different samples studied at Leeds, the plot gives an indication of the likely depth achievable with the CLSM technique as a function of fibre packing fraction and whether or not the matrix fluoresces strongly.

3.4 Application areas for CLSM in composite research

There have been relatively few papers published on the application of the confocal technique to composites and materials research since the early 1990s. Most papers are concerned with either the development of new types of CLSM or applications in the biological and physiological research fields. This section illustrates the range of phenomena of interest to researchers in polymer composites research and development that have been or which could be investigated by the CLSM technique.

Investigations at the National Physical Laboratory have revealed many potential metrological applications for the confocal scanning and optical microscope (McCormick [62]) such as ceramic wear tracks, fracture surfaces and the location of cracks due to thermal heating. NPL have used a different type of confocal system, 'tandem scanning microscope', incorporating a silicon disc which is perforated with thousands of pinholes (60 μm in diameter) and which spins at 1200 rpm. They conclude that the CLSM technique can provide surface roughness data which are more representative than those data produced by profilometers. Also, it has been argued (see p 262, Pawley [44]) that the use of piezoelectric Z drives can be used to measure intensity changes over ± 50 nm and therefore represents the potential Z precision in confocal reflection mode microscopy.

A small pilot study (unpublished) has been conducted at Leeds in collaboration with the Polymer Physics Group to illustrate how the CLSM can scan over large regions and explore the extensive crack patterns around a typical notched and stressed polycarbonate sample. The 3D topology of the cracked surface can be followed by an oil immersion objective, but an air objective is better suited to these investigations (there is no oil to leak away!).

Yeomans *et al.* [63] have reported that CLSM has been used to assess subsurface damage in a number of glass-reinforced ceramic matrix composites subjected to indentation and impact. The attraction of the CLSM is that it can be used for both surface modification studies and also for subsurface cracking, unlike alternative optical techniques.

The CLSM technique has been used successfully by Thomason and Knoester [64] to investigate the transcrystallised interphase in fibre-reinforced thermoplastic polymer composites. A transcrystallised interphase is obtained when a reinforcing fibre nucleates the polymer with sufficiently high density that the spherulites are constrained to grow in the radial direction, forming a sheath around the fibre. They observed the presence of a fluorescence signal at glass fibre–epoxy matrix interfaces (see also Section 3.4.3 and Fig. 3.36), presumably due to 'sizing agents' coating the fibres. The lateral resolution (see equation (11b)) of the CLSM technique is of the order of 200 nm, which is a factor of 1.4 better than conventional optical techniques and as the Z resolution is comparable, the volume resolution for these studies is improved by a factor of $(1.4)^3$, that is approximately three times better.

3.4.1 Low level voidage/porosity

Ultrasonic/densitometric/gravimetric techniques have been used to study voidage but there is a large scatter in any correlation of the techniques for void volume fractions $v_{\mathrm{v}} \leq 5\%$ (see for example Suarez *et al.* [65] and Fig. 3.40). There appear to be few published papers on the use of the CLSM for void determination in composite research despite the fact that the technique has great potential in this research field, which encompasses polyurethane foams through to powder impregnation processing. Recently, however, Ito *et al.* [66] have used the CLSM to investigate the time evolution (over many hours) of void structures in highly purified latex dispersions.

Increasingly research groups are deriving void volume fractions using the 2D image analyser technique and assuming that the average area fraction, A_{v}, of the voidage on a number of XY image frames is equivalent to the volume fraction, v_{v}, within the sample. The measurement errors with the 2D techniques are mainly due to illumination problems and the quality of the polishing during sample preparation in order to discriminate voids from matrix. However, the real issue is whether or not one can assume that

$$\langle A_{\mathrm{v}} \rangle = \langle v_v \rangle \tag{16}$$

It has already been pointed out that Shi and Winslow [30] give estimates for the number of frames and voids within frames which must be analysed in order to

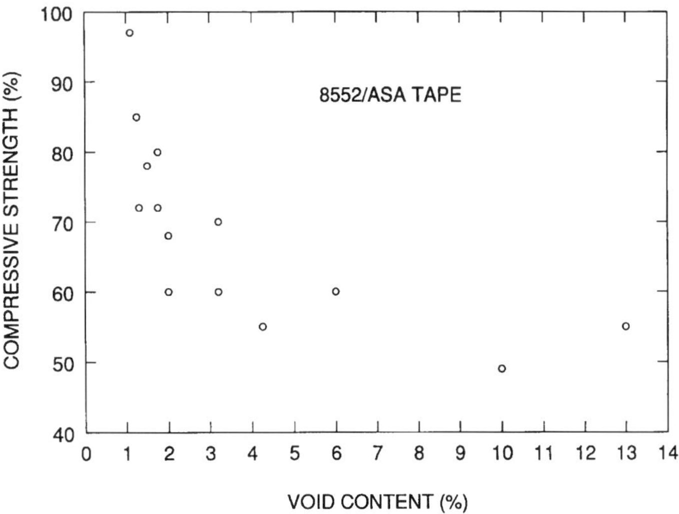

3.40 Relationship between void volume content and compressive strength, see Suarez *et al.* [65]. The void estimates were obtained by an ultrasonic technique but like all ultrasonic measurements, there is considerable scatter.

have a certain confidence in the void volume fraction estimates. Their work starts from the assumption that the voids are randomly distributed in the sample and that the void sizes are normally distributed about a mean size. Their ideas are model based and a correction factor appears in their equations which is related to the shape of the voids. A better, but admittedly more difficult, method would be to mimic the Cavalieri technique, see Section 3.1.4, in which a number of sections are taken through the objects of interest to assess their volumes.

Recently, the Leeds group has been investigating the use of both reflection and fluorescence responses to identify voids in glass reinforced epoxy composites rapidly, as shown in Fig. 3.41a and b. Voids do not fluoresce, but neither do glass fibres! However, in reflection mode, the large refractive index mismatch between void and matrix/glass fibres gives a high intensity signal. To obtain the greatest depth, a high electronic gain must be used and there is the possibility of excessive flaring of the reflection mode signal and hence an overestimation of the void regions. Therefore, by combining the two modes of operation, a better estimate of voidage may be obtained.

Finally, a robust algorithm for 3D voxel connectivity must be devised to estimate the individual void volumes and means must be found of reconstructing large volumes by overlapping high resolution subvolumes (similar to the reconstruction of fibres over large areas in the 2D image analyser) if size and spatial distributions are to be deduced. If the confocal image data collection over large regions could be automated, the assumptions of spatial randomness and

(a)

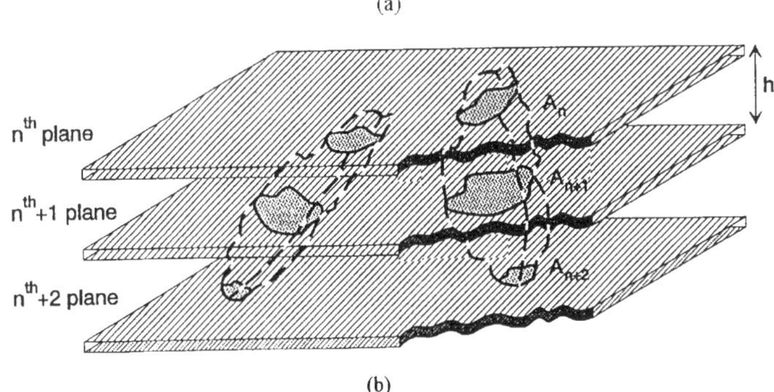

(b)

3.41 (a) An *XZ* scan into a glass fibre-reinforced liquid crystalline polymer using both fluorescence and reflection mode combined. The bright regions below the surface are due to subsurface voids. (b) Clearly, the Cavalieri approach to volume estimation can be used with the penetration depth achievable to gain an excellent estimate of void volumes with the CLSM technique.

spatially independent size distributions could be finally tested. Also spatial periodicities and short-range intercorrelations between voids could be assessed.

In Fig. 3.42a, the clarity of the reflection mode confocal *XY* image (due to the axial discrimination of the CLSM) is shown for a typical carbon fibre-reinforced sample containing voids. The frequency distribution of pixel intensities for this *XY* image frame, shown in Fig. 3.42b, illustrates that fibres and voids may be separated by an appropriate choice of pixel threshold intensity.

3.4.2 Particulate/spherical fillers in thin films, coatings and composites

Although the main thrust of this chapter is towards glass fibre-reinforced polymer composites, it is worth mentioning that the CLSM technique can be applied to

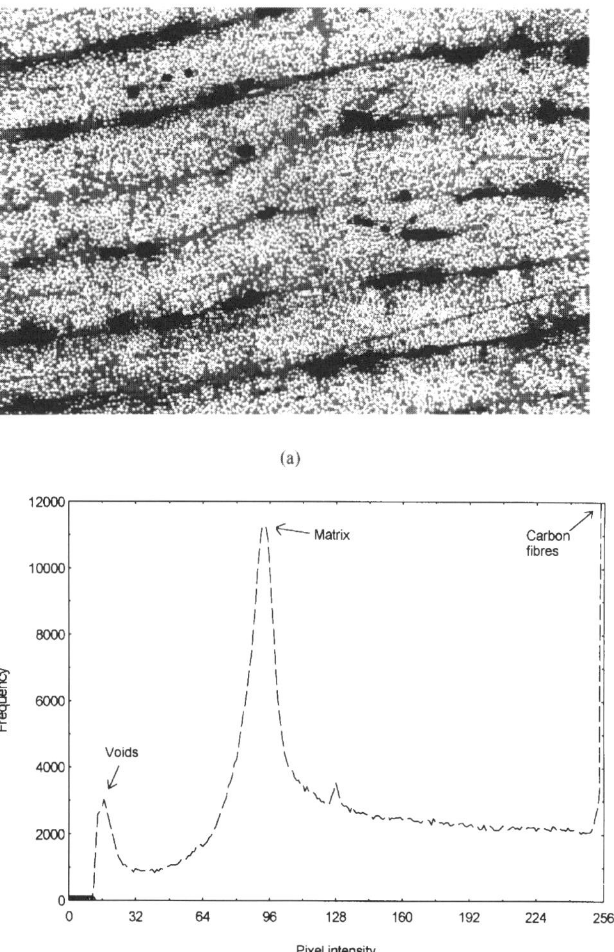

3.42 Carbon fibre-reinforced epoxy matrix viewed in reflection mode shows excellent surface detail. The voids are black, the matrix pixels are grey and the fibres are white. (b) Frequency distribution of pixel intensities shows that voidage, fibres and matrix can be differentiated easily by automatic thresholding routines.

other forms of filler too. Figure 3.43a and b shows reflection mode images of the particulates embedded in 'Melinex'* polyester film to improve the film handleability. Because of the transparency of the film and the low particulate number density, the full working range of the objective lens can be realised in this application. Comparisons between the techniques of CLSM, atomic force

* Melinex is a trade mark of the ICI group of companies.

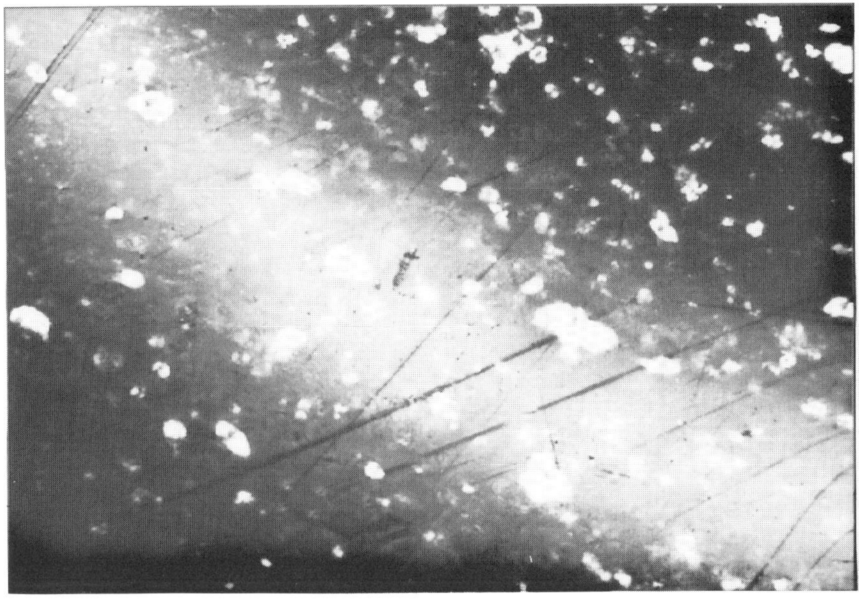

(a)

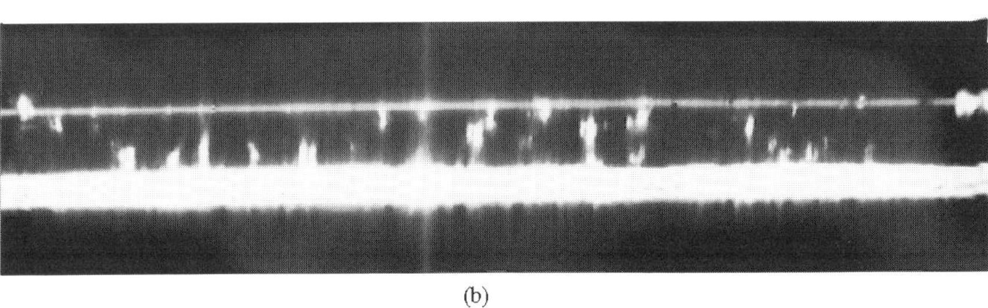

(b)

3.43 (a) An *XY* scan of a thin 'Melinex' film in reflection mode showing the particulates doping the transparent film. (b) An *XZ* scan in the same region, showing the spatial distribution of these particulates in 3D (the area of the *XZ* scan is 186 μm × 80 μm).

microscopy and scanning acoustic microscopy applied to organic coatings have been made recently by Lange *et al.* [54].

As an example of the potential of the CLSM technique for studying spherical fillers in composites, glass microspheres embedded in an MY-750 epoxy resin matrix are shown in Fig. 3.44. This image has been generated from a set of 100 *XY* sections, each separated by 0.3 μm and rendered by a depth-coded intensity to illustrate the locations of the spheres in 3D. The brightest regions are the optical sections at and near the surface and the darkest regions are deepest within the sample. The microspheres were viewed in fluorescence mode and the area of the

3.44 Intensity-coded 'extended focus' image of glass microspheres in an epoxy matrix. The brighter the intensity, the nearer the surface is that part of the image. This composite image was created from 100 *XY* frames each separated by 0.3 μm in *Z*, resulting in a kind of 3D rendering of the objects.

field of view was 186 μm by 124 μm. Despite the obviously high packing fraction, significant depths are achievable and clearly there is the possibility of novel experiments being devised, for example mapping of the 3D mesostructure before and after impacts!

Obviously the CLSM technique has potential for all kinds of coatings research, thin film packaging and substrates with particulate fillers (as well as glass fibre or low packing fraction carbon and Kevlar fibres).

3.4.3 Short glass fibre reinforcements

The Polymer Group at Leeds has been conducting experiments with glass fibre-reinforced polyoxymethylene (POM) since about 1970 (see for example Brody and Ward [67]). Deformation models have been evaluated, for example an aggregate version of the pseudoaffine model (Hope *et al.* [68]), which relate the fibre orientation distribution in an initial billet of material to the fibre orientation distribution after hydrostatic extrusion and thence to predictions of the 3D elastic moduli (Hine and co-workers [69,70]).

The most important parameter for modelling purposes is the 'draw ratio', Λ, which is given by

$$\Lambda = \text{final extrudate length/initial billet length} \tag{17}$$

The pseudoaffine model relating 3D microstructure to the extrusion process implies that the final fibre in-plane angle, Φ' (considering a plane perpendicular to the extrusion or 'draw' direction) is related to the original fibre in-plane angle, Φ, by

$$\Phi' = \Phi \tag{18}$$

whereas the angle of the fibre to the draw direction, θ', at the end of the extrusion process is related to the angle θ to the draw direction in the original billet by the expression

$$\tan \theta' = \Lambda^{-1.5} \tan \theta \tag{19}$$

Using the 3D reconstruction techniques outlined in Section 3.2.6, it has been possible to recreate the spatial positions of all glass fibres within an *XYZ* sample volume of 186 μm × 1.1 mm × 40 μm. This volume can be represented as a 'cube' on the computer screen and rotated to help visualise the fibre orientations. An example 'cube' is shown in Fig. 3.45c together with plane and side elevation views of the fibres within the sample volume. As fibre diameters are determined from the CLSM data, true aspect ratio l/d frequency distributions may be computed for fibres within the sample volume. When the fibres (as in this sample) are well aligned along the *Y* axis, fibre orientation measurements are possible to subdegree accuracy.

Note that the slightly ragged appearance of the fibres in Fig. 3.45 indicates the inherent *XZ* positional uncertainties associated with locating the fibre centres. Clearly the CLSM technique has the potential to relate flow velocities to fibre orientation and to assess the effect of fibre lengths on the final orientation states and also to investigate fibre–fibre interactions within these high packing fraction glass fibre-reinforced systems.

In Fig. 3.46 a typical view of fibres after pyrolysis of the composite sample is shown. In Fig. 3.46b two fibre length distributions are shown of one of these extruded samples taken by two different techniques. First, an *in situ* fibre length distribution from CLSM data sets like the one shown in Fig. 3.45 and, for comparison, fibre lengths (for the same extruded sample) measured by a 2D image analyser after pyrolysis of the sample. Note the good agreement between the two length distributions shown in Fig. 3.46b even though no allowance has been made for the sampling bias due to partial fibres at the edges of this finite sample volume.

In a non-fluorescing matrix, for example POM, the orientation accuracy is insensitive to the circularity of the fibre cross-section, unlike the single 2D section plane ellipticity measurements of fibre orientation. In a paper by Mattfeldt *et al.* [71] the orientations derived from a set of registered optical sections were

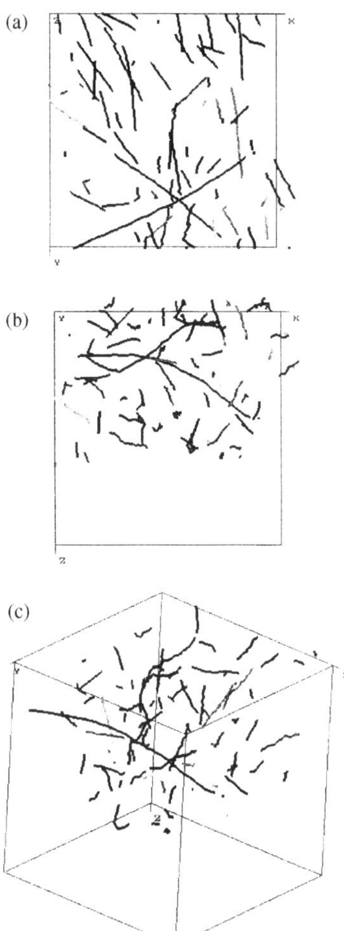

3.45 3D reconstruction of short glass fibres in POM within a sample volume of 186 μm × 1.1 mm × 40 μm in *X, Y, Z*. (a) Glass fibres in plan view. (b) Fibres in side elevation. (c) 3D representation.

used in both a new test for isotropy and also to test the pseudoaffine model predictions. It is impossible to prove that fibres are oriented isotropically from measurements made on isolated 2D sections. By taking random section planes in a sample, it might be inferred that the fibre orientations were isotropic if the orientation distributions in angle θ were identical to each other. However, as shown in Fig. 3.47, for such fibre configurations there is obviously no preferred fibre direction. The frequency distribution of angle θ shows the artificial peak in the distribution at around 15° (as discussed in Section 3.1.4) and fewer fibres at near perpendicular directions and very high orientation directions. Model-based

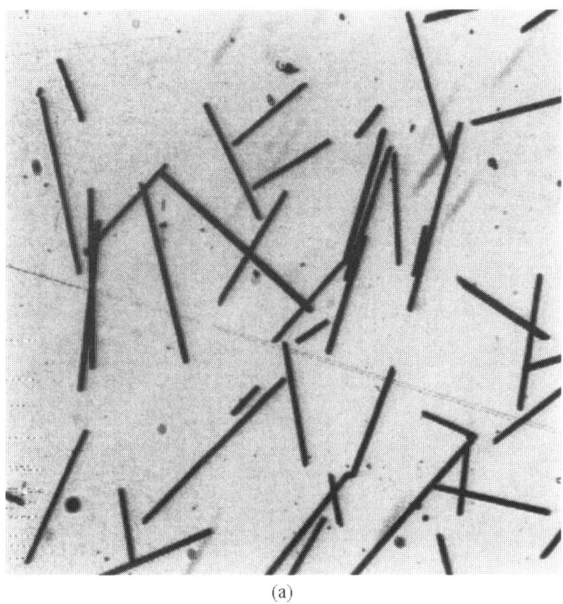

(a)

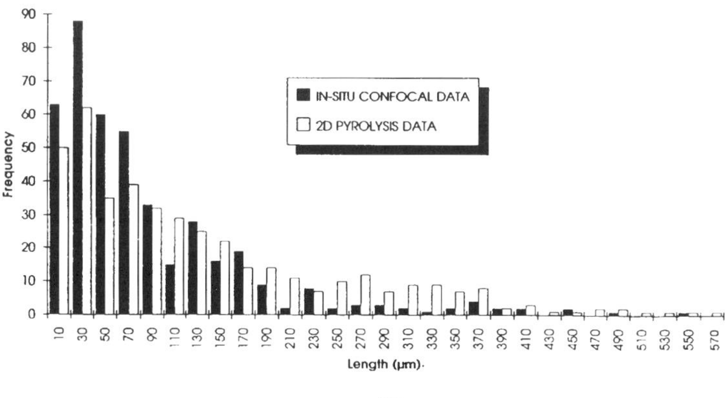

(b)

3.46 Typical *XY* frame of glass fibres remaining in a Petri dish after pyrolysis, ready for 2D fibre length determination. Note that some fibres are still connected to each other, making automatic analysis difficult. (b) Histogram of fibre lengths as determined both by 2D image analysis of pyrolysed fibres and also by *in situ* measurements of fibre lengths from CLSM volume reconstruction.

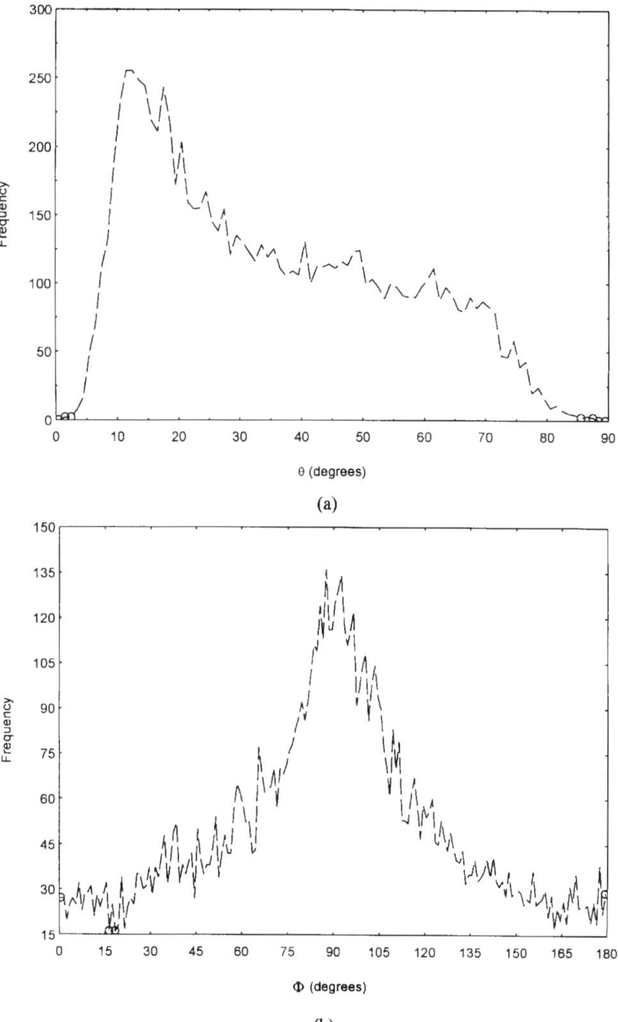

3.47 Frequency distributions of (a) the out-of-plane angle, θ, and (b) the in-plane angle, Φ, for the original billet of glass fibre-reinforced POM, as measured by the large area 2D analyser from a single section.

assumptions can be made to modify these frequency distributions, but a more reliable method is needed to test for isotropy.

Note that the CLSM technique gives the actual 3D angle, β (see Fig. 3.48a) between neighbouring fibres $\{\theta_i, \Phi_i\}$ and $\{\theta_j, \Phi_j\}$ with no ambiguity in Φ, unlike the conventional 2D measurements.

$$\beta = \cos^{-1}(\sin\theta_i \cdot \sin\theta_j \cdot \cos(\Phi_i - \Phi_j) + \cos\theta_i \cdot \cos\theta_j) \tag{20}$$

If the fibres are part of an isotropic distribution of fibre orientations, it can be shown that the expected distribution of nearest neighbour angles, β, on the unit sphere is given by

$$F(\beta) = 2 \cdot \sin^2(\beta/2) \tag{21}$$

By computer modelling, the 95% confidence limits of $F(\beta)$ for a set of 500 fibres (which corresponded to the number of fibres in each measured data set) was established and compared with the distribution derived from measurements on an original billet of glass fibres in POM (see Fig. 3.48b). Although independent ultrasonic measurements (at a scale size of many millimetres) implied that the fibre orientations in the original billet were essentially isotropic, these optical measurements of fibre orientations (on a submillimetre scale size) indicated that the material in the original billet showed 'domains' containing preferred fibre orientations, as shown in the polar plot of Fig. 3.49. The original billet was subsequently hydrostatically extruded and samples at different draw ratios were also scanned. For increasing draw ratios, the distribution of fibre orientations deviated more and more from isotropy, the fibres becoming increasingly aligned to the draw direction, and appearing to follow the pseudoaffine model description of extrusion.

3.4.4 Unidirectional glass fibre reinforcements

The CLSM technique for deducing the fibre centres of unidirectional fibre-reinforced composites has been discussed in Section 3.2.6. As it is essential to keep the fibres within the measurement window for as long as possible, the sample is best sectioned with the fibres lying parallel to the section plane and pointing along the user defined Y direction. XZ planes may then be taken every 50 μm in Y in order to plot the loci of fibre centres and to be sure that a fibre image on one XZ plane can be easily associated with a particular fibre image on the next XZ plane. Recently, a study has been made of a glass fibre-reinforced T800 epoxy which had been produced by compression moulding and the results have been discussed in Clarke *et al.* [34]. The aim of the study was to investigate fibre waviness and fibre misalignments with both the 2D large area image analyser and also the Biorad CLSM.

Early work on unidirectional fibre reinforcements has been carried out by Camponeschi [72] on ply waviness and Yurgartis [16] on fibre misalignments. The Yurgartis technique is the established standard method for determining deviations from the mean fibre direction, but this technique makes a number of assumptions about the fibres which do not seem to be valid for many composites. For example, all of the fibres in the data set are assumed to be smooth and circular in cross-section. However, although most carbon fibres have non-circular cross-sections (and an extreme example is shown in Fig. 3.50!) for glass fibres, it is reasonable to assume a circular cross-section.

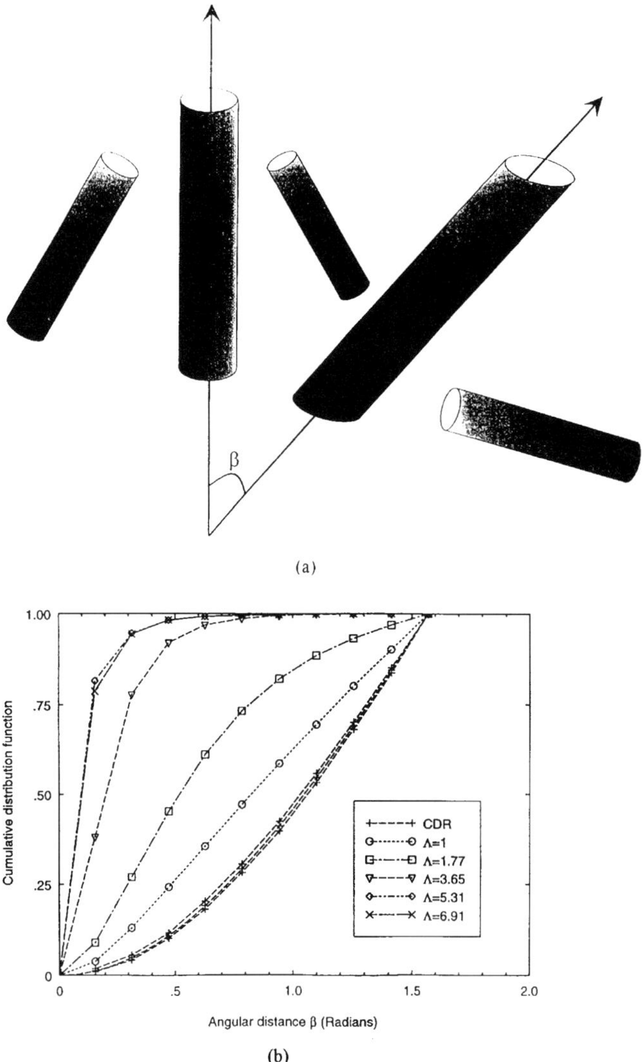

3.48 (a) Schematic view of fibres in 3D and the definition of the 3D angle between nearest neighbour fibres, β. (b) The cumulative distribution function, $F(\beta)$, deduced theoretically for an isotropically distributed ensemble of fibres – CDR (complete directional randomness) – and the actual fibre measurements for different draw ratio, Λ, hydrostatically extruded POM. Clearly, the fibres in the original billet ($\Lambda = 1$) were not isotropically distributed and at high draw ratios, the fibres become more and more aligned along the draw axis.

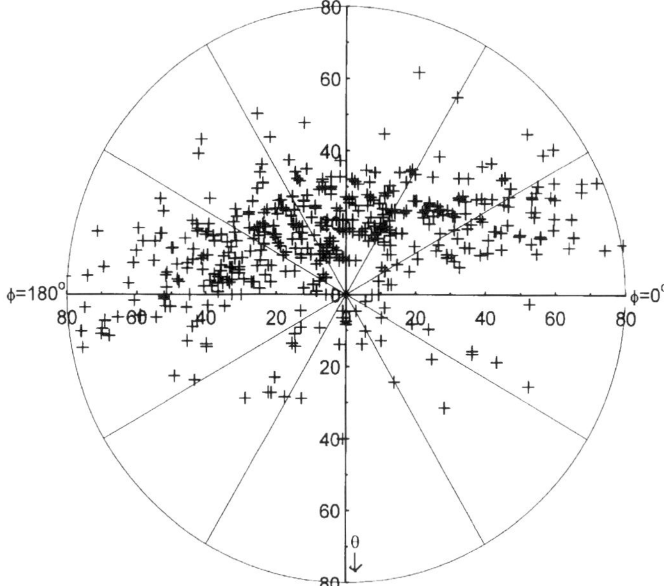

3.49 Polar plot of unambiguous fibre orientations $\{\theta, \Phi\}$ for the original billet of glass reinforced POM before extrusion, as deduced from CLSM data sets. Note that few fibres have an out-of-plane angle, θ, within the range 0° to 15° and that this is not due to the pixellation error effect discussed in Section 3.1.4 but suggests an anisotropic fibre distribution in the billet.

Another key assumption of the Yurgartis technique is that the fibres can be considered as straight rigid rods over a fibre segment length of 100–200 μm (implying that any waviness in continuous aligned fibres has a wavelength, $\lambda \gg 200\,\mu m$). If the amplitude of the fibre waviness is A, one would expect the approximate relationship in equation (22) to hold

$$\text{Misalignment angles} \leq 2A/\lambda \tag{22}$$

However, as shown in Fig. 3.51, if the true 3D movements of neighbouring fibres are plotted over fibre segments of 100 μm, significant curvature is observed and also correlated movement between fibres. The loci of fibre centres are represented by an arrow for each fibre. If the arrow degenerates into a point, this means that the fibre segment was straight and oriented along the (arbitrary) Y axis defined by the CLSM. If the arrow is straight, the length of the arrow gives an indication of that fibre's misalignment to the Y axis and the orientation of the arrow indicates the extent of the misalignment in the YZ and XY planes. However, many of the arrows are disjointed and this indicates that each fibre is curved over the 100 μm fibre segment (note that the 5° marker shows angular misalignment).

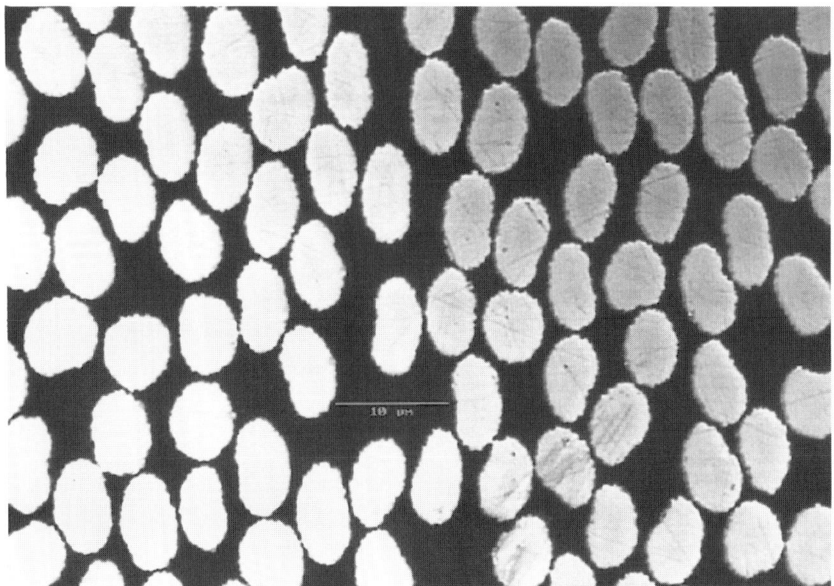

3.50 Carbon reinforced epoxy composite with the section plane at right angles to the fibre orientations showing clearly the bean-shaped cross-sections of these particular fibres. The fibre orientations cannot be determined by the 2D technique (unless one can pattern match between two or more sections to derive the shift in centres and hence the fibre orientation).

Note that the frequency distribution of fibre misalignment angles can be deduced from these sets of 100 μm fibre segments by taking the shift of centre coordinates (δx, δz) between the first and last *XZ* frame. As the pixel uncertainty for locating the fibre centres is approximately ± 0.3 μm at this magnification, the orientational error associated with each value on the frequency distribution is approximately $\pm 0.3^\circ$. There seems little point improving on this error because of the obvious local curvature of the fibres. The spread of misalignment angles in the two orthogonal planes *XY* and *YZ* are shown in Fig. 3.52a and b. The curious bimodal distribution in Fig. 3.52a is due to the fact that two neighbouring plies were within the measurement window and the plies were out of alignment by approximately 7°. Hence in multi-ply composite systems, mechanical strength will be affected not only by fibre misalignment but also by ply misalignment. Note also that angular differences of the order of 5–10° are very difficult to distinguish with conventional 2D image analysis.

The CLSM technique has tremendous advantages for unidirectional fibre research, because it can be used in two complementary ways to give useful data sets for further theoretical study, as discussed below.

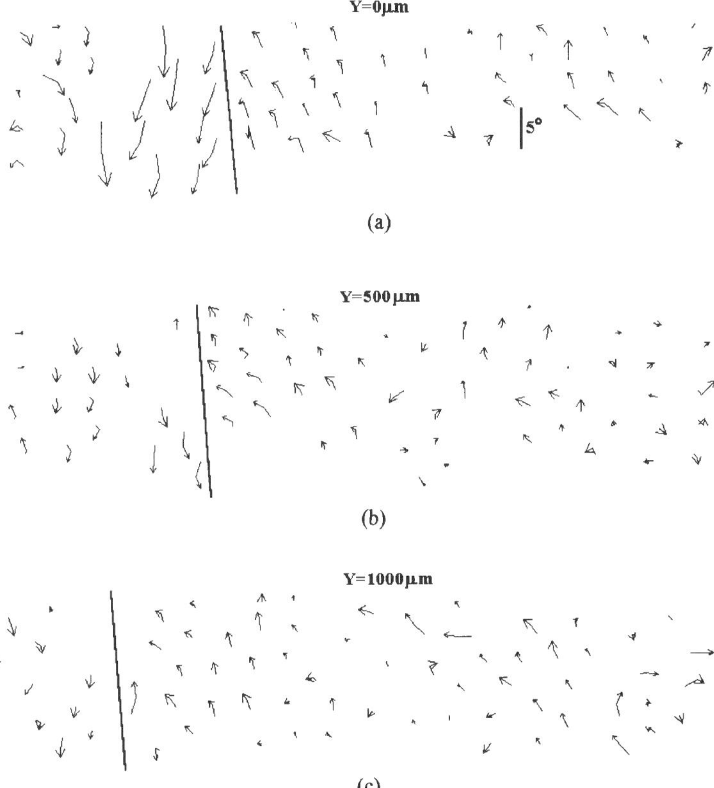

3.51 From a set of three *XZ* sections, separated from each other by 50 μm, the local fibre misalignment angles and local fibre curvatures can be visualised by arrows whose lengths indicate the degrees of misalignment and whose bent or straight nature indicates the amount of curvature. (a), (b)) and (c) are from three sets of *XZ* data each separated by 500 μm and show quite clearly the synchronism between neighbouring fibre movement and indicate the region between different plies.

3.4.4.1 Mapping of local fibre curvatures

Clearly, the misalignment angle distributions do not indicate the extent of the local fibre curvatures nor the possibility of correlated fibre movements. If the sample is sectioned with the fibres lying parallel to the surface and oriented in the user defined *Y* direction, a measurement of local fibre curvature may be obtained. A set of three *XZ* optical sections, each separated by, say, 50 μm could be taken in the *Y* direction, as shown in Fig. 3.24.

It might be the thought that local fibre curvatures could be quantified from these loci of fibre centres on three consecutive *XZ* planes. However, recent

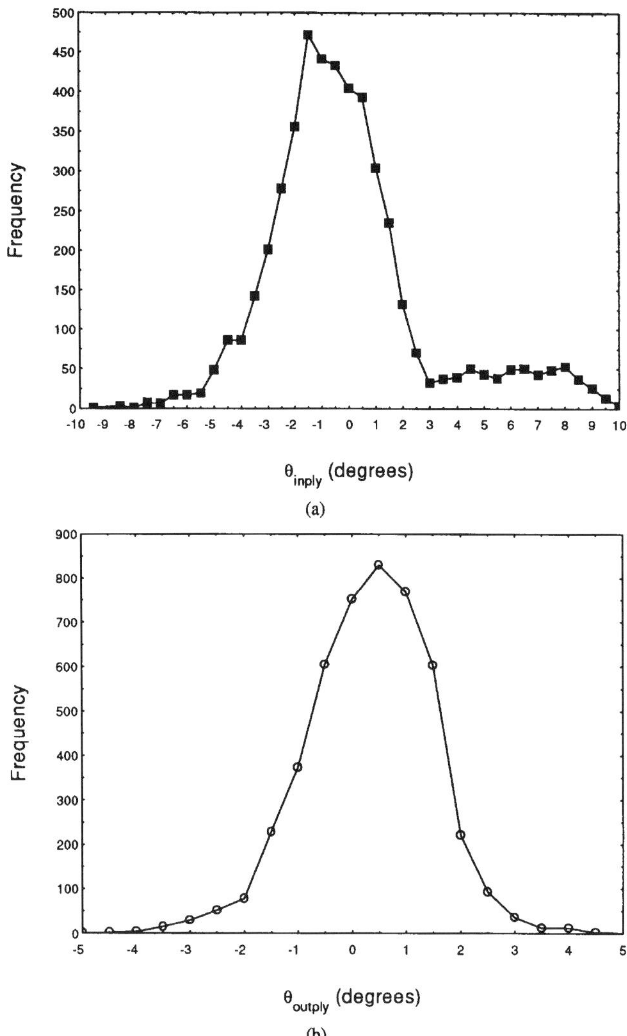

3.52 Frequency distribution of the misalignments of 100 μm glass fibre segments in a well-aligned continuous fibre reinforced epoxy are shown for (a) the plane parallel to the ply boundaries – note that the second peak implies that the fibres in one ply are misaligned by 7 or 8° to the fibres in the neighbouring ply and (b) the plane perpendicular to the ply boundaries – note that subdegree resolution in fibre orientation has been achieved.

simulations (as reported by Clarke and Eberhardt [73]) have shown that more reliable estimates of the radii of curvature can be obtained by fitting a 3D 'space curve' to the raw fibre centre data, see Fig. 3.53a. The space curve is a continuous mathematical function, $f(s)$ which is a high order polynomial in the arc length, s. In Fig. 3.53b, the frequency distribution of the local fibre radii of curvature in the T800 epoxy sample is shown together with the best-fit

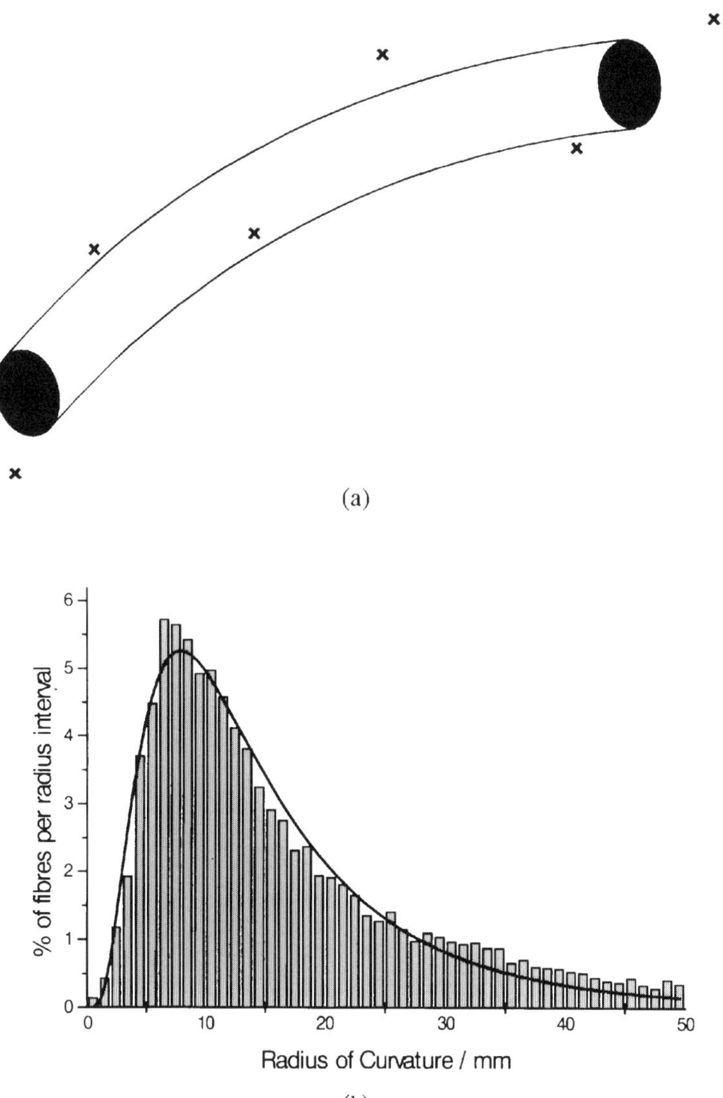

3.53 (a) Schematic view showing how the high order polynomial fit adopted in the 'space curve' representation of each fibre smooths out the intrinsic fibre centre measurement errors. Both curvature and torsion (a measure of the deviation of the fibre waviness from planarity) may be computed from the continuous function $f(s)$ of the arc length, s. (b) The frequency distribution of fibre radii of curvature for the glass fibre reinforced T800 epoxy sample is shown together with lognormal fit to the data. Note that these curvature estimates are sensitive to total fibre lengths and hence the sample volume which has been scanned, see Clarke and Eberhardt [73].

lognormal curve. Note that 32% of the fibre segments have radii of curvature less than 10 mm. As the fibres are oriented along the Y axis (within a few degrees) it is possible to represent the plane of the local curvature by one angle, χ, which can have all values between 0° and 360°. In Fig. 3.54a and b, the polar plot of the angle of local curvature is shown for two different regions of the same T800 epoxy sample. Clearly, in one region, the orientation of curvature is transversely isotropic about the Y direction but in the other region there is a clear preference for the planes of curvature to be aligned either parallel to, or perpendicular to, the ply interface. Note that further work is being undertaken at Leeds to evaluate the effect of intrinsic measurement errors on the determination of these apparent radius of curvature and plane of curvature estimates.

Whereas these data have been obtained from two small regions in the sample, significant areas of the sample could be covered efficiently by taking sets of three XZ sections at each of a large number of locations. The data do show however that care has to be exercised in the interpretation of localised structure for the modelling of global sample properties.

3.4.4.2 Characterising 3D fibre waviness

An alternative method for the characterisation of waviness is the power spectral density which has the potential to show both the amplitude and spatial frequencies of the fibre waviness within a sample. Many theoretical treatments of fibre waviness have assumed that fibre waviness is a 2D phenomenon, see for example Piggott [74]. However, Budianski and Fleck [75] and Slaughter and Fleck [76] have developed models based on a power spectral density treatment of waviness.

A pilot study to derive power spectral densities for glass fibres in T800 epoxy sample has been completed recently and is described in detail in Clarke *et al.* [77]. Once again, the sample was sectioned with the fibres lying parallel to the sectioning plane and the mean fibre direction aligned close to the defined Y direction. A set of one hundred XZ sections, each separated by 50 μm in the Y direction, yielded the true 3D waviness of approximately 40 fibres in a typical high packing fraction glass fibre unidirectional composite (by following fibre centres for fibre lengths of over 5 mm). The longest fibres in the data set are shown in Fig. 3.55.

The fibre data are treated to give a mean fibre misalignment of zero about the Y direction. As the positional fibre centre data are from discrete samples in Y, the power spectral density, $S_x(\omega)$, is defined as the discrete Fourier transform of the autocorrelation function, $C_{xx}(\tau)$, evaluated over the distance range a to b:

$$S_x(\omega) = \frac{1}{2\pi}\int_{-\infty}^{\infty} C_{xx}(\tau)\cdot \mathrm{e}^{-\mathrm{i}\omega\tau}\,\mathrm{d}\tau \tag{23}$$

$$C_{xx}(\tau) = \frac{1}{b-a}\int_{a}^{b} x(t)\cdot x(t+\tau)\mathrm{d}t \tag{24}$$

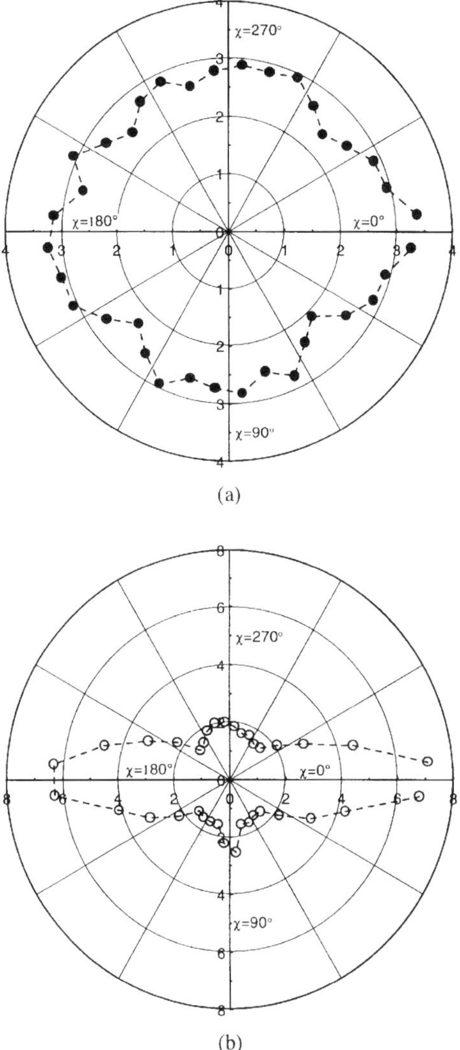

3.54 Knowing the fibre centres in 3D from CLSM measurements, the orientation of the plane in which each 100 μm fibre segment is curved (about the *Y* axis) may be computed. Interestingly, for the unidirectional glass fibre-reinforced T800 epoxy sample studied, two sample regions were analysed and two totally different sets of polar plots were derived, as shown in (a) where all orientations are equally likely and (b) where orientations of the plane of curvature are preferentially aligned either parallel or perpendicular to the ply interfaces (the radial scale is in % of fibre segments).

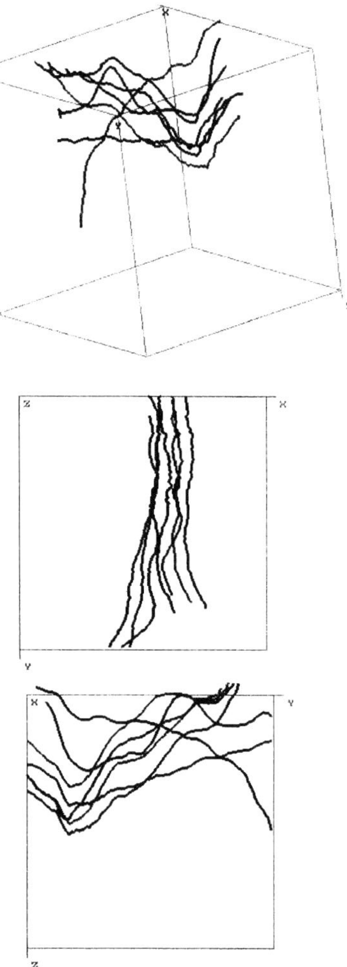

3.55 Reconstructed 3D cube containing the longest wavy fibres in the data set. The true physical dimensions of this sample space was 186 μm × 5 mm × 40 μm in *X*, *Y* and *Z*. Each fibre has been defined from fibre centre coordinates on 100 *XZ* optical sections.

The power spectral density shows the dominant spatial frequencies in the fibre waviness as well as the best estimate for the root-mean-square misalignment angles in two orthogonal planes at right angles to the mean fibre direction, as shown in Fig. 3.56. Unfortunately, the waviness wavelengths do seem to be in the range, 1.3 mm $\leq \lambda \leq$ 3 mm (i.e. most of the power is in the lowest spatial frequencies), implying that fibres must be followed for 10 mm or further in order to improve upon the power spectral density curve.

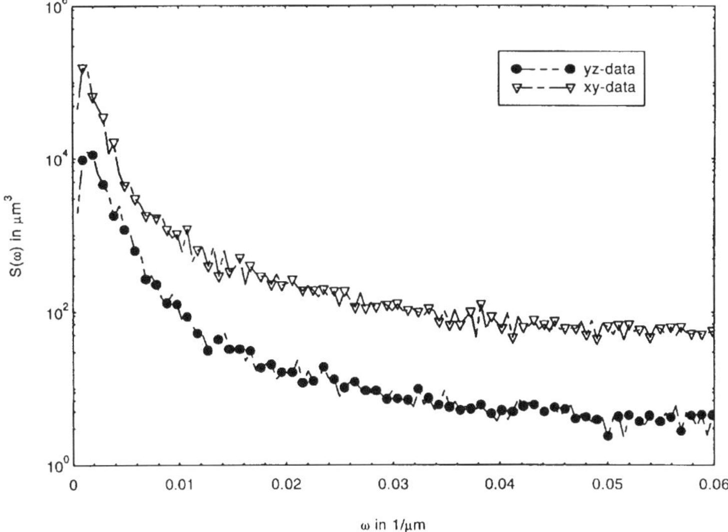

3.56 One method of characterising both amplitude and frequency information in the 3D waviness is to compute the power spectral density distributions in two orthogonal planes (a) for the *XY* plane and (b) for the *YZ* plane (for fibres oriented along the *Y* axis).

The different maximum amplitudes of the two orthogonal power spectral density plots could give an indirect indication of the orientations of the local fibre planes of curvature. If the local fibre curvatures are transversely isotropic and the measurement window in X and Z was square, the amplitudes should be approximately the same.

Unfortunately, high packing fraction carbon fibres only allow CLSM penetration to depths of the order of 5–10 μm. Hence fewer fibres are identifiable and they soon wander outside the much reduced '*XZ* measurement window' making this technique of little use for the investigation of waviness in high packing fraction carbon fibre-reinforced composites. However, the limited depth of penetration could still be useful in removing the Φ ambiguity in fibre orientation research. Robust algorithms are required for large area scanning and automatic splitting of both elliptical and non-elliptical fibre images in confocal microscopy.

3.5 Future prospects for confocal microscopy

In the previous section, the potential for using a standard, Biorad CLSM (or any other commercial CLSM) in polymer composites research has been discussed. There is a significant amount of research and development being undertaken to improve CLSM spatial resolution, imaging sensitivity and the user interface of CLSM hardware and software, as outlined by Pawley [44].

3.5.1 Automated scanning over large volumes

Reliable automated scanning over large sample subvolumes requires robust software. Most artefacts to be detected and quantified in the composite image planes will be circular or elliptical. In our 2D large area analyser, these objects have been analysed at the surface using a 'moments' algorithm which relies upon good contrast between object and matrix and an incomplete object – a split, holed or broken fibre image, that is part of a circle or part of an ellipse within the image frame – could not be fully analysed. Confocal images show excellent contrast at the sample surface, but clearly the quality of the image decreases with depth. As has been discussed in Section 3.3.3, the raw *XZ* image data can be improved by correcting for the fluorescence attenuation with depth but it is necessary to infer the circularity or ellipticity of the objects within the selected image plane (either the *XY* or the *XZ* plane) by another type of algorithm. Once the fibre images have been located, chords are sent out to find the edges of the fibre images. The list of fibre edge points $\{x_i, y_i\}$ may then be used to compute the best-fit coefficients (b, c, d, e and f) of the conic equation which describes the ellipse:

$$x^2 + bxy + cy^2 + dx + ey + f = 0 \tag{25}$$

The best-fit coefficients are computed from the matrix equation based upon the least sum of squares fitting method as discussed in Wu and Wang [78]. The conventional best-fit elliptical parameters, that is centre coordinates (x_c, y_c), in-plane angle Φ, semi-major axis (A) and semi-minor axis (B) are then derived from the conic equation coefficients, as shown below:

$$\begin{pmatrix} \sum_{i=1}^{n} x_i^2 y_i^2 & \sum_{i=1}^{n} x_i y_i^3 & \sum_{i=1}^{n} x_i^2 y_i & \sum_{i=1}^{n} x_i y_i^2 & \sum_{i=1}^{n} x_i y_i \\ \sum_{i=1}^{n} x_i y_i^3 & \sum_{i=1}^{n} y_i^4 & \sum_{i=1}^{n} x_i y_i^2 & \sum_{i=1}^{n} y_i^3 & \sum_{i=1}^{n} y_i^2 \\ \sum_{i=1}^{n} x_i^2 y_i & \sum_{i=1}^{n} x_i y_i^2 & \sum_{i=1}^{n} x_i^2 & \sum_{i=1}^{n} x_i y_i & \sum_{i=1}^{n} x_i \\ \sum_{i=1}^{n} x_i y_i^2 & \sum_{i=1}^{n} y_i^3 & \sum_{i=1}^{n} x_i y_i & \sum_{i=1}^{n} y_i^2 & \sum_{i=1}^{n} y_i \\ \sum_{i=1}^{n} x_i y_i & \sum_{i=1}^{n} y_i^2 & \sum_{i=1}^{n} x_i & \sum_{i=1}^{n} y_i & n \end{pmatrix} \begin{pmatrix} b \\ c \\ d \\ e \\ f \end{pmatrix} = \begin{pmatrix} -\sum_{i=1}^{n} x_i^3 y_i \\ -\sum_{i=1}^{n} x_i^2 y_i^2 \\ -\sum_{i=1}^{n} x_i^3 \\ -\sum_{i=1}^{n} x_i^2 y_i \\ -\sum_{i=1}^{n} x_i^2 \end{pmatrix} \tag{26}$$

where

$$x_c = \frac{-2cd + be}{4c - b^2}, \qquad y_c = \frac{-2e + bd}{4c - b^2}, \qquad \Phi = \frac{1}{2}\tan^{-1}\left(\frac{b}{1-c}\right),$$

$$A = \left(\frac{-f'}{a'}\right)^{1/2}, \qquad B = \left(\frac{-f'}{c'}\right)^{1/2}$$

$$a' = \frac{1}{2}(1 + c + \sqrt{(1-c)^2 + b^2}) \quad \text{if } b \geq 0$$

or

$$a' = \frac{1}{2}(1 + c - \sqrt{(1-c)^2 + b^2}) \quad \text{if } b < 0$$

$$f' = f + \frac{1}{2}(dx_c + ey_c)$$

and

$$c' = (1 + c) - a'$$

A major goal is to process the raw image data as fast as possible, on the fly, so that data storage requirements are minimised. In Fig. 3.57, a typical confocal *XY* frame and the elliptical fits to all fibre images using the above equations are shown. The fibre elliptical parameters may be computed using the above

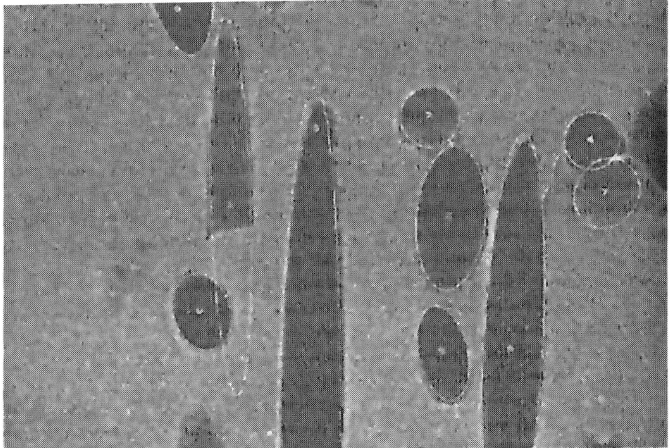

3.57 Typical surface fluorescence mode CLSM image of long glass fibres in a fluorescing matrix showing both complete fibre cross-sections and also incomplete fibre cross-sections. The best-fit ellipses (which are used to determine the fibre orientations) have been generated by the least squares algorithm. Note that the moments algorithm could not be used to deduce the orientations of the partial fibre images.

equations within the time taken to Kalman filter the next *XY* (or *XZ*) image frame (i.e. within a few seconds). Therefore only a few hundred bytes of memory storage may be needed to save the relevant image information from an image frame (instead of the 1/4 Mbyte or greater memory requirements of a complete image frame). Hence, at the end of a large sample volume run, the fibre orientation distributions and associated fibre length distributions could also be computed within seconds.

For particulate research, or void volume determination studies, similar efficient 3D voxel connectivity routines must be developed. Eberhardt [79] has described a technique employed at Leeds in the derivation of object volumes and the object 3D central coordinates for size and spatial distribution analysis. At Leeds, work is ongoing to optimise these software algorithms for CLSM systems in a standard PC environment (rather than a workstation environment).

3.5.2 Miscellaneous ideas

3.5.2.1 New CLSM designs

There is considerable research activity in optimising confocal designs, for example fibre optic designs (Wilson [80]) and special Hadamard type masks to improve spatial resolution by a factor ×2 (Pike [48]). At Massachussetts Institute of Technology (MIT) in the USA, a more sensitive confocal system is being developed as described by Bell [81]. The OCT (optical coherence tomography) and OCM (optical coherence microscopy) techniques promise better depth penetration using a combination of confocal and interferometric concepts. At the NPL laboratory in the UK, a compact 'palmtop' design of CLSM, incorporating optical fibres terminated with microlenses is being investigated (Gee [82]). A consortium involving the University of Amsterdam, Dornier and ESTEC (the European Space Centre in Nordwijk) have developed a small robust lightweight CLSM system ultimately for inclusion into sounding rockets (Schiller [83]). New generations of confocal systems are now available and most are discussed in detail by Pawley [84].

3.5.2.2 New research opportunities

Rheological studies with Noran Instruments full spatial resolution videorate or the reduced spatial resolution supervideorate CLSM will be possible and should improve significantly the spatial resolution of McGrath and Wille's optical sectioning method, described earlier in Section 3.1.6.

Future work could investigate the change of 3D mesostructure *in situ* with applied stress on the sample, instead of the secondhand interpretation through micromechanical modelling. This idea has been investigated using a scanning electron microscope and a stress stage for polycrystalline sample stressing (Viaris [85]). A mechanical stress stage attached to the CLSM microscope could follow 3D fibre movements *in situ*, that is the 3D fibre waviness in an unstressed unidirectional sample could be scanned and after stressing the sample, what happens to the fibre waviness could be observed (effects at strains at the few % levels should be readily observed). Similarly, the possibility certainly exists for the CLSM to investigate the 3D deformation of woven or crimped glass fibres in thermoplastics due to processing route or external impact. In order to give an idea of these possibilities, a typical *XZ* and three *XY* frames are shown for a resin transfer moulded, glass fibre-reinforced composite in Fig. 3.58a, b, c and d.

3.5.2.3 New measurement standards

Although not quite the conventional idea of a fibre-filled composite, there are real possibilities for the CLSM to have an impact in the automated measurement of asbestos, manmade mineral fibres (MMMF) and organic fibres in cellulose filters for health and safety quality control.

The traditional method for asbestos fibre monitoring is to use the phase contrast microscope and to scan in *X* and *Y* systematically with the operator recording the number of fibres within a Walton and Beckett graticule, subject to established counting rules. This is an incredibly tedious and hence unreliable method of establishing asbestos fibre number density. Around 1980, attempts were made to automate the counting of the phase contrast microscope images (see for example Baron and Shulman [86]), but with limited success because of the complexity of the fibre images. In Fig. 3.59 a typical CLSM image field is shown together with a simulated Walton–Beckett graticule for manual counting of asbestos fibres.

A pilot study is in progress to show the superb clarity of the asbestos fibre images with a CLSM operating in reflection mode and to compare the manual counting of fibres with automated measurements. Different types of asbestos fibre are being investigated, that is crocidolite, amosite and chrysotile forms. Not only number densities but also fibre length distributions as a function of depth within the collapsed filters are being measured using a Biorad CLSM and an oil immersion objective, *NA* 1.4, $\times 60$. In Fig. 3.60, the typical length distribution is shown for chrysotile fibres. Note that the criterion for fibres to be selected by the manual counter is that their lengths should be greater than 5 μm. Unfortunately, the frequency distribution is increasing rapidly at this fibre length and hence the final fibre count is very sensitive to interoperator bias. The software challenge is to scan over significant sample volumes in *X*, *Y* and *Z* and automatically to differentiate the fibres from trapped particulates, as discussed in Burdett *et al.* [87].

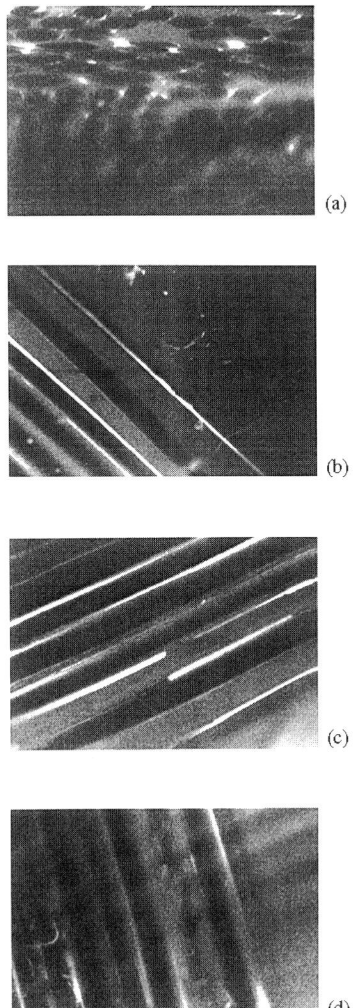

3.58 Illustration of the ability of the CLSM to section through complex glass fibre-reinforced composites optically. It is clear that the complete 3D structure of the resin transfer moulded sample could be reconstructed from sets of *XZ* or *XY* planes. An *XZ* section (186 μm × 150 μm) is shown in (a) and three *XY* sections taken at: (b) the surface, (c) 50 μm depth and (d) 100 μm depth.

3.5.3 Concluding remarks

The CLSM technique has been little used to date because it has been perceived as an exotic expensive measurement technique looking for suitable applications in the field of materials science and composites research. It is true that the cost of these systems ranges from £65 000 to £150 000 at the moment and obviously

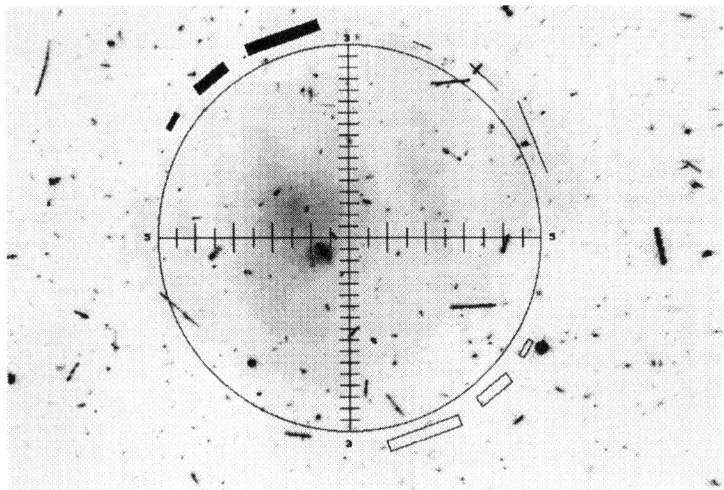

3.59 Simulated Walton–Beckett graticule (used for manual fibre counting) superimposed over a typical CLSM image of asbestos fibres within compacted cellulose film (by courtesy of the Health & Safety Executive, UK).

represents a serious investment for any potential user. However, if the number of new application areas for this technique steadily increases over the next few years and novel designs become established (e.g. McEntee [88]) the overall cost of a confocal system should be reduced significantly.

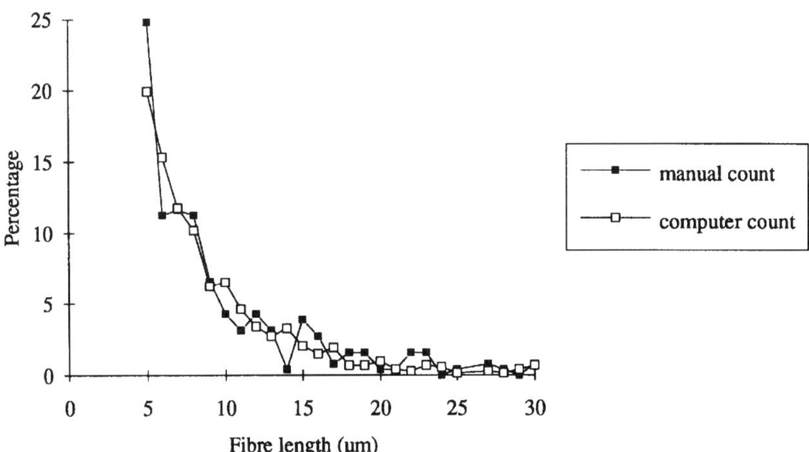

3.60 Typical plot of the fibre length distribution of asbestos fibres in cellulose film as determined automatically by the CLSM technique and, for comparison, by manually identifying, counting and measuring the length of the fibres.

It has been shown that the CLSM technique is capable of subdegree angular orientation errors and *in situ* fibre length determination. Hence it could be used to lay to rest the question of correlations between fibre orientations and fibre lengths in real high packing fraction industrial composites. The CLSM has been shown to be ideal for 3D fibre waviness investigations and has great potential for a range of other studies, not least the determination of surface topology, low void volume fractions and the 3D spatial monitoring of any fibrous or particulate material in thin films and coatings. Without doubt, the confocal laser scanning microscope has a great future as an indispensable tool for high quality measurements of 3D mesostructure in material science research and especially in polymer composites research.

Acknowledgements

We would like to thank a number of colleagues for their encouragement. Our special thanks to Dr Vyvyan Howard for illustrating the concepts of stereology to one of the authors (ARC) in 1992 and suggesting that the CLSM technique should be considered for our fibre orientation studies; Dr Andrew Dixon at Biorad Microscience Ltd, for his support with the loan of the MRC600 confocal system; Paul Smith at Noran Instruments (UK) for his recent support; David Atkinson at Nikon (UK), for the loan of numerous objective lenses when we needed to explore different possibilities; Prof Paul Curtis and Dr Mike Pitkethly at DERA (Farnborough), for crucial financial assistance at a difficult time and also for introducing us to the problem of fibre waviness in unidirectional glass fibre-reinforced composites; Dr Paul Mills and Dr Simon Allen at ICI Films, Wilton for their recent CASE studentship and interest in our work; Dr Gunther Fischer and Dr Christian Ludwig at IKP, Stuttgart and our other partners (Dr Michel Vincent, Dr Einar Hinrichsen and Dr Tom Berland) in our BRITE/EurAM consortium for many fruitful meetings; Prof Klaus Friedrich and Dr Veronika Klinkmuller at the IVW, Kaiserslautern with whom we have a British Council/DAAD project grant; PhD student Peter From and his supervisor Prof Ryszard Pyrz at the Institute of Mechanical Engineering, Aalborg University for their collaborative work on 3D simulations; Dr Norman Fleck at Cambridge University, Engineering Department and Dr Will Slaughter at the Mechanical Engineering Department, University of Pittsburgh, for their assistance with the power spectral density calculations; Dr Garry Burdett at HSE Research Laboratories, Sheffield for a recent MSc studentship and for his financial assistance to purchase a Noran Odyssey CLSM system; Keith Norris, Dr Steven Wire, Dr Pete Hine, Dr Alan Duckett and Prof Ian Ward of the Polymer IRC at the University of Leeds, for many stimulating discussions and their collaboration on the glass fibre-reinforced POM; Dr Mike Wisnom at the Aerospace Engineering Department, University of Bristol for the unidirectional glass fibre-reinforced epoxy samples. Last, but not least, thanks to the rest of our small group: PhD students Mike Enderby and Colin Eberhardt and MSc student David Hunter for their constant input and support.

ND and GA acknowledge the financial assistance from the EU initiative, BRITE EurAM project BE-8081: 'Push-pull Processing of Liquid Crystalline Polymers'.

References

1. G K Haritos, J W Hager, A K Amos, M J Salkind and A S D Wang, Mesomechanics – the microstructure mechanics connection, *Internat. J. Solids Structures* 1988 **24**(11) 1081–1096.
2. P C M Grim and G Hadziioannou, Imaging and characterisation of materials by the new scanning probe techniques (STM/AFM), In *Characterisation of Polymer Composites*, ed H Ishida, Butterworth-Heinemann, 1994, Chap. 7, pp 129–146. ISBN 0-7506-9386-X.
3. M W Darlington, P L McGinley and G R Smith, Structure and anisotropy of stiffness in glass fibre-reinforced thermoplastics, *J. Mater. Sci.* 1976 **11** 877–886.
4. M H Geier, *Quality handbook for composite materials*, Chapman & Hall, 1994, pp 225–232. ISBN 0-412-43120-3.
5. M Maisl, T Scherer, H Reiter and S Hirsekorn, Nondestructive investigation of fibre reinforced composites by x-ray computed tomography, In *Nondestructive Characterisation of Materials*, eds P Holler, V Hauk, G Dobmann, C Rudd and R Green, Springer-Verlag, 1988, pp 147–154. ISBN 0-387-51856-8.
6. B Blumich and P Blumler, NMR imaging of polymer composites, *Makromol.Chem.* 1993 **194** 2133–2161.
7. A L Segre, D Capitani, M Malinconico, E Martuscelli, D Gross and V Leheman, Nuclear magnetic resonance microscopy of multicomponent polymeric materials, *J. Mater. Sci. Lett.* 1993 **12** 728–731.
8. F Lisy, A Hiltner, E Baer, J L Katz and A Meunier, Application of scanning acoustic microscopy to polymeric materials' *J. Appl. Polym. Sci.* 1994 **52** 329–352.
9. K Urabe and S Yomoda, A nondestructive testing method of fiber orientation by microwave, *Adv. Comp. Mater.* 1991 **1**(3) 193–208.
10. S H McGee and R L McCullough, *An optical technique for measuring fiber orientation in short fiber composites*, University of Delaware report no. CCM-81-09, 1981.
11. F Polato, P Parrini and G Gianotti, A new technique for the measurement of glass fibre orientations in composite materials, *Adv. Comp. Mater.* 1980 **2** 1050–1058.
12. R J Young, Evaluation of composite interfaces using Raman spectroscopy, *Key Eng. Mater.* 1996 **116** 173–192.
13. C Galiotis, Micromechanics of reinforcement using laser Raman spectroscopy, Chapter 8 of this book.
14. F J Guild and J Summerscales, Microstructural image analysis applied to fibre composite materials: a review, *Composites* 1993 **24** 383–393.

15. S Fakirov and C Fakirova, Direct determination of the orientation of short glass fibres in an injection moulded polyethylene terephthalate system, *Polym. Composites* 1985 **6**(1) 41–46.
16. S W Yurgartis, Measurement of small angle fibre misalignments in continuous fibre composites, *Comp. Sci. Technol.* 1987 **30** 279–293.
17. S Toll and P-O Andersson, Microstructural characterisation of injection moulded composites using image analysis, *Composites* 1991 **22**(4) 298–306.
18. G Fischer and P Eyerer, Measuring spatial orientation of short fibre reinforced thermoplastics by image analysis, *Polym. Composites* 1988 **9**(4) 297–304.
19. S W Yurgartis and K Morey, Measurement of yarn shape and nesting in plain-weave composites, *Comp. Sci. Technol.* 1993 **46** 39–50.
20. A R Clarke, N C Davidson and G Archenhold, A multitransputer image analyser for 3D fibre orientation studies in composites, *Transactions of the Royal Microscopical Society*, ed H Y Elder, Adam Hilger, 1990, Vol. 1, pp 305–309.
21. A R Clarke, N C Davidson and G Archenhold, A large area, high resolution, image analyser for polymer research, *Proc. Int. Conf. Transputing '91*, eds P Welch and A Bakkers, IOS Press, 1991, Vol. 1, pp 31–47.
22. A R Clarke, N C Davidson and G Archenhold, Image analyser of carbon fibre orientations in composite materials, *Proc. Int. Conf. on Applications of Transputers, TA90* IOS Press, University of Southampton, 1991, pp 248–255.
23. P J Hine, R A Duckett, N C Davidson and A R Clarke, Modelling of the elastic properties of fibre reinforced composites, 1: orientation measurement, *Comp. Sci. Technol.* 1993 **47** 65–73.
24. A R Clarke, N C Davidson and G Archenhold, Mesostructural characterisation of aligned fiber composites, *Flow-Induced Alignment in Composite Materials*, eds D C Guell and T D Papathanasiou, Woodhead Publishers, 1997, Chap. 7, pp 230–292.
25. S G Advani and C L Tucker III, The use of tensors to describe and predict fibre orientation in short fibre composites, *J. Rheology* 1987 **31** 751–784.
26. I M Ward, Optical and mechanical anisotropy in crystalline polymers, *Proc. Phys. Soc.* 1962 **80** 1176–1188.
27. M Taya, K Muramatsu, D J Lloyd and R Watanabe, Determination of distribution patterns of fillers in composites by micromorphological parameters, *JSME Internat. J.* 1991 **34**(2) 198–206.
28. R Pyrz, Correlation between microstructure variability and local stress field in two-phase materials, *Mater. Sci. Eng.* 1994 **A177** 253–259.
29. A R Clarke, N C Davidson and G Archenhold, The measurement and modelling of fibre directions in composites, *IUTAM Symposium on Microstructure–Property Interactions in Composite Materials*, ed R Pyrz, Kluwer Academic, 1995, pp 77–88.
30. D Shi and D Winslow, Accuracy of a volume fraction measurement using areal image analysis, *J. Testing Evaluation* 1991 **19**(3) 210–213.
31. N C Davidson, G Archenhold and A R Clarke, Measurements of fibre length and fibre curvature with a large area, high spatial resolution 2D image analyser, *Proc. PPS10, Akron, OH* 1994 287–288.
32. B Moginger and P Eyerer, Determination of the weighting function $g(b, r, v)$ for the fibre orientation analysis of the short fibre reinforced composites, *Composites* 1991 **22** 394–398.
33. R S Bay and C L Tucker, Stereological measurement and error estimates for three dimensional fibre orientation, *Polym. Eng. Sci.* 1992 **32**(4) 240–253.

34. A R Clarke, G Archenhold and N C Davidson, A novel technique for determining the 3D spatial distribution of glass fibres in polymer composites, *Comp. Sci. Technol.* 1995 **55** 75–91.
35. H J G Gundersen, P Bagger, T F Bendtsen, S M Evans, L Korbo, N Marcussen, A Moller, K Nielsen, J R Nyengaard, P Pakkenberg, F B Sorensen, A Vesterby and M J West, The new stereological tools: disector, fractionator, nucleator and point sampled intercepts and their use in pathological research and diagnosis, *APMIS 96*, 1988, 857–881.
36. L M Cruz-Orive and E R Weibel, Recent stereological methods for cell biology: brief survey, *Amer. J. Physiol.* 1990 **258** L148–L156.
37. V Howard, Stereological techniques in biological electron microscopy, *Biophysical Electron Microscopy*, Academic Press, Chap. 13, 1990. ISBN 0-12-333355-5.
38. V Howard, The confocal microscope as an instrument for measuring microstructural geometry, *Confocal Microscopy*, Academic Press, Chap. 10, 1990. ISBN 0-12-757270-8.
39. B Paluch, Analysis of geometric imperfections in unidirectionally reinforced composites, *Proceedings European Conference Composite Materials, ECCM6*, Bordeaux, 1993, pp 305–310.
40. M Merickel, 3D reconstruction: the registration problem, *Computer Vision, Graphics and Image Processing* 1987 **42** 206–219.
41. N C Davidson, A R Clarke and G Archenhold, Large area, high resolution image analysis of composite materials, *J. Microscopy* 1997 **185**(2) 233–242.
42. J J McGrath and J M Wille, Determination of 3D fiber orientation distribution in thermoplastic injection moulding, *Comp. Sci. Technol.* 1995 **53** 133–143.
43. A R Clarke, N C Davidson and G Archenhold, Measurements of fibre directions in reinforced polymer composites, *J. Microscopy* 1993 **171** 69–79.
44. J B Pawley, *Handbook of Biological Confocal Microscopy*, 2nd edn, Plenum Press, New York, 1995.
45. J B Pawley, W B Amos, A Dixon and T C Brelje, Simultaneous, non-interfering, collection of optimal fluorescent and backscattered light signals on the MRC-500/600, *Proc. Microsc. Soc. Amer.* 1993 **51** 156–157.
46. G Archenhold, *The Physical and Optical Sectioning of Polymer Composites*, M.Sc. Thesis, Department of Physics, University of Leeds, 1993.
47. K Carlsson, The influence of specimen refractive index, detector signal integration and non-uniform scan speed on the imaging properties in confocal microscopy, *J. Microscopy* 1991 **163**(2) 167–178.
48. E R Pike, Inverse problems in confocal microscopy, In *Inverse Problems in Scattering and Imaging*, eds M Bertero and E R Pike, Adam Hilger, Bristol, 1992.
49. S W Hell and E H K Stelzer, Fundamental improvement of resolution with a 4π-confocal fluorescence microscope using two photon excitation, *Optic Commun.* 1992 **93** 277–282.
50. J B Pawley, Fundamental limits in confocal microscopy, In *Handbook of Biological Confocal Microscopy*, 2nd edn, Plenum Press, 1995, Chap. 2, pp 27–28.
51. J L Thomason and A Knoester, Application of confocal scanning optical microscopy to the study of fibre reinforced composites, *J. Mater. Sci. Lett.* 1990 **9** 258–262.
52. G J Brakenhoff, H van der Voort, M W Baarslag, J L Oud, R Zwart and R van Driel, Visualisation and analysis techniques for 3D information acquired by confocal microscopy, *Scanning Microscopy* 1988 **2**(4) 1831–1838.

53. R G King and P M Delaney, Confocal microscopy, *Mater. Forum* 1994 **18** 21–29.
54. J Lange, J-AE Månson and A Hult, Defects in solvent-free organic coatings studied by atomic force microscopy, scanning acoustic microscopy and confocal laser microscopy, *J. Coatings Technol.* 1994 **66**(838) 19–26.
55. T D Visser, J L Oud and G J Brakenhoff, Refractive index and axial distance measurements in 3D microscopy, *Optik* 1992 **90**(1) 17–19.
56. S W Hell and E H K Stelzer, Lens aberrations in confocal fluorescence microscopy, In *Handbook of Biological Confocal Microscopy*, 2nd edn, ed J B Pawley, Plenum Press, Chap. 20, 1995.
57. C Hall, *Polymer Materials – An Introduction for Technologists and Scientists*, Macmillan Education 1989. ISBN 0-333-46379-X.
58. T D Visser, F C A Groen and G J Brakenhoff, Absorption and scattering correction in fluorescence confocal microscopy, *J. Microscopy* 1991 **163**(2) 189–200.
59. A Kriete, Undesirable phenomena in 3D image cytometry, In *Visualisation in Biomedical Microscopies – 3D Imaging and Computer Applications* VCH, 1992, Chap. 9, pp 214–218. ISBN 1-56081-222-2.
60. J C Russ, *Computer-assisted Microscopy*, Plenum Press, 1992. ISBN 0-306-43410-5.
61. R Y Tsien and A Waggoner, Fluorophores for confocal microscopy: photophysics and photochemistry, In *Handbook of Biological Confocal Microscopy*, 2nd edn, ed J B Pawley, Plenum Press, Chap. 16, 1995.
62. N J McCormick, Confocal scanning optical microscopy in material science, *Metals Mater.* 1992 May 274–275.
63. J A Yeomans, A J Winn and K L Powell, Confocal scanning laser microscopy of ceramic matrix composites, Paper 4a, Abstracts of Microscopy of Composite Materials II, Oxford, *J. Microscopy* 1994 **29**(2) 89.
64. J L Thomason and A Knoester, Application of confocal scanning optical microscopy to the study of fibre-reinforced polymer composites, *J. Mater. Sci. Lett.* 1990 **9** 258–262.
65. J C Suarez, F Molleda and A Guemes, Void content in carbon fibre/epoxy resin composites and its effects on compressive properties, *Proceedings ICCM9, Madrid*, ed A Miravete, University of Zaragoza, Woodhead 1993, Vol. VI, pp 589–596.
66. K Ito, H Yoshida and N Ise, Void structure in colloidal dispersions, *Science* 1994 **263** 66–68.
67. H Brody and I M Ward, Modulus of short carbon and glass fibre reinforced composites, *Polym. Eng. Sci.* 1971 **11**(2) 139–151.
68. P S Hope, A Richardson and I M Ward, The hydrostatic extrusion and die drawing of glass fibre reinforced polyoxymethylene, *Polym. Eng. Sci.* 1982 **22** 307–315.
69. P J Hine, R A Duckett, A R Clarke and I M Ward, Orientation measurement in hydrostatically extruded, glass fibre reinforced POM, *Proceedings Deformation and Fracture of Composites II, PPS8*, Manchester, 1993, Paper 27, pp 1–9.
70. P J Hine, N C Davidson, R A Duckett, A R Clarke and I M Ward, Hydrostatically extruded, glass fibre reinforced polyoxymethylene: I The development of fibre and matrix orientation, *Polym. Composites* 1996 **17**(5) 720–729.
71. T Mattfeldt, A R Clarke and G Archenhold, Estimation of the directional distribution of spatial fibre processes using stereology and confocal scanning laser microscopy, *J. Microscopy* 1994 **173**(2) 87–101.

72. E T Camponeschi, Lamina waviness levels in thick composites and its effect on their compression strength, *Proceedings International Conference Composite Materials, ICCM8*, Honolulu, 1991, 30-E-1 to 30-E-13.
73. A R Clarke and C N Eberhardt, Automated acquisition and 3D reconstruction of glass fibre misalignments and local curvatures using confocal microscopy. Proceedings ECCM-8, Naples, 1998, in press.
74. M R Piggott, The effect of fibre waviness on the mechanical properties of unidirectional fibre composites, *Proceedings Conference Mesostructures and Mesomechanics in Fibre Composites* 1994 pp 145–158.
75. B Budianski and N A Fleck, Compressive failure of fiber composites, *J. Mech. Phys. Solids* 1993 **41** 183–211.
76. W S Slaughter and N A Fleck, Microbuckling of fiber composites with random initial fiber waviness, *J. Mech. Phys. Solids* 1994 **42**(11) 1743–1766.
77. A R Clarke, G Archenhold, N C Davidson, W S Slaughter and N A Fleck, Determining the power spectral density of the waviness of unidirectional glass fibres in polymer composites, *Appl. Comp. Mater.* 1995 **2** 233–243.
78. W-Y Wu and M-J J Wang, Elliptical object detection by using its geometric properties, *Pattern Recognition* 1993 **26**(10) 1499–1509.
79. C N Eberhardt, 3D voxel connectivity algorithms for large volume CLSM studies, University of Leeds internal report MPI 98-1, 1998.
80. T Wilson, *Confocal Microscopy*, Academic Press, London, 1993.
81. J Bell, Interferometry reveals eye's microstructure, *Opto Laser Europe* 1994 **12** 29–31.
82. M Gee, *Parallel Confocal Microscopy*, National Physics Laboratory, Materials Research Booklet, 1994.
83. P Schiller, Confocal laser scanning microscope, *European Space Agency Bulletin*, Summer 1994, pp 8–9.
84. J B Pawley, Appendix 2: Light paths of current commercial confocal light microscopes for biology, In *Handbook of Biological Confocal Microscopy*, 2nd edn, ed J B Pawley, Plenum Press, 1995.
85. P Viaris, *In-situ Deformation Experiments and Measurements on Materials*, Royal Microscopical Society, Digital Imaging Special Interest Group, Ecole des Mines, Paris, 7 July 1995.
86. P A Baron and S A Shulman, Evaluation of the Magiscan image analyser for asbestos fibre counting, *Amer. Ind. Hyg. Assoc. J.* 1987 **48**(1) 39–46.
87. G Burdett, G Archenhold, A R Clarke and D M Hunter, The use of scanning confocal microscopy to measure the penetration of asbestos into membrane filters, *Advances in Environmental Measurement Methods for Asbestos*, ASTM STP 1342, eds M E Beard and H L Rook, American Society for Testing Materials, in press.
88. J McEntee, Encoded pixels enhance microscope's 3D images, *Opto Laser Europe* 1996 **32** 17–21.

4
Geometric modelling of yarn and fiber assemblies

MICHAEL KEEFE

4.1 Introduction

The rapid increase in computer hardware and technology has created a powerful tool for scientists and engineers. One particular computational field of computer graphics, called 'solid modelling' (but more appropriately named geometric modelling) has become quite important in engineering analysis applications. To this end, there are many commercial packages available that give the user powerful creation ability and computer manipulation of the geometry of objects. This chapter gives a brief overview of the modelling techniques behind the current commercial packages and then provides an indication of how this technology is being used in areas related to the study of yarn and fiber assemblies. It is hoped that this introduction to solid modelling will serve as an invitation to expand and develop more approaches in the study of fiber-reinforced composites that take advantage of this technology.

It could easily be argued that the current wave of scientific research on the mechanics and behavior of textile assemblies began with Peirce [1] and his study of the geometry of cloth structure. It is perhaps revealing, but not surprising, that geometry initiated that scientific modelling. Although geometry is not sufficient to describe the behavior of yarn and fiber assemblies, it is necessary to have at least statistical assumptions on the geometrical aspects in order to develop a non-empirical model.

However, the irregularity and sheer number of interacting elements in any practical structure make the precise geometric representation of any specific assembly of questionable use and, certainly for the case of an actual composite, beyond the capability of even a super computer. Yet, it is surprising that the growth in computer technology relating to graphics, geometric engines and visualization has seemed to have little effect on the scientific understanding of the mechanics of yarn and fiber assemblies. This is certainly not true with the aesthetic nor large-scale geometric aspects of these structures [2, 3].

4.2 Model

The aspect of scientific visualization that deals with the computer representation of an actual physical object is known as 'solid modelling'. The key to solid modelling is the idea that the computer model used to generate the graphic visualization is a complete and unambiguous representation of the three-dimensional physical object. Therefore, in theory, all information required to categorize, identify and even differentiate a unique individual object is an integral part of the computer model.

The development of this computer technology is closely tied to the engineering community's application of computer-aided design and manufacturing. When trying to create a computer-integrated product development environment, attempts simply to automate the existing engineering documentation and drafting process highlighted the inadequacies (inefficiency and ambiguity) of the then-current two-dimensional wireframe representations. There has been tremendous progress in this field as computational availability, speed and efficiency have grown, and a number of commercial products using this technology are available. This chapter does not attempt to present a comprehensive review of the field but instead provides a basic overview of the technology. The interested reader is directed toward the detailed history and survey papers for a more complete report on this field of research [4–6].

4.2.1 Techniques

Although solid modelling promises a complete computer representation of an actual physical object, in practice, the commercial modellers are limited to providing a complete and unambiguous computer representation of the geometry of that object. This restriction can be partially overcome in an integrated development environment by using the integrating database functionality provided by the geometric model; for example, by adding material properties as part of a finite-element analysis component. (This will be discussed in Section 4.3, Applications.) As mentioned previously, researchers are also active in areas such as texture and colour mapping in order to give the computer model of an object a realistic but computer-representable 'look' and 'feel' for non-technical users [7]. However, only the geometric aspect has been formalized into a general methodology. Therefore, the reader should recognize that the current generation of commercially available 'solid modellers' are restricted to unambiguous handling of only the geometry of a physical object. Therefore, we will limit the discussion that follows to this geometric modelling.

4.2.2 Representation schemes

Although there are many possible mathematical ways of defining a three-dimensional object, three representation techniques have become the *de facto* standards in geometric modelling. We will call them boundary, construction and enumeration. Because each of the techniques has advantages and disadvantages, commercial modellers often provide (at least at the user–interface level) a hybrid scheme, seemingly able to switch between techniques depending on the specific geometric task. However, a fundamental bias in the software package can often be discerned depending on how it handles (or, with naive programs, mishandles) certain tasks.

4.2.2.1 Enumeration

The enumeration technique basically approximates the geometry of the object by subdividing three-dimensional space into some predetermined arrangement of cells. The object is simply that collection of cells occupied by the object, see Fig. 4.1. The validity of the computer model is guaranteed by the underlying structure imposed on three-dimensional space. Although the figure shows (and systems use) Cartesian enumeration, cylindrical and/or spherical schemes could be similarly developed.

This approach is popular because of its simplicity and its potential for parallelism. Computer software and hardware are being developed that take advantage of this approach, i.e. 'voxels' where the fundamental unit is a three-dimensional *vo*lume *el*ement instead of traditional raster-graphics where the fundamental unit is a two-dimensional box or 'pixel' [8].

Because the cell elements have a predetermined shape, this approach can only approximate an arbitrary shape. In theory, one could always decrease the cell size and increase the number of cells, to approximate an arbitrary shape to any predefined required accuracy. However, the global decrease in the cell size often leads to fine resolution in relatively featureless areas. More complex, but more efficient cell creation techniques can be employed. For example, a 'quad-tree' decomposition on a planar object would divide any partially filled cell into four subcells. This would continue until a predefined accuracy limit was reached. A similar three-dimensional approach (octree) would divide each target cell into eight subcells. A Cartesian implementation of this is shown in Fig. 4.2.

Enumeration is closely related to certain applications such as finite-element analysis where the discretization of the domain is fundamentally a cell-like decomposition process, or stereolithography [9] where the object is approximated by stacks of slices made up of two-dimensional cross-sections. (Note: although we present each representation technique separately, a hybrid representation could be created; for example, one of the other techniques could be used to create the enumeration cells or enumeration could be used to create two-dimensional cross-sections that are then pieced together by one of the other techniques.)

4.2.2.2 Construction

This representation technique defines an object by a sequence of construction operations performed on a set of predetermined primitive objects. (The primitives

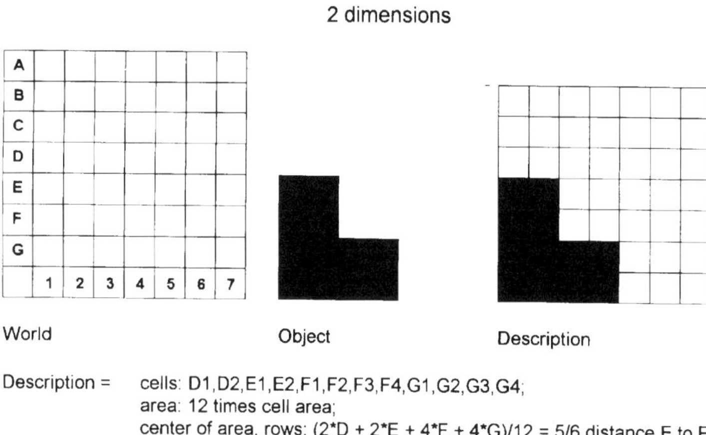

Description = cells: D1,D2,E1,E2,F1,F2,F3,F4,G1,G2,G3,G4;
area: 12 times cell area;
center of area, rows: (2*D + 2*E + 4*F + 4*G)/12 = 5/6 distance E to F
center of area, columns: (4*1 + 4*2 + 2*3 + 2*4)/12 = 13/6; etc.

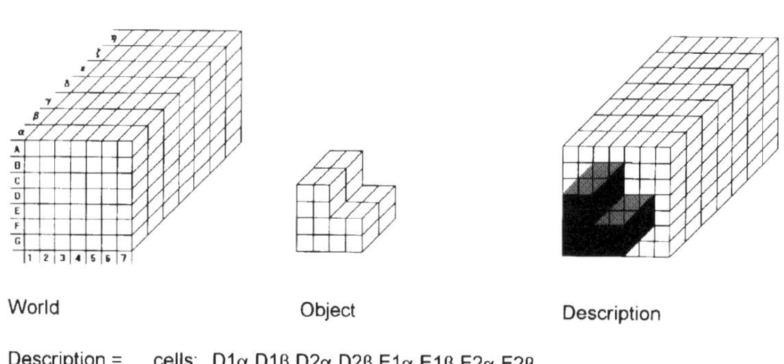

Description = cells: D1α,D1β,D2α,D2β,E1α,E1β,E2α,E2β,
F1α,F1β,F2α,F2β,F3α,F3β,F4α,F4β,
G1α,G1β,G2α,G2β,G3α,G3β,G4α,G4β; etc.

4.1 Spatial enumeration.

are parameterized so that the user may control the size of the features on the primitive; that is, a 'box' primitive would be defined with user-entered lengths, 'height', 'width' and 'depth' rather than defined with specific numbers. The more recent 'parametric' modellers go one step further and instead of just allowing the user to input various dimensions, actually construct the system symbolically, allowing powerful interactive editing of system features during analysis. Of course the software must always check for geometric validity with the actual specific dimensions.) The validity of the object is mathematically tracked through set theoretic algebra, and the construction operations are fundamentally the Boolean operators: union, intersection, addition and subtraction, see Fig. 4.3. These operators are 'regularized' which means that the set resulting from the operator must be of the same dimensionality as the two operands.

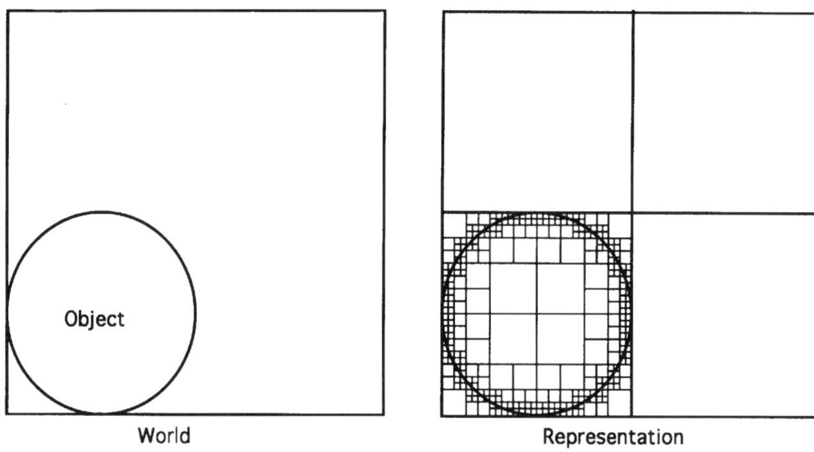

4.2 Spatial enumeration: subdivision technique.

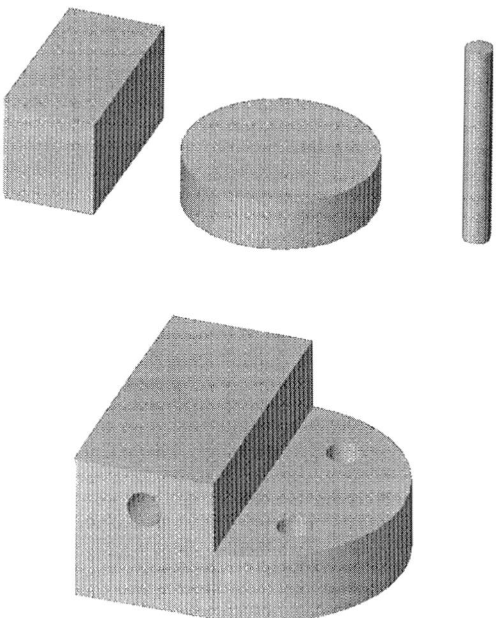

4.3 Construction representation.

From an engineering aspect, the popularity of this approach is due to its similarity to the traditional manufacturing process. The primitives can be thought of as the stock material available to the machinist, and a direct relationship can be seen between the Boolean operators and welding, drilling, milling, etc. However, the ability to represent any arbitrary curved object is restricted by the objects that have been placed in the finite set of available primitives.

Another interesting point is the question of comparing two phyical objects. Consider a simple staircase consisting of a single step. That object can be constructed in many ways. For example, it could be made by 'adding' two appropriate parallelepipeds together or by 'subtracting' one parallelepiped from a larger one. Since construction is not unique, deciding when two physical objects are geometrically identical is not trivial, given only the construction representation technique.

4.2.2.3 Boundary

This final representation technique defines an object by the topology of the surface that separates the 'inside' from the 'outside' of the object. Euler operators are used to manipulate boundary geometry and to verify that the model is valid. This approach, unlike the other two, does not specifically model the interior of the object. However, it does provide a structural hierarchy for increasing dimensionality of the generated object: moving from spatial locations (points, 0-dimension), to edges (curves, 1-dimension) and then to faces (patch, 2-dimension).

This type of representation has been well studied since it is necessary for calculating intersections with the objects and is needed for graphical display on current monitors. For that reason, all commercial modellers either carry along a boundary representation data structure of the object or possess the ability to convert from their representation scheme to a boundary representation. Moreover, because of its ability to define curves directly, a boundary representation is the best mathematical choice for modelling an object with complex shapes.

4.2.3 Curved objects

Whether one is trying to determine a cell size to ensure a required model accuracy or creating an appropriate primitive object or developing the boundary of an object, it would not be feasible to try to model a yarn and fiber assembly without having some ability to describe a curved object. Although quadratic curves have frequently been used to create textile geometry, the current standard for complex curvature in geometric modellers is to use more general parametric curves called non-uniform rational B-splines (NURBS). For a more complete mathematical treatment of NURBS, the interested reader should look through the literature [10,11] or any of the many textbooks on the subject [12,13].

As with any mathematical treatment of an actual object, the trade off in choosing a model is between the generality of treatment (that is, the computer model must have algorithmic approaches for calculating intersections and interpolating) and the exactness of representing the desired object. The NURBS have become the accepted compromise in the geometric modellers.

4.2.3.1 Splines

Splines get their name from the ancient drafting technique of using a thin flexible rod and altering its shape by hanging various weights along its length. Elasticity allows the shape to be approximated by assuming a linear-continuous moment curve, relating that to the curvature and then integrating twice to obtain the displacements in the plane containing the weights. Alternatively, a polynomial representation of the desired final shape can be assumed and then position, slope and curvature continuity can be demanded along the length of the spline. In either approach, the result is a set of cubic polynomials representing the thin flexible rod.

Polynomial approximations are useful in computer models because their evaluation is straightforward given the present computer architecture: they do not require transcendental or trigonometric evaluation and rely on addition, multiplication and their inverses. The calculation of derivatives and integrals is also simple with polynomials. Since the generalized polynomial representation will not be able to match all curves exactly, there must be a way to control the deviation of the mathematical curve from some desired curve. If one wanted to increase the level of control possible for a single polynomial representation, the only way would be to increase the degree of the polynomial and thus the number of coefficients. However, high degree polynomials will suffer from an often undesirable oscillatory behavior. Therefore, the model does not attempt to create a single polynomial representation and instead creates a piecewise set of polynomials with appropriate continuity constraints. Besides the obvious link to the physical spline model, cubic polynomials are also considered a reasonable compromise since they are the smallest degree polynomial that will allow an inflection point.

4.2.3.2 Rational

However, polynomials cannot exactly represent a conic section, and the conics (especially the circle) are important and ubiquitous curves. Therefore the model uses rational polynomials, that is, the division of two polynomials. For example, consider the first-quadrant arc on the unit circle: $x^2 + y^2 = 1$. Rather than using a functional representation for the curve C, i.e. $C = F(x, y)$, everything will be defined in a parametric manner; for example, $C = X(u)$, $y(u)$ with $X(u) = \cos(u)$, $Y(u) = \sin(u)$; u being the parameter and $0 \leq u \leq 90°$. The parametric form clearly indicates that this is a curve (a one-dimensional object) since only a single parameter is needed. Secondly, the parametric representation makes it easier to distinguish the invariant form of the object from the particular coordinate system that is being used for representation.

Unfortunately, that parametric representation contains trigonometric functions, and it is not possible to use simple polynomials such as:

$$X(t) = \sum_{k=0}^{n} a_k \cdot t^k \qquad Y(t) = \sum_{k=0}^{n} b_k \cdot t^k$$

to represent a circle. Putting those parametric polynomials into the equation for the unit circle ($x^2 + y^2 = 1$) and then equating polynomial coefficients on each side of the equality (since the relationship must be true for all values of the parameter of variation t) will allow for only a single point solution of $x = a_0, y = b_0$ (with $a_0^2 + b_0^2 = 1$) and will not generate a curve.

However, with careful choice of rational quadratic polynomials, exact circular arcs can be generated, e.g. the first-quadrant arc of the unit circle [14]:

$$X(t) = \frac{(2 \cdot t - 2 \cdot t^2) \cdot \frac{\sqrt{2}}{2} + t^2}{(1 - t)^2 + (2 \cdot t - 2 \cdot t^2) \cdot \frac{\sqrt{2}}{2} + t^2}$$

$$Y(t) = \frac{(1 - t)^2 + (2 \cdot t - 2 \cdot t^2) \cdot \frac{\sqrt{2}}{2}}{(1 - t)^2 + (2 \cdot t - 2 \cdot t^2) \cdot \frac{\sqrt{2}}{2} + t^2} \qquad 0 \leq t \leq 1$$

4.2.3.3 B-Spline

Recognizing the actual model will contain quotients of the curves, we now turn to defining a single B-Spline curve. The B-Spline formulation can be presented as a power-based expansion, also called the analytic form:

$$\text{B-Spline curve } \boldsymbol{C} = \boldsymbol{C}(t) = \sum_{k=0}^{n} \boldsymbol{A}_k \cdot t^k$$

where $\boldsymbol{A}$ and $\boldsymbol{C}$ are vectors whose elements represent the individual spatial coordinates. However, this does not give the user an intuitive feel for how to modify the curve. Therefore, a more common development uses the geometric form that relies on a set of control points and a set of blending functions:

$$\boldsymbol{C}(t) = \sum_{m=1}^{M} N_{m,k}(t) \cdot \boldsymbol{P}_m$$

where $\boldsymbol{P}$ represents a set of M spatial locations called control points or vertices, the Ns are scalar-valued polynomial functions called the blending or basis functions and k is an integer greater than zero called the order of the B-Spline. The resulting polynomial will have degree $k-1$ and, given the previous discussion, most models use a fourth-order B-Spline in order to generate a piecewise continuous set of cubic polynomials. Note the shape of the curve can now be more intuitively modified by adjusting the spatial locations of the control points.

Many different curves can be generated through various choices for the blending functions. The B-Spline is a generalization of the Bezier (Bernstein polynomial blending functions) curve and it can be defined in the following recursive formulation (the recursive formulation is also algorithmically useful given the repetitive nature of the computer hardware):

$$N_{i,1}(t) = 1 \qquad \text{if } T_i \leq t < T_{i+1}$$

$$N_{i,1}(t) = 0 \qquad \text{for all other values of } t \text{, and}$$

(note from the inequality that if $T_i = T_{i+1}$, then $N_{i,1}(t) = 0$ for all t)

$$N_{i,j}(t) = (t - T_i) \cdot \frac{N_{i(j-1)}(t)}{T_{(i+j-1)} - T_i} + (T_{(i+j)} - t) \cdot \frac{N_{(i+1),(j-1)}(t)}{T_{(i+j)} - T_{(i+1)}}$$

where $\boldsymbol{T} = (T_1, T_2, \ldots)$ is a finite, non-decreasing sequence of values called the knot vector.

The resulting curve is called ‘uniform’ if the knot values satisfy the following relationship:

$$T_{i+1} - T_i = 1 \quad \text{or} \quad T_{i+1} - T_i = 0$$

For mathematical completeness, we allow t to attain the final value of $T_{\max}$ and we agree that $\frac{0}{0} = 0$ for the coefficients involving the blending functions and knot vector values.

One can develop a better understanding of the B-Spline curve by looking at the tree-like structure resulting from the definition:

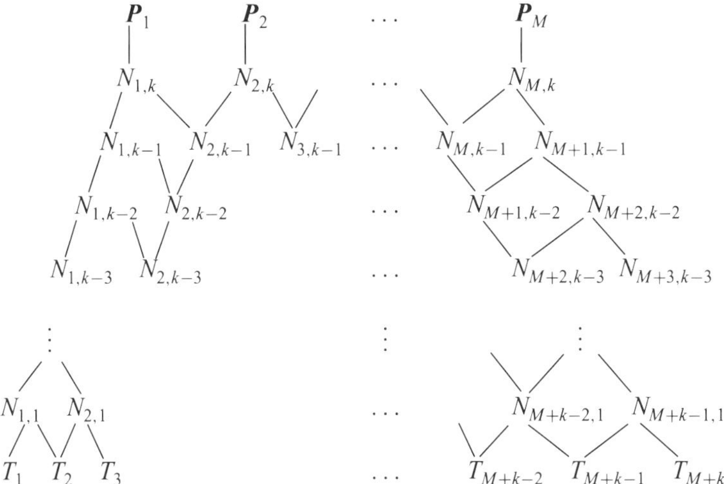

where direct dependence is indicated vertically by the solid lines and summations take place along the horizontal rows. One can see by following up the tree for a certain value range of the parameter t that the resulting B-Spline curve depends only on, typically, a smaller subset of the control points. This gives the curve what is called a 'local' control property. This means that changing the location of a control point will modify only a local portion of the B-Spline curve (only a finite subrange of the parameter t is affected by any single control point) leaving the rest unchanged. The B-Spline blending curves ($N_{m,k}$) are developed as a weighting function for the control points:

$$\sum_{i=1}^{M+(k-j)} N_{i,j}(t) = 1 \quad \text{(for all } t\text{)},$$

and one can show that the B-Spline curve is completely contained within the convex hull formed by the control points. These features are particularly important for predictable behavior when the user is adjusting the control points to design a free-form curve.

When creating a curve, the designer uses the control points for general shaping and the B-Spline curve is not required to pass through any of the control points. However, since these curves will often be used as pieces that ultimately must fit together, the designer typically wants to have direct control over the starting and ending point of the curve. This can be achieved by creating a k-tuple duplicate knot at each end of the knot vector, or:

$$T_1 = T_2 = \cdots = T_k = 0 \quad \text{and} \quad T_{M+1} = \cdots = T_{M+k-1} = T_{M+k} = T_{\max}$$

This forces the B-Spline curve to pass through the first and last control point and will create what is called the open or non-periodic B-Spline. If one has this k-tuple at each end of the knot vector and no other duplicate knots and is generating a uniform B-Spline, then $T_{\max} = M - k + 1$.

The user can modify the curve in many ways. Adding and moving control points are the easiest intuitively, but the curve can also be pulled closer to a particular control point by adding duplicate control points. As was hinted at in the previous paragraph, adding duplicate knots also has the effect of pulling the curve towards the control points but the two approaches are not the same. Multiple knots actually reduce the continuity of the B-Spline bending functions. To increase the potential level of control, without changing the physical shape of the current B-Spline curve, it is also possible to calculate appropriate new control points or to calculate and insert knots (and control points) to create a new B-Spline with the same shape as the previous one [15].

4.2.3.4 Expansion

Through the techniques described above, the B-Spline can be manipulated to create a wide variety of curves. However, manipulating the knot values has moved the user away from a more intuitive approach, and certain generalizations of the B-Spline add user-controlled parameters that (it is hoped) have a more intuitive feel to 'pushing' and 'pulling' the curve. One example, called the Beta-Spline, relaxes the first and second derivative continuity constraints by requiring only directional continuity and allowing the user to control the value of the ratio of the derivatives' magnitudes at the piecewise boundaries [16].

As can be seen in the previous discussion, the B-Spline curve need not actually pass through the controlling vertices. However, in many applications, an object is to be generated such that the user wants the model to pass through a set of user-defined points. In this case, there are algorithms for calculating an appropriate set of control vertices such that the resulting curve does interpolate a given set of data points [17]. The current geometric modellers typically leave the object-generating control-vertex mesh hidden from the user.

Once there is a model for curves, there is a 'recursive' approach for creating surfaces. The basic approach for increasing the dimensionality is to extrude the results into another dimension by substituting some previously created objects for the control points in the curve model. For example, start with a model that defines a curve, $\boldsymbol{C}$, as a blended function on M points: $\boldsymbol{C} = F(\boldsymbol{P}_m,\ m = 1, 2, \ldots, M)$. Now suppose one has enough data to create J different curves, $\boldsymbol{C}_k, k = 1, 2, \ldots, J$, where $\boldsymbol{C}_k = F(\boldsymbol{P}_{m,k}, m = 1, 2, \ldots, M)$. To create a two-dimensional surface, one simply replaces the points in the curve definition with the created curves: $\boldsymbol{S} = F(\boldsymbol{C}_k,\ k = 1, 2, \ldots, J)$.

Although current geometric modellers stop at B-Spline surfaces, it is mathematically possible to continue and extrude surfaces along curves to create a solid. However, in general, users are satisfied with assuming a homogeneous interior and thus are not interested in the ability to control the interior of the object; the bounding surface is all that is needed.

4.2.4 Manipulation

The object can now be thought of as a mathematically connected sequence of spatial locations. To visualize and to locate one object with respect to another, for example, to position or duplicate a feature on a primitive, the user will want to manipulate the original object. The movement of the object in space is represented as a sequence of transformations operating on the set of coordinates defining the object. This can be demonstrated through a simple example.

Suppose one has a pyramid: coordinates A, B, C, D, E with respective $[x, y, z]$ coordinates [0, 0 ,0], [1, 0, 0], [1, 1, 0], [0, 1, 0], [0.5, 0.5, 1] and with straight line segments between AB, BC, CD, DA, AE, BE, CE, DE. To rotate that pyramid 90 degrees counterclockwise about the x axis, one transforms the original coordinate matrix by multiplying with the appropriate rotation matrix (this example treats spatial coordinates as column vectors; transpose operations can be applied to treat coordinates as row vectors):

$$\text{3 by 3 rotation matrix} * \text{3 by } n \ (n = 5 \text{ in example) coordinates matrix} = \text{transformed 3 by } n$$

$$\begin{vmatrix} 1 & 0 & 0 \\ 0 & \cos(90) & -\sin(90) \\ 0 & \sin(90) & \cos(90) \end{vmatrix} * \begin{vmatrix} 0 & 1 & 1 & 0 & 0.5 \\ 0 & 0 & 1 & 1 & 0.5 \\ 0 & 0 & 0 & 0 & 1 \end{vmatrix}$$

$$= \begin{vmatrix} 0 & 1 & 1 & 0 & 0.5 \\ 0 & 0 & 0 & 0 & -1 \\ 0 & 0 & 1 & 1 & 0.5 \end{vmatrix}$$

If locations remain 3-tuples, then translation would require matrix addition compared to rotations being matrix multiplication. Introducing homogeneous coordinates (described in more detail in any book on computer graphics [18]) will keep all operations matrix multiplication by making the coordinate a 4-tuple: the fourth component of a spatial location is always one; if, after transformation, the fourth component is not one, the x, y, z values must be divided by the value of the fourth component prior to the coordinate becoming a spatial location. To expand the previous example: rotate the original pyramid by 90 degrees counterclockwise about the x axis and then translate by $[-1, 0, 0]$:

4 by 4 translation $*$ 4 by 4 rotation $*$ 4 by n ($n = 5$) coordinates
= transformed 4 by n

$$\begin{vmatrix} 1 & 0 & 0 & -1 \\ 0 & 1 & 0 & 0 \\ 0 & 0 & 1 & 0 \\ 0 & 0 & 0 & 1 \end{vmatrix} * \begin{vmatrix} 1 & 0 & 0 & 0 \\ 0 & \cos(90) & -\sin(90) & 0 \\ 0 & \sin(90) & \cos(90) & 0 \\ 0 & 0 & 0 & 1 \end{vmatrix} * \begin{vmatrix} 0 & 1 & 1 & 0 & 0.5 \\ 0 & 0 & 1 & 1 & 0.5 \\ 0 & 0 & 0 & 0 & 1 \\ 1 & 1 & 1 & 1 & 1 \end{vmatrix}$$

$$= \begin{vmatrix} -1 & 0 & 0 & -1 & -0.5 \\ 0 & 0 & 0 & 0 & -1 \\ 0 & 0 & 1 & 1 & 0.5 \\ 1 & 1 & 1 & 1 & 1 \end{vmatrix}$$

Since matrix multiplication is associative, the total transformation matrix becomes the matrix multiplication of the individual transformations. A generic transformation matrix, with spatial locations as column vectors, can then be analyzed as the upper left 3 by 3 submatrix representating rotations and reflections, the upper right 3 by 1 submatrix representing translations, the lower left 1 by 3 submatrix representing perspective modifications and the lower right 1 by 1 submatrix representing global scaling (the last two are used primarily for display on the two-dimensional (2D) monitor screens). Objects are generally considered rigid bodies thus spatial transformations should preserve Euclidean distance.

Current computer hardware for graphic visualization (the graphics 'engine') includes special hardware designed for matrix multiplication. This allows the user to manipulate the object on the monitor interactively without long computation delays.

Although matrix multiplication is associative, it is not generally commutative. Thus the order of multiplication will make a difference. Suppose one has m 4 by 4 transformation matrices, $[T_1], [T_2], \ldots, [T_m]$. If $[P]$ represents the matrix of coordinates, then $[T_m][T_{m-1}]\ldots[T_2][T_1][P]$ will create the set of coordinates representing the motion, in the global reference frame, of the object transformed by $[T_1]$ then $[T_2]\ldots$ through $[T_m]$ – as in the previous pyramid example. However, the reverse: $[T_1][T_2]\ldots[T_{m-1}][T_m][P]$, can be thought of as moving a local object set of coordinate axes by $[T_1]$ then $[T_2]\ldots$ through $[T_m]$ and then generating the object represented by $[P]$ in that local coordinate system. Alternatively, one could imagine the object remaining fixed but the viewer moving by $[T_1]$ then $[T_2]\ldots$ through $[T_m]$.

4.2.5 Yarn and fiber assemblies

As mentioned previously, the most appropriate model for the complex curves of a yarn and fiber assembly is to develop a surface boundary representation. For our structures, the best approach is to move, 'sweep,' a closed cross-section curve along a centerline path [19]. To locate the cross-section properly, one also will need the tangent and normal to the centerline path:

$$[\boldsymbol{C}] = \begin{vmatrix} C_X(t) \\ C_Y(t) \\ C_Z(t) \end{vmatrix} \text{ centerline curve; e.g. } \begin{vmatrix} \alpha_X & \beta_X & \gamma_X & \delta_X \\ \alpha_Y & \beta_Y & \gamma_Y & \delta_Y \\ \alpha_Z & \beta_Z & \gamma_Z & \delta_Z \end{vmatrix} * \begin{vmatrix} t^3 \\ t^2 \\ t \\ 1 \end{vmatrix}$$

for cubic polynomial;

$$[\widehat{\mathbf{tan}}] = \text{tangent} = \frac{\mathrm{d}[\boldsymbol{C}]}{\mathrm{d}s} = \frac{\mathrm{d}[\boldsymbol{C}]}{\mathrm{d}t}\frac{\mathrm{d}t}{\mathrm{d}s}, \text{ where } \mathrm{d}s = \sqrt{\mathrm{d}x^2 + \mathrm{d}y^2 + \mathrm{d}z^2};$$

$$[\widehat{\mathbf{norm}}] = \text{normal} = \frac{\mathrm{d}[\widehat{\mathbf{tan}}]}{\mathrm{d}s} \Bigg/ \left\| \frac{\mathrm{d}[\widehat{\mathbf{tan}}]}{\mathrm{d}s} \right\|$$

$$[\hat{\mathbf{b}}] = \text{binormal} = [\widehat{\mathbf{tan}}]\mathbf{x}[\widehat{\mathbf{norm}}]$$

The cross-section will be a closed curve defined in terms of its own parameter:

$$[\boldsymbol{X}_{\mathbf{sect}}] = \begin{vmatrix} X_{Sx}(u) \\ X_{Sy}(u) \\ X_{Sz}(u) \end{vmatrix}$$

The surface is then generated by moving the cross-section along the centerline path; assume that one wants the x, y, z direction of the cross-section to correspond respectively to the binormal, normal and tangent directions of the centerline:

$$[\boldsymbol{S}] = \begin{vmatrix} S_X(u,t) \\ S_Y(u,t) \\ S_Z(u,t) \end{vmatrix} = [\boldsymbol{C}] + X_{S_X}(u)\cdot[\hat{b}] + X_{S_Y}(u)\cdot[\widehat{\mathbf{norm}}] + X_{S_Z}(u)\cdot[\widehat{\mathbf{tan}}]$$

(Depending on the difficulty in obtaining ds, approximations could be used for the normal, binormal and tangent by applying rotation transformations along the centerline [19]).

As a concrete example of this construction, let us sweep an elliptical cross-section along a helical path.

$$\text{Helix, amplitude } A\text{, period } 2\pi P = \begin{vmatrix} C_X(t) \\ C_Y(t) \\ C_Z(t) \end{vmatrix} = \begin{vmatrix} A\cos\left(\frac{t}{P}\right) \\ A\sin\left(\frac{t}{P}\right) \\ \frac{t}{P} \end{vmatrix}$$

with $\mathrm{d}s = \mathrm{d}t\dfrac{\sqrt{A^2+1}}{P}$

$$\text{tangent} = \frac{1}{\sqrt{A^2+1}} \cdot \begin{vmatrix} -A\sin\left(\frac{t}{P}\right) \\ A\cos\left(\frac{t}{P}\right) \\ 1 \end{vmatrix}, \qquad \text{normal} = \begin{vmatrix} -\cos\left(\frac{t}{P}\right) \\ -\sin\left(\frac{t}{P}\right) \\ 0 \end{vmatrix},$$

$$\text{binormal} = \frac{1}{\sqrt{A^2+1}} \cdot \begin{vmatrix} \sin\left(\frac{t}{P}\right) \\ -\cos\left(\frac{t}{P}\right) \\ A \end{vmatrix}$$

If we assume the ellipse has minor axis of length 'b' and major axis 'a' and further agree that the minor axis will be placed along the centerline normal, the major axis along the binormal (most cross-sections are planar and are meant to lie in the plane perpendicular to the tangent), a parametric definition of our ellipse is:

$$x_{\text{ellipse}} = \sqrt{\frac{a^2b^2}{a^2\sin^2(u) + b^2\cos^2(u)}} \cdot \cos(u),$$

$$y_{\text{ellipse}} = \sqrt{\frac{a^2b^2}{a^2\sin^2(u) + b^2\cos^2(u)}} \cdot \sin(u), \; -\pi < u \leq \pi$$

Thus we end up with the surface:

$$\begin{vmatrix} A\cos\left(\frac{t}{P}\right) \\ A\sin\left(\frac{t}{P}\right) \\ \frac{t}{P} \end{vmatrix} + \sqrt{\frac{a^2b^2}{a^2\sin^2(u) + b^2\cos^2(u)}} \cdot \frac{\cos(u)}{\sqrt{A^2+1}} \cdot \begin{vmatrix} \sin\left(\frac{t}{P}\right) \\ -\cos\left(\frac{t}{P}\right) \\ A \end{vmatrix}$$

$$+ \sqrt{\frac{a^2b^2}{a^2\sin^2(u) + b^2\cos^2(u)}} \cdot \sin(u) \cdot \begin{vmatrix} -\cos\left(\frac{t}{P}\right) \\ -\sin\left(\frac{t}{P}\right) \\ 0 \end{vmatrix},$$

or

$$\begin{vmatrix} [A - K(u)\cdot\sin(u)]\cdot\cos\left(\frac{t}{P}\right) + K(u)\cdot\frac{\cos(u)}{\sqrt{A^2+1}}\cdot\sin\left(\frac{t}{P}\right) \\ [A - K(u)\cdot\sin(u)]\cdot\sin\left(\frac{t}{P}\right) - K(u)\cdot\frac{\cos(u)}{\sqrt{A^2+1}}\cdot\cos\left(\frac{t}{P}\right) \\ \frac{t}{P} + K(u)\cdot\frac{A * \cos(u)}{\sqrt{A^2+1}} \end{vmatrix}$$

where

$$K(u) = \sqrt{\frac{a^2b^2}{a^2\sin^2(u) + b^2\cos^2(u)}}$$

Although topologically correct as a swept surface, the final surface must still be checked for geometric validity. For example, if $a > A$ then the surface would self intersect and therefore could not be a valid solid model.

4.3 Applications

4.3.1 Visualization

Scientific visualization was the original engineering use of computer graphics. Computer hardware and software will allow the user to see, manipulate and move through a defined object. Of course, if interactive computer manipulation is required, the object being modelled will be treated as a rigid body in the graphic manipulation. Computer power is still too slow to permit an interactively manipulatable simulation although is is certainly being discussed with super computers.

Visualization can provide insights into a structure that would be impossible to develop otherwise. Figure 4.4 shows a visualization of an analytic model for a structure created by twisting an already twisted bundle of elements. By using the graphics to traverse the length of the object interactively or by varying the parameters in the model, the user can quickly determine cross-sections, verify tightness of the structure, and so on. Colour coding, translucency and other display techniques can be used to aid in visualization of complex structures.

Two obvious applications would be to visualize a real object by obtaining experimental information about the structure [20] and using that to generate the solid model of the object, or to use the visualization to aid in verifying the appropriateness of an analytic representation or user-created model of the structure [19]. With the creation technique of sweeping cross-sections along a centerline curve, a basic template is available for putting the experimental or analytic frame information in a form suitable for a geometric modelling package.

4.3.2 Integrating database

Another powerful outcome of the solid model is its ability to become the information database that defines the object. Since a true solid model would contain all the necessary information to define uniquely and completely the object, it is by definition the natural core for driving other digital applications. There are meshing interface algorithms to create a finite-element discretization

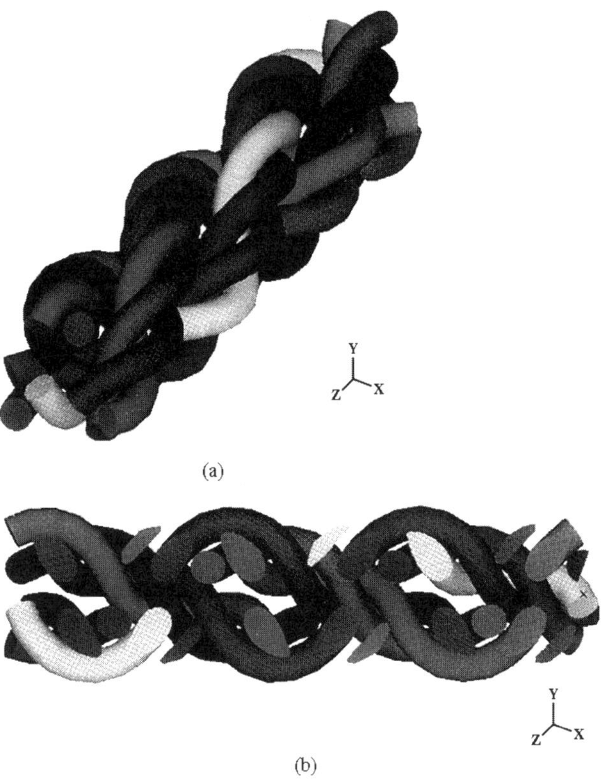

4.4 (a) Visualization, 3 by 3 twisted ply. (b) Visualization, 3 by 3 twisted ply, longitudinal section.

(or boundary-element surface characterization) directly from the geometric computer description of the object. Manufacturers are able to use the geometric model to generate the code necessary to drive numerically controlled machines or to generate a stack of cross-sections needed for rapid prototyping applications.

The current model for computer-integrated product development typically has the solid model as the database hub around which each of the types of interface described previously would be the spoke leading to its particular application. Once a geometric model is available, the researcher can take advantage of these links to perform other analyses. However, the differences between yarn and fiber assemblies and typical engineering structures will need to be addressed.

For example, it would not be reasonable to model an entire piece of cloth as a collection of a large number of yarns, each yarn made up of a large number of filaments, at the filament level. By using modelling on one scale, it might be possible to generate a continuum set of material properties, perhaps varying over the structure, that would allow a larger scale of analyses. However, in a finite-element analysis of a textile structure modelled as a collection of yarn elements,

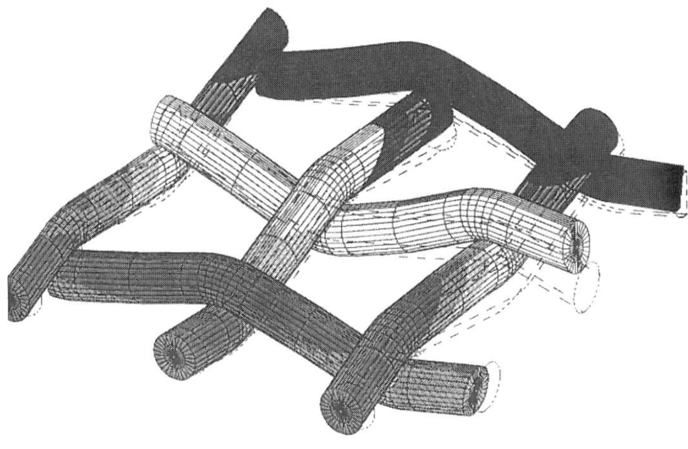

4.5 Finite-element simulation of plain weave cell under uniaxial tension. Dashed lines represent original geometry, shading indicates *Z* movement.

even with appropriate material properties, it is not a trivial problem to generate the appropriate boundary conditions, to generate realistic loading or to handle the interfaces between the interacting elements. Yet, these techniques can provide approximations to textile mechanical behavior, see Fig. 4.5 [21].

4.3.3 Developing models

One final use of this technology would be to use solid modelling techniques to develop fabric models. For example, there has been work studying the geometric limits to twist in a yarn bundle [22] and shear in a woven fabric [23]. Compression was modelled by assuming an ellipical cross-section for the yarn and then varying the ellipse's aspect ratio, b/a, where ellipse b is in the direction normal to the path of the centerline of the yarn. The effect on fiber volume in a unit piece was then analyzed.

If the twist is modelled essentially as a helix, there is a theoretical maximum volume fraction for an elliptical cross-section yarn bundle. This maximum is about 87% and occurs when there is no twist, Fig. 4.6. Twist can help lock the bundle together but at the cost of reducing the volume fraction. For multiyarn bundles, a reasonable approximation for calculating volume fraction is to use the straight-yarn simplification for aspect ratios up to one and then asymptotically approach a volume fraction of $\pi/4$. By choosing a trigonometric model for yarns

3-Yarn Bundle

ellipse axes ratio, $\frac{b}{a} = \frac{1}{\sqrt{3}}$

bundle center to yarn center = $\frac{2\sqrt{ab}}{\sqrt{2\sqrt{3}}}$

volume fraction = 0.866

4-Yarn Bundle

ellipse axes ratio, $\frac{b}{a} = 0.432$

bundle center to elliptical yarn center $= 0.829 * 2\sqrt{ab}$

radius of circular yarn = $\sqrt{ab}$

volume fraction = 0.878

4.6 Yarn bundles with maximum volume fraction.

in a plain weave, maximum shear angle $(90 - \theta)^\circ$ can be determined, where θ° is the acute angle between warp and fill yarns, see Fig. 4.7.

4.4 Conclusions

Geometric modelling technology is beginning to have an impact on the scientific modelling of composites. Most of the work to date has been in visualization of the complex structures although some of the traditional engineering analyses that are integrated around the geometric database have also been used, e.g. finite-element analysis. Current geometric modelling technology seems to have untapped potential available for advancing the study of fiber-reinforced composites:

- The fundamental mathematics of weighting the curve based on control parameters has physical similarities to some of the manufacturing process parameters, for example, weaving.
- Computer-graphic mapping techniques are digital methods for storing pattern information.
- B-Spline-like definitions may prove to be appropriate for defining and storing information on a particular topology and structure.

In addition, this particular approach automatically links the developed research model to the computer-graphics visualization world and to computer-aided engineering analysis packages and applications.

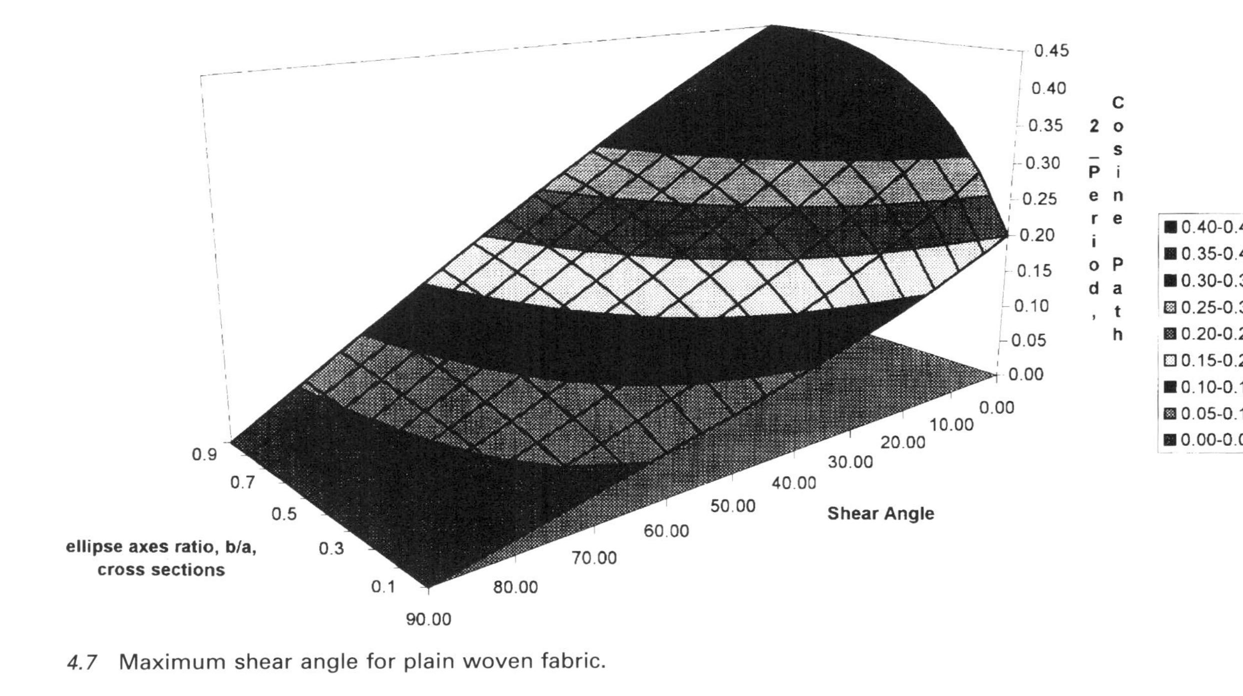

4.7 Maximum shear angle for plain woven fabric.

References

1. F T Peirce, The geometry of cloth structure, *J. Text. Inst.* 1937 **28** T45–T96.
2. T Yasuda, S Yokoi, J Toriwaki and K Inagaki, A shading model for cloth objects, *IEEE Computer Graphics Appl.* 1992 **12**(6) 15–24.
3. J Weil, The synthesis of cloth objects, *Computer Graphics* 1986 **20**(4) 49–54.
4. A A G Requicha and H B Voelcker, Solid modeling: a historical summary and contemporary assessment, *IEEE Computer Graphics Applic.* 1982 **2**(2) 9–24.
5. A A G Requicha and H B Voelcker, Solid modeling: current status and research directions, *IEEE Computer Graphics Applic.* 1983 **3**(7) 25–37.
6. A A G Requicha and J R Rossignac, Solid modeling and beyond, *IEEE Computer Graphics Applic.* 1992 **12**(5) 31–44.
7. M Neal, Keep it simple, *IEEE Computer Graphics Applic.* 1989 **9**(5) 3–5.
8. C Upson and M Keeler, V-buffer: visible volume rendering, *Computer Graphics* 1988 **22**(4) 59–64.
9. P F Jacobs, *Rapid Prototyping & Manufacturing: Fundamentals of Stereolithography*, Society of Manufacturing Engineers, Dearborn, MI, 1992. ISBN 0872634256.
10. B A Barsky, Computer-aided geometric design: a bibliography with keywords & classified index, *IEEE Computer Graphics Applic.* 1981 **1**(3) 67–108.
11. B A Barsky, A description and evaluation of various 3-D models, *IEEE Computer Graphics Applic.* 1984 **4**(1) 38–52.
12. D F Rogers and J A Adams, *Mathematical Elements for Computer Graphics*, 2nd edn, McGraw Hill, New York, 1990. ISBN 0070535299.
13. M Mortenson, *Geometric Modelling*, John Wiley, New York, 1985. ISBN 0471882798.
14. G Farin, *Curves and Surfaces for Computer Aided Geometric Design: A Practical Approach*, 3rd edn, Academic Press, Boston, 1993, p233. ISBN 0122490525.
15. W Boehm, Inserting new knots into B-spline curves, *Computer Aided Design* 1980 **12**(4) 199–201.
16. B A Barsky and J C Beatty, Local control of bias and tension in beta-splines, *Computer Graphics* 1983 **17**(3) 193–218.
17. B A Barsky and D P Greenberg, Determining a set of B-spline control vertices to generate an interpolating surface, *Computer Graphics Image Processing* 1980 **14** 203–226.
18. J Foley, A van Dam, S Feiner and J Hughes, *Computer Graphics, Principles and Practice*, 2nd edn, Addison-Wesley, 1990. ISBN 0201121107.
19. M Keefe, D C Edwards and J Yang, Solid modeling of yarn and fiber assemblies, *J. Text. Inst.* 1992, **83**(2) 185–196.
20. J E Masters, R L Foye, C M Pastore and Y A Gowayed, Mechanical properties of triaxially braided composites: experimental and analytical results, *J. Composites Technol. Res.* 1993 **15**(2) 112–122.
21. S Means and M Keefe, Finite-element simulation of deformations in woven structures under tensile loading, *Proceedings of the Second International Conference on Composites Engineering*, University of New Orleans, New Orleans, August 21–24, 1995.
22. M Keefe, Solid modeling applied to fibrous assemblies. Part I: Twisted yarns, *J. Text. Inst.* 1994 **85**(3) 338–349.
23. M Keefe, Solid modeling applied to fibrous assemblies. Part II: Woven fabrics, *J. Text. Inst.* 1994 **85**(3) 350–358.

5

Characterisation of yarn shape in woven fabric composites

STEVEN W YURGARTIS AND JULIUS JORTNER

5.1 Introduction

Properties of fiber-reinforced composites are affected by the orientations of the fibers. When textile fabrics are used as reinforcement, the fibers typically are organized in more-or-less collimated bundles called yarns. By yarn shape we mean the path the yarn's axis follows in space and the geometry of the yarn's cross-sections, which may vary along its length. To a first approximation fibers within a yarn are oriented parallel to the yarn's axial path; thus the study of yarn shape may be viewed as a shortcut to the characterization of fiber orientations.

Yarns are used to make a wide range of fabric architectures including braids, knits, mats and weaves. In this chapter we focus on laminates made with woven cloth, which is the most common form of woven reinforcement in laminated composites. Even within the restricted scope of two-dimensional (2D) orthogonally woven cloth there is a surprising variety of cloth architectures, as will be shown below.

In a composite laminate the stack of cloth layers is made rigid by a solid matrix. In such laminates, yarn shape is determined not only by the factors operating in the as-woven cloth, which are relatively well understood, but also by the spatial relationships among cloth layers and the various distortions caused by relative movements (compaction and sometimes shear) that are part of composite fabrication [1]; this point is illustrated in Fig. 5.1. The shape of a yarn segment reconstructed from a series of cross-sections is shown in Fig. 5.2; note the complex geometry of yarns which are part of a composite.

This chapter introduces some of the terminology used in describing cloth geometry, defines measures of yarn shape in laminate composites, describes techniques for the measurement of several yarn shape parameters, briefly reviews the known and suspected roles of yarn shape in composite material properties and concludes with some discussion of areas where more understanding is needed.

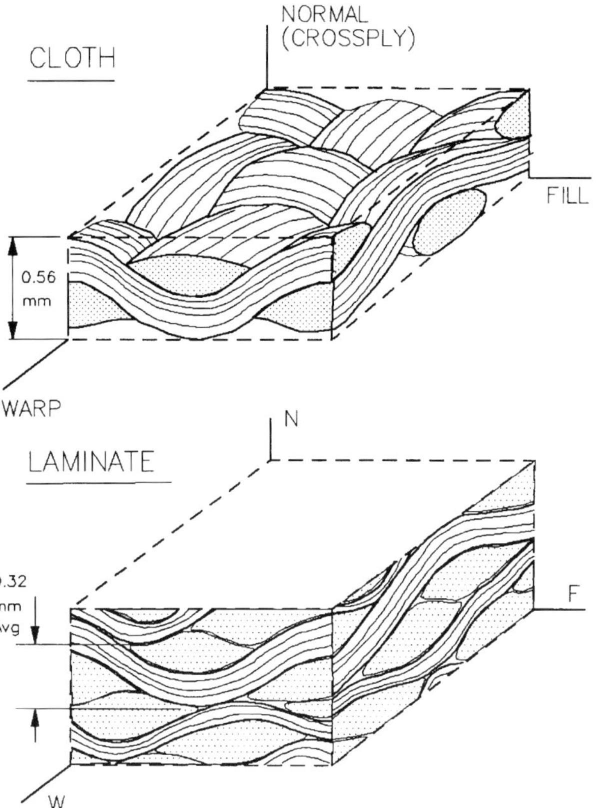

5.1 Contrast between yarn shapes in a plain-weave cloth and in a composite, from Jortner [1].

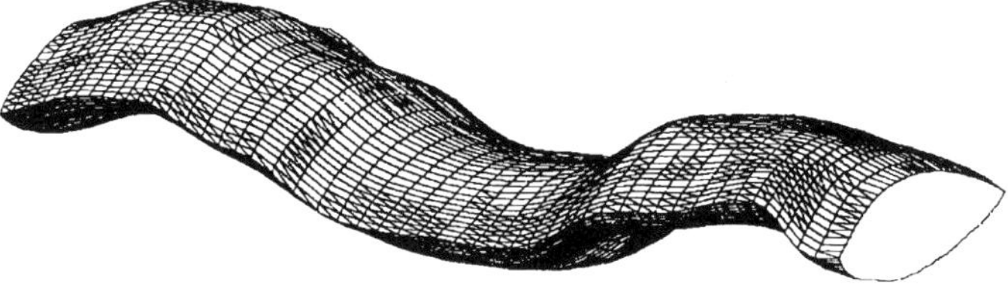

5.2 Three-dimensional reconstruction of a yarn segment in a composite like that of Fig. 5.1.

5.2 Yarn and cloth terminology

As expected, much of the terminology for composite material cloth reinforcement comes from the textile industry. Hearle *et al.* [2] give an excellent review of textile yarn and fabric terminology, as do Schwartz *et al.* [3]. The following discussion is taken largely from their works.

5.2.1 Yarns

Yarns used in reinforcement of composite materials are generally composed of continuous filaments (fibers). Yarns can also be composed of interlocked short fibers, in which case they are called spun yarns.

The most common textile descriptions of yarn geometry are yarn twist and yarn count. Twist is often measured in turns-per-inch (tpi), while count is the mass per unit length or a linear density. Typical units of yarn count are the tex, which is g km^{-1}, and the denier, which is g $(9000\ m)^{-1}$. The advantage of using yarn count as a measure of size (dimension) is that the yarn radius is often difficult to define while count is an easy quantity to measure. The surface twist angle is defined as the angle between the axis of the yarn and the axis of a surface fiber; note that fiber angle is necessarily a function of the distance from the center of the yarn. Yarns are twisted to improve their structural coherence. Since yarns in composite reinforcement are soon to be surrounded by a supportive matrix, they are often found with minimal or no twist. Untwisted yarns are sometimes called tows. Yarns are also commonly described by the number of filaments per yarn, which of course is related to yarn size.

5.2.2 Fabrics

Yarns are assembled into fabrics. The primary purpose of the assembly is to make a structure that has good in-plane properties but a very low bending stiffness. Indeed, these properties are the primary attractions of fabric as a reinforcement for composite materials. Often composite reinforcement fabrics are chosen for their ability to 'drape' [4].

There are many ways to assemble yarns into fabrics. For example, knits interlock yarns by looping around neighbors and braids have each yarn interwoven with every other yarn, usually at an acute angle. Weaves, the most common textile geometry, are produced on a loom that weaves yarns at right angles to one another. Continuous yarns laid out longitudinally on the loom are the warps (also called ends), while the yarns that are shuttled transversely across the loom perpendicular to the warps are called fills (also picks or the weft).

The weave type is determined by the way that the warps and fills are interlaced [3]. A good way to represent weave patterns is on square grid paper; columns represent the warps while the rows are the fills. At the location where a warp yarn weaves over a fill the square on the grid paper is filled; additional squares are filled in if need be until the warp again passes under the fills. Three of the more common weave types used in composite reinforcements, the plain weave, twill weave and the satin weave, are illustrated in Fig. 5.3. An interlacing point is often described (loosely) as a place in the cloth where a warp yarn crosses over then under a fill yarn. We refine that definition with the term interlacing region and a description of the 'order' of the interlace. The order of an interlace region is the

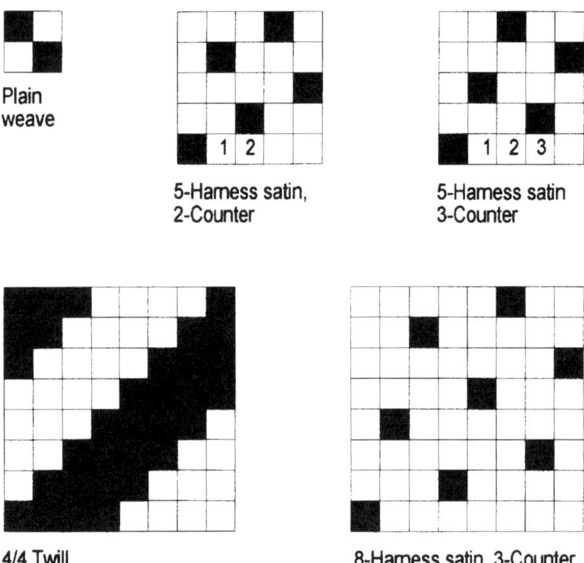

5.3 Grid pattern showing plain, twill and satin weaves.

number of transverse yarns crossed by a single yarn between the point it passes from the bottom to the top of the cloth and the point it passes again to the bottom, or vice versa. Thus, a plain weave has interlaces all of order I_1, and the 5-harness satin illustrated in Fig. 5.3 has interlaces of order I_1 and I_4.

The twill weave is represented as a fraction, with the top number being the number of fills that a warp yarn crosses over and the bottom number being those it crosses under. A characteristic of a twill weave is that the adjoining yarns have diagonally adjacent interlacing points. If top and bottom numbers are the same, the twill is called 'balanced'.

Satin weaves are distinguished by each warp passing over a single fill before floating under several fills, then repeating the pattern, with the restriction that I_1 interlaces on adjacent warps do not touch. The number of harnesses used to weave a satin weave is the number of warp yarns, or squares on a row of grid paper, that are needed to define one fill wavelength. For example, a 5-harness satin has a yarn that passes over one and then under four more yarns before repeating. (Notice that a simple plain weave, where warps interlace with every fill, is a 2-harness weave.) There are variations, often overlooked, that are possible within a particular satin weave. To distinguish between different types of satins with the same harness number the term counter is used. On grid paper, the counter of a satin is the number of squares from one interlacing point to the next on adjacent fills. From the formal definition of satin weaves, which prohibits adjacent interlacing points on adjoining yarns, counters of 1 are forbidden; an attempt to use a counter of 1 in a 5-harness satin would produce, formally, a 1/4 twill weave. By now the reader new to this subject should already be gaining an appreciation for the bewildering range of cloth geometries.

Chou [5] has used a single parameter, n_g, the interlace frequency, as a simplified representation of weave geometry. While a useful simplification for mechanics modeling, a single parameter cannot accurately describe weave architecture.

Also worth noting are the variations in geometry possible within a single fabric style. For example, an 8-harness satin can be made with variety of yarn sizes and a variety of yarn spacings, and the warp yarns need not be the same as the fill yarns! Yarn spacing is described by 'ends per inch' for warps and 'picks per inch' for fills (the odd terms come from loom mechanisms). Taken together, these spacings are the count of the fabric.

Fabrics with equal warp and fill counts are called balanced. Each variation gives a significantly different cloth that may perform quite differently as a composite reinforcement. The 'tightness' of the yarns in the fabric is sometimes defined by the cover factor. When yarns are woven together so tightly that they cannot slide relative to one another this is called the jammed condition. In composite applications a common (but weak) measure of yarn size and yarn cover is the areal density of the fabric, typically in units of g m^{-2}.

Finally, woven fabrics can have a 'top' and 'bottom', also commonly referred to as the face and back of the cloth, based on the face-up position of the cloth in the loom. For many purposes the definition of which side is top is arbitrary, all that is required is consistency. The ply normal is defined here as pointing in the direction of the 'top' of the ply. When cloth lamina are laminated with both normals in the same direction, the resulting sequence has been defined as stacked; when they are opposite, the sequence is called folded [6]. Differences between stacking and folding are important in laminates of satin weaves because, for example, the face may be mostly warps while the back is mostly fills. Then, except where I_1 interlacings occur, stacked satins laminate with fills against warps, whereas folded satins laminate with fills against fills and warps against warps. Unbalanced twill cloth laminates also would be influenced by the choice of stacking or folding whereas plain weave laminates would not.

5.2.3 Remarks about fabric geometry

Many details of fabric geometry have been left out of this brief description. Nonetheless, the description should alert the reader to the potential complexity of fabric geometry. There are many, many variations of fabric architectures. Indeed, one of the continuing challenges for designers of composite materials is to invent fabric architectures that provide reinforcement in the desired directions and magnitudes for each structural situation. Some variations in fabric architectures that are being investigated for composite reinforcement, to which we could not do any justice in this chapter, include mats, triaxial weaves, three-dimensional (3D) interlock fabrics, knits and non-planar braids [7–10].

5.3 Descriptors of yarn shape

5.3.1 Yarn path

The yarn path is defined by the yarn centerline. Several geometric features of yarn path, as seen on polished sections of composites like Fig. 5.4, have been defined [11], and are illustrated in Fig. 5.5.

Inclination angle is the angle between the tangent to the yarn centerline path and the mean yarn direction (which is taken to be in the plane of the cloth). The distribution of inclination angles gives a statistical representation of yarn shape with respect to orientation.

Crimp angle is the maximum absolute value of inclination angle between an adjacent yarn path peak and trough. As it represents the extremes of reinforcement deviation from the lamination plane, the crimp angle may be expected to influence strength and stiffness. Notice the distinction between the definition of crimp angle, given here, and the textile term yarn crimp, which is the percent by which the length of the yarn before weaving (essentially equal to the length of the straight yarn) exceeds the length of the woven cloth.

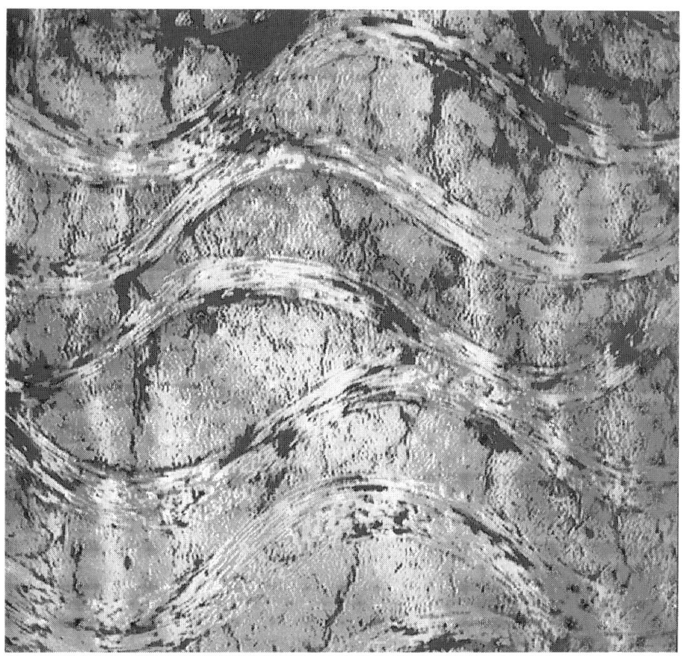

5.4 Micrograph of a plain weave composite section.

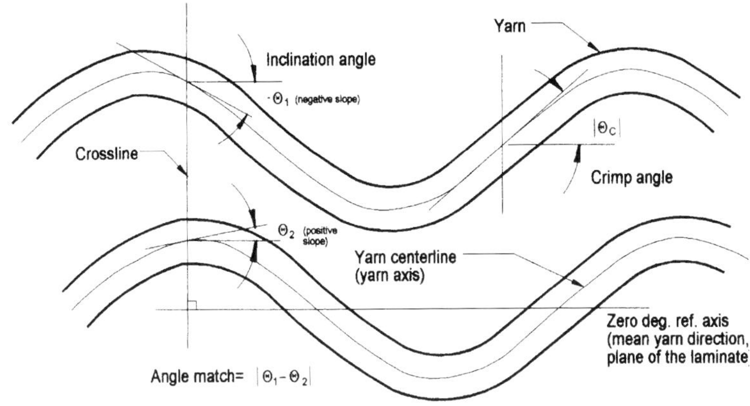

5.5 Sketch showing definitions of yarn shape parameters.

Wavelength is the distance, along the mean yarn path, between two nearest peak or two nearest trough points. Wavelength is a variable quantity that can have a range of values for a single reinforcement fabric.

Ramp amplitude is the height of the yarn path between adjacent peak and trough points. The ramp is the portion of the yarn path between an adjacent peak and trough. From this definition easily follow definitions of ramp length and ramp angle. Notice that the ramp amplitude as defined here is twice the usually defined amplitude of a sinusoidal wave; the difference is necessary because yarn paths are often not symmetrical about the midplane of the cloth.

5.3.2 Internal geometry

The internal geometry of yarns includes twist, packing fraction and cross-section shape. Twist and packing fraction have already been defined.

A description of yarn cross-section shape can be approached by measuring the cross-section area and the major and minor caliper dimensions. Since yarns in a composite are rarely circular in cross-section it is not often of much value to describe yarn diameter. Previous efforts to characterize yarn cross-section shape can be found in work by Du *et al.* [12] and Vas *et al.* [13] with the most detailed work done by Xu *et al.* [14].

The shape of individual fibers which make up a yarn may also be of interest. The cross-sectional shape of the fibers and the axial path of the fibers will have a direct effect on the packing reaction and related properties such as laminate compaction and cure shrinkage [15]. Another potentially interesting feature is the distribution of fibers within a yarn, which is not necessarily uniform. However, little work has been done to quantify such details of internal yarn structure.

5.3.3 Nesting

Nesting refers to the settling together of cloth layers so the boundary between two layers is not a plane, but rather a complex surface, as suggested in Fig. 5.1. Nesting can be expected to have an effect on interlaminar shear properties of composites.

Varieties of nesting and some potential effects have been discussed by Jortner [1,16]. For plain weave laminates there are several basic nesting possibilities (shown as a two-dimensional schematic in Fig. 5.6). The best laminate compaction probably is achieved with the cooperative nesting implied in Fig. 5.7. However, variations in yarn counts are likely to cause alterations of nesting and collimation even in otherwise well nested laminates.

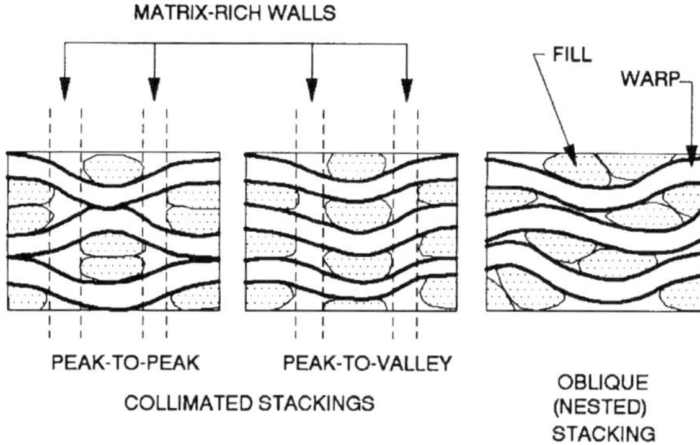

5.6 Sketches of basic nesting possibilities, from Jortner [1].

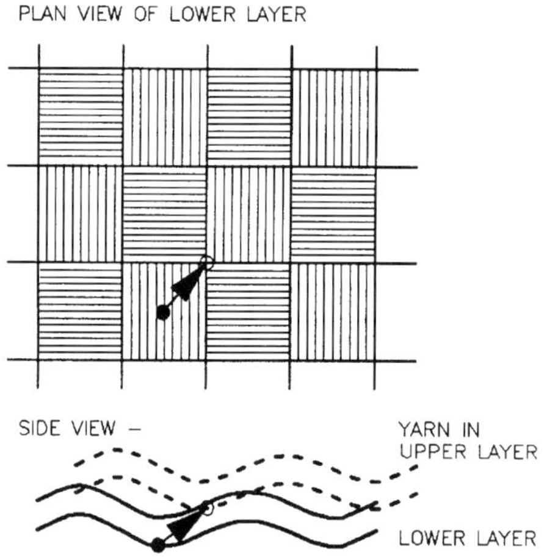

5.7 Displacement of one plain weave layer over another for optimal nesting (shown by arrow).

Unfortunately there does not yet appear to be a good quantitative description of nesting in relation to yarn shape. Yurgartis *et al.* [11] defined angle match as a first attempt. Angle match is the absolute value of the difference between inclination angles of adjacent yarns. If yarns were periodic and uniform in shape, angle match would be derivable from the phase match between paths of adjacent yarns. An advantage to using angle match rather than phase angle is that a periodic yarn shape is not required. However, this metric is rather difficult to interpret; the interested reader is referred to Yurgartis *et al.* [11] for details.

5.4 Techniques for the measurement of yarn shape

5.4.1 Manual methods

The simplest approach for measuring yarn shape is to observe a polished specimen through a microscope fitted with an rotatable reticle and angular scale, or an angular reticle, or a rotational stage. Alternatively, micrographs can be examined with ruler and protractor in hand. Such hand-and-eye methods are laborious, but can be quite effective if done carefully [17, 18]. The advantage of hand–eye methods is their accessibility and conceptual ease of implementation. No sophisticated hardware or software is required. New measures are readily implemented on a small scale. However, statistical confidence is limited by the time and patience of the investigator, and the measures are usually the simpler ones, like crimp angles, amplitudes and wavelengths of the yarn paths. The reader is referred to the discussion in the following section for advice on experimental details that should be considered in any method.

5.4.2 Computer-aided image processing

The statistical nature of microstructural measurements makes large sample sizes desirable, and when coupled to the complexity of the microstructure, the use of computer-aided image processing is attractive. Yurgartis *et al.* [11] describe an approach to characterizing yarn shape which is the basis of the following discussion. A similar approach for dry cloth was independently developed by Xu *et al.* [19].

The measurement of inclination angle, crimp angle and angle match proceeds in three stages:

- obtaining a micrograph which reveals the morphology of interest,
- identifying the yarn boundaries and fitting a function to match the boundaries, and
- extracting information from the yarn boundary data.

The technique is illustrated here with data collected from plain weave carbon–carbon composites, but is readily adaptable to other composite systems.

5.4.2.1 *Obtaining a micrograph for yarn shape measurements*

Obtaining a micrograph for the purpose of these measurements is quite straightforward. Magnifications available from standard reflected light microscopes are suitable. Care must be taken with low magnification stereo microscopes to avoid the complications from the stereo optics. Accurate dimensional measurements require that a stage micrometer micrograph be prepared in parallel with specimen micrographs. Readying the specimen requires typical metallographic preparation techniques. The process is necessarily destructive. A specimen must be cut from the material and here care is required to make this cut as close to a principal material direction as possible; that is, parallel to the warp or fill yarns. The specimen is potted in a polymer resin (often epoxy that has low cure shrinkage), polished and photographed so as to obtain a digital micrograph.

In collecting these micrographs two important considerations must be kept in mind. First, a decision must be made on how many micrographs must be collected for analysis and how these will be collected. In all such collections of data, the sampling of areas to be viewed, the magnification at which observations are gathered and the number of measurements taken, must all be appropriate to the use that will be made of the data. Our experience shows that sample sizes of about 100 crimp angles are desirable for good statistical representation. Images are collected from random non-overlapping fields around the available specimen area. The second requirement is that the specimen orientation be such that the average longitudinal yarn direction is parallel to an edge of the digital micrograph. The extraction of angle measurements force this condition on the data collection. Once the specimen orientation is aligned, it must be maintained constant for all subsequent micrographs within a data set. Notice that this requirement makes interpretation of data from multiple specimens difficult and not advisable. Figure 5.4 is an example of a micrograph ready for the next step in the measurements.

5.4.2.2 *Identification of yarn boundaries*

The method of the second step, identifying the yarn boundaries, depends on the nature of the material. If there is sufficient contrast between the longitudinal yarns (yarns in the plane of the section) compared with transverse yarns (yarns normal to the plane of the section), then simple image processing techniques such as thresholding and edge detection can be used [20]. Often there is not sufficient contrast between longitudinal and transverse yarns. In this case a possible approach is to segment the image using a texture mapping technique [20]. Yet another approach is to use a semi-manual technique in which the operator uses a mouse-and-cursor setup to identify points along the yarn boundaries. We have found this latter approach to be most generally efficient, so additional description of the technique is given.

In the semi-manual technique, points defining the edges of the yarns (about 12 per yarn) are identified by the operator using a mouse-and-cursor setup. An equation is smoothed through the points to approximate the yarn boundary. A seventh order polynomial was chosen for convenience and was found to work well. Good fits to yarn boundaries are obtained. Figure 5.8 shows the functional fits to the yarn boundaries superposed on the original image; the white lines are the polynomial fits to the yarn boundaries. Although operator input tends to slow the measurement rate, by making the input simple the time investment and accuracy prove to be efficient compared with the alternative computational investment. Operator input takes advantage of the superb pattern recognition capability of the human vision system. It can identify points on the yarn boundaries when there is little contrast between longitudinal and transverse yarns, as in the images of the carbon–carbon composites illustrated here, and can readily avoid computer-aided image processing complications encountered in distinguishing yarns that touch each other. An important goal is to minimize the sensitivity of the measurements to the operator. In this case the strategy of smoothing a function through the operator-selected points makes the location of a particular point uncritical. In a variation of this approach we have let the computer

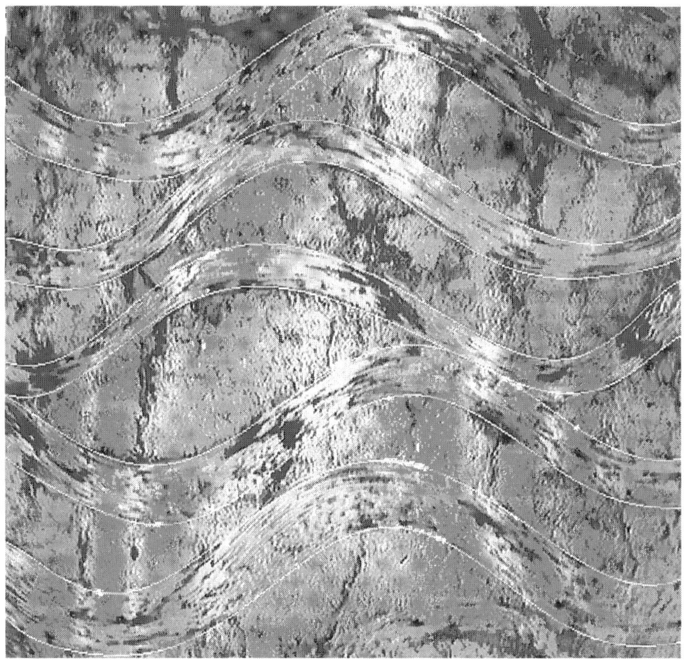

5.8 Functional fits to the yarn boundaries micrograph.

attempt to locate points on the yarn boundary using a texture recognition algorithm and then the human operator has corrected any bad guesses. This speeds up the process by a factor of about 5. The typical operator time is 1 min to identify the yarn boundaries in an image such as Fig. 5.4.

5.4.2.3 Extraction of yarn shape measures from the boundary functions

The third stage of the process takes the yarn boundary fits and extracts information. For convenience of discussion, we define the x-axis as parallel to the bottom edge of the image and the y-axis parallel to the left edge of the image. Definition of yarn boundaries as described above results in an output file containing the polynomial coefficients for each yarn boundary segment, $F(y(x))$. This information can be used to collect the desired distributions of yarn shape measures. Both the top and bottom boundaries of the yarns are recorded since yarn thickness is not always uniform. As it seems desirable to have data reflect the path of the yarn centerline, for each yarn segment the midpoint along the y-axis between the upper and lower boundaries is determined at 15 uniformly spaced intervals along the yarn. Another seventh order polynomial is fitted to these midpoints and becomes the functional representation of the estimated yarn centerline.

Inclination angles are easily found from the first derivative of the polynomial fit. Samples of the inclination angles along a yarn are taken at the intersections of 'crosslines' and yarns. Crosslines are lines parallel to the y-axis and traverse the complete image field. Placement of crosslines near the edge of the image is to be avoided; inaccuracy near the edges results from misbehavior of the polynomial fit which has no constraints beyond the image boundaries. The number of inclination angle measurements depends on the number of digital images analyzed and the number of crosslines chosen; the data presented in Fig. 5.9 were collected from 25 crosslines randomly spaced per image field.

There now arises a question of the accuracy of the absolute value of the angles measured. Since the specimen was probably initially aligned by eye on the microscope stage, the average yarn direction (defined as the zero-degree direction) is not accurately known *a priori*. However, the definition that the average yarn direction is at zero degrees provides a partial solution. Thus, the inclination angle data are shifted until the average inclination angle is zero. This is not a completely satisfactory solution, unfortunately, since it assumes yarn shapes have symmetric inclination angle distributions. If this is not the case, the shifting to a mean of zero will give a false representation of the angle magnitudes, although the distribution shape will not be affected. Consideration of a saw-tooth waveform will make this clear. In our experience the uncertainty is small, shifts were always less than 1.6 degrees, but for unusual weaves this issue might become important.

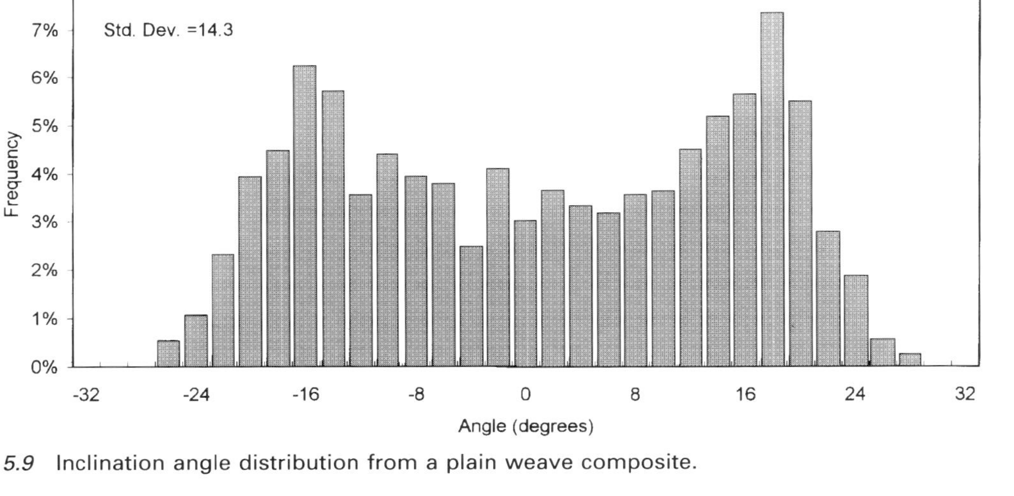

5.9 Inclination angle distribution from a plain weave composite.

Crimp angles are found by searching each yarn shape function for inflection points. Points along each yarn centerline are polled for the first and second derivative at each point. Peaks or troughs are identified by the first derivative changing sign. Between a peak and a trough the inclination angle is measured wherever there is an inflection point revealed by the second derivative changing sign. Consistent with the definition of crimp angle, the largest inclination angle in each peak-to-trough interval is recorded as the crimp angle. To compensate for initial specimen misalignment, crimp angle distributions are shifted by the same shift required to bring the inclination angle distribution to a mean of zero degrees. An example of crimp angle distribution taken from the material of Fig. 5.5 is shown in Fig. 5.10.

Angle match distribution is calculated using a slight variation of the crossline method. Along a single crossline, inclination angle is recorded along with the angle difference between adjacent yarns. The data consist of two columns, the first containing the inclination angle of the first yarn, and the second containing the inclination angle difference between this first yarn and its lower neighboring yarn. Example data are given by Yurgartis *et al.* [11].

Based on test cases involving known angles and using the 512×480 pixel digital images shown in the figures, the accuracy of the angle measurements is estimated to be within $\pm 1^\circ$. From a series of repeated measurements, in which the operator starts from the same image but redefines the yarn boundaries each time, reproducibility of the angle measurements is estimated to be about $\pm 0.5^\circ$. Similar reproducibility is found between experienced operators.

5.4.2.4 Other measures from boundary functions

Other morphological measures are available from the functional fits to the yarn shape. Of particular interest to modeling efforts is approximating the yarn shape with a sinusoid, perhaps using a least squares method. Yarn volume fraction measurements are readily available from the first stage data using established methods [21]. Information about yarn wavelength and amplitude distribution can also be extracted from the yarn shape functions. These measures are not illustrated here. The two-stage approach – fitting functions to the yarn boundaries and then extracting information from the functions – has proved to be efficient; it allows the stored yarn–boundary fits to be used several times to extract a variety of information.

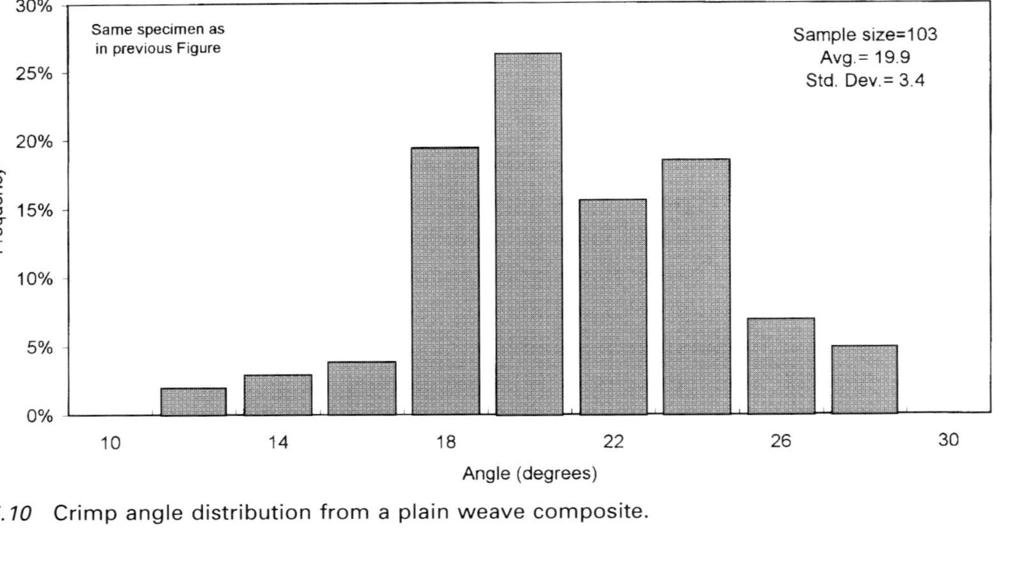

5.10 Crimp angle distribution from a plain weave composite.

5.5 Role of yarn shape in composite properties

We have focused in this chapter primarily on yarn path. The other essential aspect of yarn shape, the bounding surface of the yarn, which is the totality of its interfaces with other yarns, pores and matrix pockets, has been touched on only briefly, mainly in connection with descriptions of nesting. Both the path and the bounding surface are believed to affect composite behavior. In this section we point out some of the effects and their representation in some of the available analytical models of composite behavior.

Yarn shape affects composite behavior directly through the local orientations of the fibers (e.g. Jortner [22]). Usually, however, attempts to model cloth composite properties only approximate fiber orientations. That is, the fiber orientations are not measured directly but assumed to be parallel to the yarn path. Most modeling ignores yarn twist and variations in cross-section along a yarn, which could cause fiber orientations to deviate from the yarn inclination angle; cross-sectional variation also may cause variations of fiber-volume fraction along a yarn.

To provide input to most models, the yarn path may be measured via inclination angles as suggested in this chapter or, as frequently seems to be the case, simply to be a sinusoid of appropriate amplitude and wavelength.

In effect, such first-order approximate models idealize each in-composite yarn as locally unidirectional composite (LUC), which includes the matrix material within the yarn boundaries whose orientation varies along the yarn path. The properties of the LUC can be measured or estimated (as equal, perhaps, to the properties of a unidirectional model composite of the same fiber volume fraction made with the same fiber and matrix). Then, the properties of the cloth-reinforced laminate may be estimated from the LUC properties, as rotated into laminate coordinates through angles implicit in the inclination-angle distribution, and the relative volume fractions of warp and fill yarns. Usually, uniform-strain models are used to integrate (volume average) the rotated LUC properties [5,23]; such volume averaging models usually give reasonably good approximations to elastic constants of cloth-reinforced laminates.

Attempts have been made [5,23] to predict some aspects of non-linear stress–strain behavior from similar volume averaging models. A failure criterion, expressed in LUC coordinates, is applied to the state of stress or strain predicted to occur in a volume element. When the composite stress state reaches the point that any volume element fails, new properties are assigned to that volume element and new composite elastic constants estimated by volume averaging. Such approaches, although they provide useful insights, cannot be reliably accurate, inasmuch as uniform-strain assumptions inevitably are accompanied by local stresses that are discontinuous and inaccurate. However, there are enough loosely defined input parameters to enable such models to become fairly accurate and intuitively appealing correlators of non-linear behavior for specific composites.

Sometimes, even less information than the full distribution of inclination angles can prove useful. Crimp angle information has been found to correlate with variability in tensile strengths of cloth laminates [17,18]. Strength in tension along one set of yarns is decreased when the average crimp angle of the specimen, in that set of yarns, is increased. The mechanisms are not solidly understood. Pollock [18] concludes that the bundles behave like curved beams and failure initiates by shear across them at the points of maximum inclination angle. Jortner [17] advanced the hypothesis that failure initiates by composite-level shear on 'planes of weakness' that are inclined at the average crimp angle to the laminate plane. It is still an open question why, in view of the extreme-value weakest-link nature of fracture, the average crimp angle appears to be a better correlator than the maximum crimp angle.

Within the general framework of volume averaging treatments of the inclination angles, it is sometimes possible to include some nesting information. For example, in Jortner [16], application of a volume averaging model to four-layer stacks of cloth variously nested has provided some initial estimates of the effects of nesting variations on laminate elastic properties. The model of Jortner [16] can do so because it combines uniform-strain averaging over certain volume subelements with uniform-stress averaging of subelements over the composite as a whole.

In all such volume averaging models, changes in yarn shape do not affect properties unless the shape changes are manifest in the distribution of inclination angles. The amount of information ignored may be appreciated when we realize that, when analysts say they assume the yarn path is sinusoidal, strictly speaking, they really assume only that it has the same inclination-angle distribution as a sinusoid. Models that make use of additional information, such as continuity of fibers, the actual shape of yarns and the spatial relationship among contacting yarns (e.g. nesting), would necessarily be more complex and appear to be relatively rare.

We believe the actual shape of yarns and the spatial relationships among them are important to an understanding of non-linear stress–strain responses and strengths, some of which may be controlled by damage at specific locations within a cloth or within certain nesting configurations. Thus, Williams and Yurgartis [24] are finding that certain forms of fatigue damage in satin weave laminates tend to initiate at I_1 interlace regions where yarn curvature is greatest. We may speculate that details of yarn shape and nesting can affect non-linear behaviours and strength through, at least:

- inclination angle distribution,
- radius of curvature of the yarn path (because, among other possibilities, a small radius of curvature may mean some fiber damage has occurred during weaving by bending of fibers, or because curved phases may cause local stress gradients during service), and
- angular orientations, size (dimension) and locations of interyarn interfaces, which might, perhaps, be areas of weakness in the composite.

Finite-element analysis (FEA) of representative volumes of cloth laminates has been attempted (e.g. Whitcomb [25]). The additional complexity appears unwarranted to estimate elastic responses. However, the approach shows great promise for finding locations of initial damage. In principle, FEA can make use of almost all the yarn shape information that we can imagine measuring. There are, however, difficulties. Input properties for various regions of the cloth would be needed to make maximal use of FEA; currently, however, we do not have enough information to justify more than application of LUC properties, as above. Another difficulty relates to the definition of the representative volume to be analyzed. Clearly, nesting information is required to define the geometry. Also, it seems difficult to define the right boundary conditions for the representative volume, given that details of nesting do not necessarily repeat within a laminate. Finally, rather large and expensive FE models are implied if we wish to model a suitably extensive and complex representative volume. Some combination of small FEA models and insightful heuristic analyses seems to be indicated if we are to gain understanding efficiently.

5.6 Unresolved issues

Much remains to be done in the characterization of yarn shape. Some of the remaining unresolved issues are noted here.

The yarn shape characterization discussed above has generally assumed a two-dimensional framework. However, the path of the centerline of a yarn section, as seen on the micrographic plane (e.g. Fig. 5.4), is not necessarily equal to the path of the full yarn (cf. Fig. 5.2). The distortions implied in Fig. 5.1 and 5.2 suggest inclination angles may differ on the various possible polishing sections. Furthermore, the true yarn path may be three dimensional, not capable of being fully described on any one plane. This issue has not received much attention to our knowledge.

Another limitation to the work described is its restriction to a particularly simple type of laminate, in which the cloth layers are stacked so their warp yarns are aligned. This warp-aligned construction simplifies the microstructure in comparison with the more general laminate in which each layer may be oriented at some non-zero angle to the layer below. Stacking sequence is defined as the position, orientation and sequence of how multiple plies of cloth reinforcement are stacked to make a laminate. In principle, the methods described above should apply fairly readily, if micrographic sections are prepared in each of the yarn directions. However, there may be some subtle difficulties relating to the greater likelihood of non-planar yarn paths due to more complex nestings in angle-plied laminates. Yurgartis and Maurer [6] have explored some aspects of stacking sequence, but much more needs to be done.

Efforts to model composite behavior have tended to rely on the definition of a unit cell [5,26]. An attempt is made to choose the unit cell geometry such that the salient (or manageable) features of yarn shape are defined. However, real composites reinforced with cloth have a range of yarn shape. When the properties of interest move beyond 'averaged' properties, such as modulus, and attention turns to locally controlled properties, such as strength, then it will be necessary to invent better ways to describe the variability of yarn shapes and the interaction of the various yarn shape features. Pastore [27] has made useful first steps on this problem. As a final note along these lines, it must be said that despite the frequency of assumptions, actual yarn shape is rarely 'sinusoidal'.

As noted in the previous discussion, internal features of yarn shape, such as fiber packing and fiber path, have received only minimal study. It is quite possible that these features have an important role in composite manufacturing (e.g. resin infiltration, cure shrinkage).

The morphological description of yarns in composites made from other textile structures, such as mats, knits and braids, remains largely uncharted territory, with some exceptions (see for example, references [10,12,28–31]). The vast array of textile geometries and their complexity make this a challenging problem.

Perhaps the biggest issue remaining is the continued development of the connections between processing, yarn shape and properties. Until these connections are better established the characterization of yarn shape will remain an academic exercise, albeit a requisite one in the search for these connections.

References

1. J Jortner, Microstructure of cloth-reinforced carbon–carbon laminates, *Carbon* 1992 **30**(2) 153–163.
2. J W S Hearle, P Grosberg and S Backer (eds), *Structural Mechanics of Fibers, Yarns, and Fabrics, Vol. 1*, Wiley-Interscience, New York, 1969.
3. P Schwartz, T Rhodes and M Mohamed, *Fabric Forming Systems*, Noyes Publications, Park Ridge, NJ, 1982.
4. J Z Yu, Z Cai and F K Ko, Formability of textile preforms for composite applications, Parts 1 and 2, *Composites Manuf.* 1994 **5**(2) 113–132.
5. T-W Chou, *Microstructural Design of Fiber Composites*, Cambridge University Press, New York, 1992.
6. S W Yurgartis and J P Maurer, Modelling weave and stacking configuration effects on interlaminar shear stresses in fabric laminates, *Composites* 1993 **24**(8) 651–658.
7. J Summerscales, High-performance reinforcement fabrics, *Prog. Rubber Plastics Technol.* 1987 **3**(3) 20–32.
8. J W S Hearle, Textiles for composites, *Textile Horizons* 1994 **14**(6) 12–15.
9. B J Hill, R McIlhagger and P McLaughlin, Weaving multilayer fabrics for reinforcement of engineering components, *Composites Manuf.* 1993 **4**(4) 227–232.

10. Y Q Wang and A S D Wang, On the topological yarn structure of 3D rectangular and tubular braided preforms, *Composites Sci. Technol.* 1994 **51**(4) 575–586.
11. S W Yurgartis, K Morey and J Jortner, Measurements of yarn shape and nesting in plain-weave composites, *Composites Sci. Technol.* 1993 **46**(1) 39–50.
12. G-W Du, T-W Chou and P Popper, Analysis of 3D textile preforms for multidirectional reinforcement of composites, *J. Materials Sci.* 1991 **26** 3438–3448.
13. L M Vas, G Halasz, M Takacs, I Eordogh and K Szasz, Measurement of yarn diameter and twist angle with image processing system, *Periodica Polytechnica Mech. Eng.* 1994 **38**(4) 277–296.
14. B Xu, B Pourdeyhimi and J Sobus, Fiber cross sectional shape analysis using image processing techniques, *Textile Res. J.* 1993 **63** 717–730.
15. J Jortner, Shrinkage of fiber bundles during carbonization, *Extended Abstracts 20th Biennial Conference Carbon*, American Carbon Society, 1991, pp 402–403.
16. J Jortner, Fabric nesting and some effects on constitutive behavior of plain-weave cloth-reinforced laminates, *Proceedings 6th Japan–US Conference on Composite Materials*, Technomic, 1992, pp. 464–473.
17. J Jortner, Effects of crimp angle on the tensile strength of a carbon–carbon laminate, *Proceedings Symposium High-Temperature Composites*, American Society Composites, Technomic, 1989, pp 243–251.
18. P Pollock, Tensile failure in 2D carbon–carbon composites, *Carbon* 1990 **28** 717–732.
19. B Xu, B Pourdeyhimi and J Sobus, Characterizing fiber crimp by image analysis: definitions, algorithms, and techniques, *Textile Res. J.* 1992 **62**(2) 73–80.
20. J C Russ, *Computer-Assisted Microscopy: The Measurement and Analysis of Images*, Plenum Press, New York, 1990.
21. R T de Hoff and E T Rhines, *Quantitative Microscopy*, McGraw Hill, New York, 1968.
22. J Jortner, A model for prediction of thermal and elastic constants of wrinkled regions in composite materials, In *Effects of Defects in Composite Materials*, ASTM STP 836, American Society for Testing and Materials, 1984, pp 217–236.
23. J Jortner, A model for nonlinear stress–strain behavior of 2D composites with brittle matrices and wavy yarns, In *Advances in Composite Materials and Structures*, eds S S Wang and Y D S Rajapakse, American Society Mechanical Engineers, AMD-Vol. 82, 1986, pp 135–146.
24. J C Williams and S W Yurgartis, unpublished results.
25. J D Whitcomb, *Three-Dimensional Stress Analysis of Plain Weave Composites*, Report NASA-TM-101672, NASA Langley Research Center, 1989.
26. N K Naik, *Woven Fabric Composites*, Technomic, Lancaster, PA, 1994.
27. C M Pastore, Quantification of processing artifacts in textile composites, *Composites Manuf.* 1993 **4**(4) 217–226.
28. X Bugao and Y-L Ting, Measuring structural characteristics of fiber segments in non-woven fabric, *Textile Res. J.* 1995 **65** 41–48.
29. L V Smith and S R Swanson, Micro-mechanics parameters controlling the strength of braided composites, *Composites Sci. Technol.* 1995 **54**(2) 177–184.
30. O A Kallmes, A comprehensive view of the structure of paper, In *Theory and Design of Wood and Fiber Composite Materials*, ed. B A Jayne, Syracuse University Press, New York, 1972, p 157.
31. D Guang-Wu and F K Ko, Unit cell geometry of 3D braided structures, *J. Reinforced Plastics Composites* 1993 **12** 752–768.

6
Quantitative microstructural analysis for continuous fibre composites

FELICITY J GUILD AND JOHN SUMMERSCALES

6.1 Introduction

Current definition of the microstructure of fibre-reinforced composite materials is usually limited to the definition of the materials used and the proportion and orientation of the fibres. Significant differences can be achieved within such a definition, particularly in respect of the spatial distribution of fibres within the material. The microstructure–mechanics relationship is gaining importance as the discipline of mesomechanics. Recent advances in computer hardware, and in software for image processing and analysis, have permitted the rapid definition of spatial microstructural parameters. This chapter presents a review of the techniques available for quantitative definition of the microstructure of continuous fibre composites. The importance of such definition is demonstrated by some examples of relationships between properties and microstructure.

The usual description of fibre composite materials consists of definition of the materials used for the fibre and matrix, the proportion of fibres contained and their orientation. Such description is incomplete; identical materials, according to this description, could contain very different arrangements of fibres. Continuous fibres may be all aligned as a single parallel array (unidirectional), random swirl, cross-plied (bidirectional), woven (plain-weave, twill or satin), knitted or stitch-interlocked fabrics.

There is growing evidence in the literature that fibre arrangement may be a critical factor in the performance of composite materials. The importance of the microstructure of composite materials has recently been highlighted [1, 2]; the subject of 'mesomechanics' has been defined as the relationship between microstructure and material properties [3]. It is clear that composite materials are now becoming accepted for use in critical applications, so their proper microstructural definition and control must become increasingly important. Such microstructures can now be easily defined using automatic image analysis.

Techniques of automatic image analysis have been generally available for about 20 years. Recent developments, in cost effective hardware and high performance software, have made their use increasingly accessible. Current developments in methods of image analysis have become available from increasing computer power both within the systems and for further off-line analysis. Edge-recognition algorithms [4–6] may be used to enhance boundaries. Useful image enhancements for composite materials include the elimination of scratches as detected features and the separation of neighbouring fibres which may have been detected as a single feature; such enhancements are carried out via definitions of morphology. An example of the separation of fibres allowing their detection as separate features is shown in Fig. 6.1. Jeulin [7] has surveyed useful morphological measurements applicable to composite materials.

It is clear that successful application of image analysis to fibre composite materials is critically related to the quality of the image. The most usual source of image for these materials is an optical microscope; other possible sources include electron microscopes, X-ray micrographs and computed tomography. Techniques of polishing and etching to achieve good contrast must be considered. Automatic image analysis has been successfully used to generate the usual description of these materials, namely the volume fraction occupied by the fibres and their orientation; most analysis of fibre orientation has been carried out for short fibre composites. Modern techniques of automatic image analysis may now be leading to the goal of proper association of material behaviour with fibre arrangement.

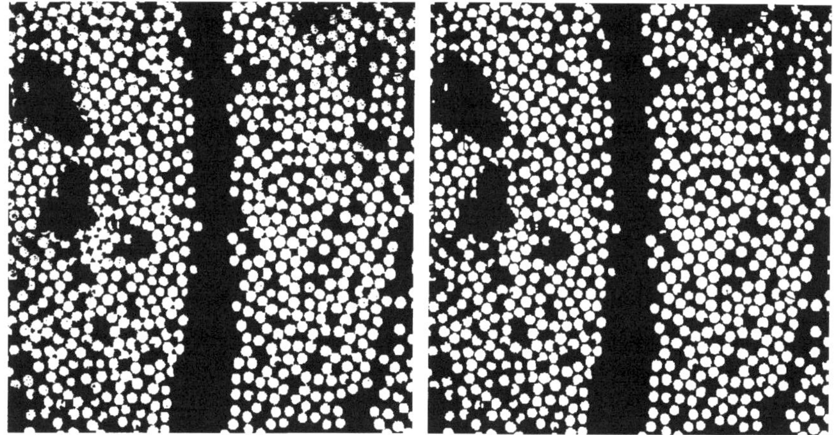

6.1 Binary image of section of unidirectional continuous carbon fibre/epoxy laminate. The left hand image is as acquired, before image processing (the white areas are detected pixels). The right hand image has been subjected to segmentation: image processing has been used to separate the individual features. The images were acquired on the University of Plymouth Quantimet 570.

6.2 Alternatives to microscopy

The volume fraction of fibres in a fibre-reinforced composite can be determined via several different 'bulk' methods. The simplest method is via the composite density; a density bottle is used [8]. The volume fraction determined in this way is accurate provided the densities of the constituent materials are accurately known, and there is negligible void content. These conditions may not always be satisfied and a method based on resin removal may be preferred.

Methods of resin removal include acid digestion and resin burn-off. The weight fraction of the fibres is determined directly; if the densities of the fibre and the composite or resin are known, the volume fraction can be calculated. Acid digestion can be performed using sulphuric acid and hydrogen peroxide, or nitric acid. The procedures have been fully described [9]. An alternative method of acid attack combined with use of a microwave oven, designated an acid digestion bomb, was found to give consistent results for a carbon fibre/epoxy composite [10].

Resin burn-off may be a suitable method for composites but cannot be used for carbon fibre composites. Typically the sample is placed in an oven preheated to about 600°C; the resin is allowed to burn off for about an hour until all traces of blackness have disappeared [9]. The fibre content of random glass mat/polypropylene has been measured by introducing a weighed sample into a furnace at 550°C for 1 h; the sample was reweighed and the content calculated [11].

The methods of resin removal are all destructive. Non-destructive methods include the density method, described above, and methods involving radiation. Hendron *et al.* [12] have shown that the β-ray backscatter method is capable of accuracies of approximately 1.5% in the determination of epoxy resin–glass content, when measured statistically. The tests were referenced to the results of burn-off tests. Hofer and Gayer [13] have reported on the applicability of non-destructive measurement of the glass content and void fraction of GFRP (glass fibre-reinforced plastics) using absorption measurements of high and low energy gamma radiation. Resolution of 20 μm was achieved with an accuracy of 1.5% at an average glass content of 15%. Martin [14] calculated the mass absorption coefficients for both X-rays and neutron beams, on the assumption that the material is homogeneous. The value of the mass absorption coefficient for a given change in resin content was significantly greater for the neutron beam. This method may be useful for the non-destructive measurement of composite resin content.

This brief survey of methods of volume fraction determination demonstrates that many methods, both non-destructive and destructive, are available. However, the methods generate values of volume fraction alone; no spatial information can be deduced. Techniques of image analysis lead to both accurate measurement of volume fraction and further microstructural definition based on the spatial distribution of the fibres.

6.3 Specimen preparation and examination

The quality of the original image is the most crucial factor in determining the quality of the results of image analysis. The preparation of good polished sections of composite materials for optical examination is dominated by the difficulties arising from the different hardnesses of the constituent materials; successful polishing routines are generally based on long polishing times at small abrasive sizes. The preparation of polymeric based samples for optical light examination has been fully described elsewhere [15,16].

The different hardness values of the constituents of a composite material generally lead to incomplete flatness in the finished section. This topographical information can be used to enhance the image. Nomarski interference techniques [17] can be used to create contrast arising from the slopes at the interface between the constituents. A section of unidirectional continuous glass fibre/polyester laminate examined using Nomarski interference is shown in Fig. 6.2. Differences in topography can also be highlighted using the technique of confocal microscopy [18]. This technique involves collection of information from within the depth of focus alone; the specimen is scanned and an image can be reconstructed. Contrast can be obtained by careful selection of the focus depth [19].

Further methods of contrast enhancement are based on surface treatment of the section. Flaws, including porosity and interlaminar splits, can be revealed by staining [20]. Etching techniques may be applicable for dissimilar chemical constituents, such as a polymer matrix and glass filler; for example, hydrofluoric acid can be used to introduce contrast in glass fibre/resin composites [21]. Careful etching can also be used to introduce contrast in the carbon fibre/carbon composite; however, coating may be a simpler and more consistent technique for these difficult materials [22]. Selective coating may also be a valuable technique

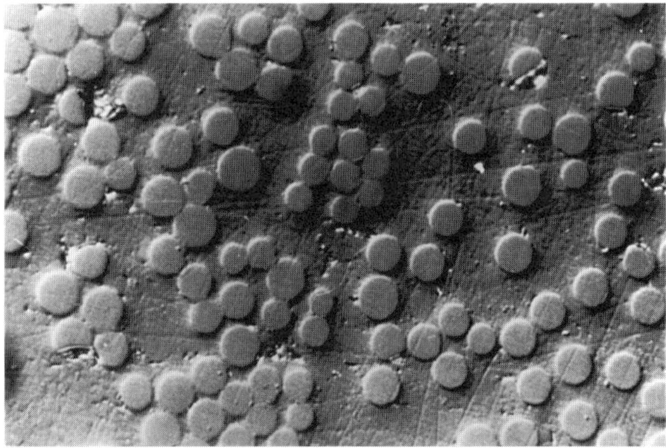

6.2 Enhanced contrast in the microscopy of a unidirectional glass fibre/polyester composite using Normarski interference.

for a wide range of composite materials containing constituents of epoxy resin, glass fibres and metal [23,24]. The use of thin oxide films to produce interference film colour contrast in metal matrix composites has been shown to be a useful technique [25]. However, this technique relies on careful attention to specimen preparation.

The description of preparation techniques above assumes examination in reflected light; this is the more usual illumination. However, transmitted light through thin sections can be used [26]; careful preparation methods are essential to prevent the introduction of defects.

The specimen is then ready for examination. As described above, the most usual method of examination uses an optical microscope. However, stored images, such as that recorded from examination using a confocal microscope, can be examined. Other stored images can be analysed including those from an electron microscope, scanning acoustic microscope, X-ray transmission and computed tomography (X-radiation, γ-radiation or nuclear magnetic resonance).

6.4 Large scale features

6.4.1 Stereology

The analysis of microstructure, whether using direct methods or using automatic image analysis, is essentially observation in two dimensions. The microstructure of a continuous fibre composite may be fully defined in two dimensions, by the plane perpendicular to the fibre direction. This definition is based on the assumptions that the fibres are perfectly straight and invariant along their length. The determination of three-dimensional structures via measurements in two dimensions is the subject of stereology; the fundamentals of this science are contained in various textbooks (see, for example, Underwood [27] and de Hoff and Rhines [28]).

The 'classical' methods of quantitative microscopy may be broadly divided into three categories, namely lineal and areal analysis and point counting. The most simple use of areal analysis is for continuous unidirectional fibre composites viewed perpendicular to the fibre direction, where the area fraction is identical to the volume fraction. Lineal analysis and point counting allow quantitative measurements of complex microstructures, such as volume fraction and grain size, to be measured from intercept lengths or fractional distribution of an array of points. These methods were preferred for manual measurements, but areal analysis is generally more applicable to automatic image analysis. The measurement of grain size in advanced ceramics using a point-sampled intercept technique, developed from the 'classical' methods, has recently been presented [29]; these measurements were made manually without automation.

The expected values of variance for volume fraction measurement by point counting, lineal analysis and areal analysis have been compared [30]. Systematic point counting in two dimensions was found to be the most accurate method. The comparative accuracy of the different methods of measurement has recently been considered by Shi and Winslow [31]. For areal analysis, expressions were derived for the required number of frames to be examined for a given confidence level; the results were confirmed by simulation. The availability of automatic image analysis allows the examination of sufficient frames for sensible confidence limits.

The role of the science of stereology in materials science was initiated by the application of quantitative microscopy before automatic image analysis was available. The role of quantitative microscopy in materials engineering was reviewed in 1972 [32]. The application of statistical methods in stereology was described by Nicholson [33]; some methods for analysis of data derived by hand may be applicable to results derived today from automatic image analysis. Comparison of data derived by hand and from an early Quantimet automatic image analyser show how such early systems could introduce constraints in the measuring system; such constraints should not be introduced in systems available today.

6.4.2 Determination of volume fraction

The measurement of volume fraction of continuous fibre composites from area fraction may be directly compared with measurements via other methods such as density measurements. Inaccuracies in areal measurements arise from sampling, sample preparation, the quality of the image and the choice of the threshold. Ideally, maximum contrast between the two phases is introduced in the preparation (Section 6.3 above). The quality of the image, including the focal depth of the microscope, may often introduce a gradual change in grey level between the two phases. The exact choice of threshold thus determines the size of the detected features, and thus the areal measurement [34]. This problem is largely solved in modern automatic image analyser systems which allow the use of an algorithm to convert the gradual change in grey level between the phases to a step change; measured volume fraction is not then affected by the precise threshold chosen. Use of such features can introduce consistency in areal measurement of volume fraction; however the precise numerical accuracy may still be questioned. Advanced imaging systems permit the elimination of rogue features on the basis of detected feature morphology (size, roundness, etc.).

These inherent inaccuracies in areal measurement were recognised by Waterbury and Drzal [35] who described an alternative method of volume fraction determination using an automatic image analyser system to count the number of fibres observed. The results were in good agreement with volume fraction results obtained from acid digestion; no systematic errors were observed. However, the application of this method may be limited since it relies on known consistent fibre radius. This method was used, in conjunction with a template to determine the presence of a fibre, by Yang and Colton [36]. The overall volume fraction of a thermoplastic composite reinforced with AS-4 carbon fibres of known constant radius was measured. The results were compared with results from conventional area fraction measurements using the threshold method; these measurements required higher magnification, so more samples were required. Good agreement was found between the two methods. Saidpour [37] used transmitted light microscopy to determine the void content and distribution from the shadows cast by the voids. The signals were processed in a Quantimet 800 system which enabled the measurement of void length, area and concentration to be performed with good repeatability and accuracy.

6.4.3 Measurement of fibre orientation

The orientation of continuous fibre composites is generally well known. Specimens for microstructural observation are cut perpendicular to the fibre direction. Assuming that the fibre has a circular cross-section, the viewed cross-section is therefore a distribution of discs. Orientation of the fibres at angles other than perfectly perpendicular to the plane is indicated by the presence of ellipses instead of circular discs. Measurement of fibre orientation is based on comparative measurements of the axes of these ellipses. This technique is particularly important for short-fibre composites, when orientation is unknown and must be determined [38]. A similar technique may be applied to continuous fibre composites. Yurgartis [39] has described this technique for continuous fibre composites and measured misalignment in a carbon fibre composite. Most accurate measurements are obtained from sectioning at a shallow angle to the fibre perpendicular direction. The fibres were found to lie within 3° of the mean fibre direction. Measurements using the orthogonal plane of view are described by Yang and Colton [36]. In this view, perfectly aligned fibres should form continuous lines. The alignment was measured from the length of the lines observed from measurements in the two orthogonal planes, after sophisticated simplification of the image.

6.4.4 Hybrid composites

Marshall [40] reported that in composites formed from a hybrid carbon/glass tape, the tows of carbon fibre adopted a trapezoidal cross-section in order to interlock more efficiently with the glass fibres. The problems of microstructure definition are compounded when the composite is reinforced with two or more types of fibre (a hybrid composite). There is now significant scope for microstructural changes through variation of the amount, the orientation and the relative positions of the reinforcements. Kulkarni *et al.* [41] identified two major types of patterning in hybrids. In 'intimate' hybrids each fibre is surrounded by fibres of the other type (intraply). Examples of intimate hybrids are shown in Fig. 6.3. In 'zebra' or 'discrete' hybrids, fibres of the same type occur in aggregates (interply).

Bader and Manders [42] defined the dispersion of a hybrid composite microstructure as the reciprocal of the thickness, in metres, of the smallest representative repeat unit of the laminate. In the case of a simple sandwich laminate this repeat distance will be the total laminate thickness. Similarly, this

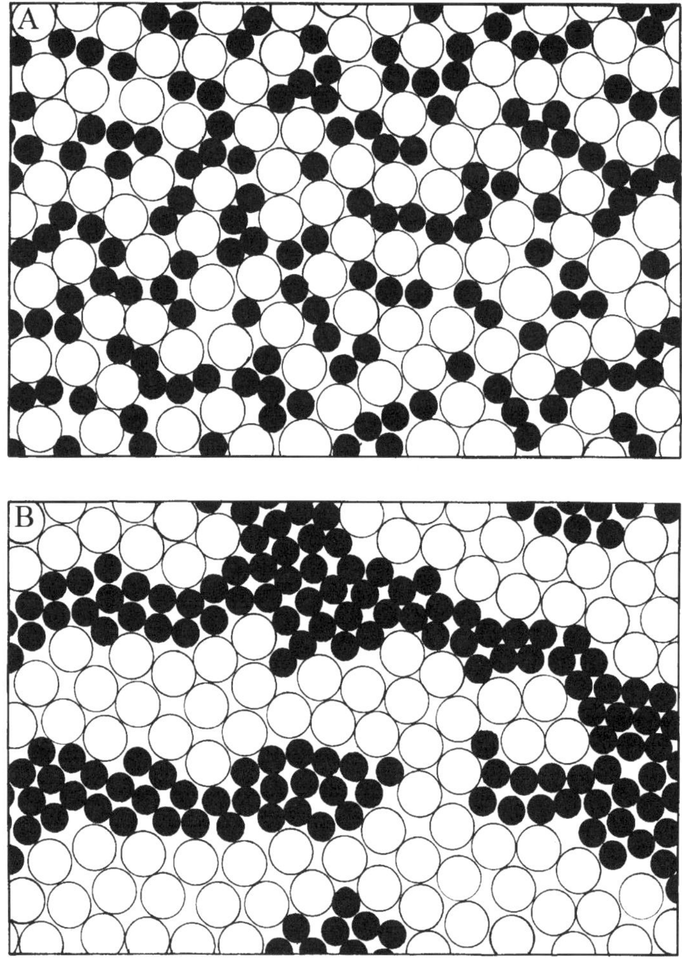

6.3 Schematic representation of the microstructure of intimate hybrids, A: typical of short fibre reinforcement, B: typical of continuous fibre reinforcement.

parameter will provide little useful information in the definition of unbalanced laminates.

Aveston and Kelly [43] examined the case of a bicomponent composite of constant volume fraction. A decrease in the size of one component automatically involves a decrease in the size of the other component. For uniaxial fibres the separation of the surfaces can be written in terms of the fibre diameter, volume fraction and a parameter with a value of 0.912 for a hexagonal array or 0.785 for a square array. This method of definition is clearly limited to highly organised structural composites.

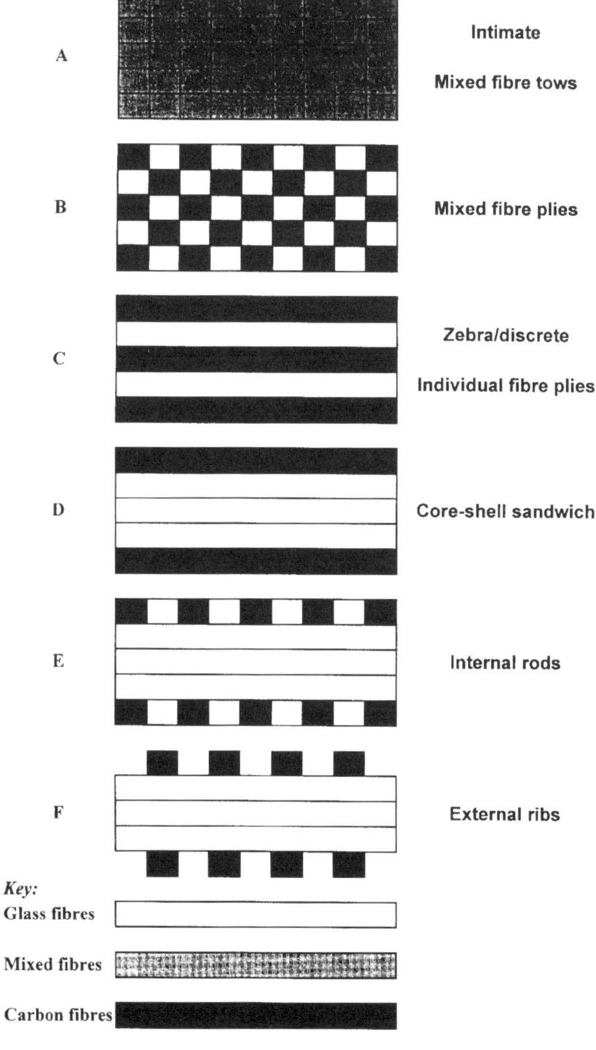

6.4 Schematic representation of the microstructure of hybrid composites.

Short and Summerscales [44] proposed a similar scheme to Kulkarni *et al.* [41] but with six levels of organisation (shown in Fig. 6.4). This was further quantified [45] to define more precisely intimately-mixed composites. Three parameters were considered, nearest-neighbour index, chi-squared index and contiguity index. Of these, the latter is the most simple conceptually and was found to be the most consistent. The contiguity index is the ratio of the number of changes of fibre type (along a transect of the micrograph) to the number of fibre–fibre spaces. When fibres occur alternately along the transept the index will have a value of 1. When the fibres are clustered in large groups of a single type the value will approach zero. In layered composites the index will, of course, be direction dependent.

The use of hybrid fibre composite materials is increasing, particularly to realise either/both property tailoring and cost savings. The accurate definition of such materials may be realistically achieved using automatic image analysis.

6.5 Measurement of fibre arrangement

Quantitative methods of description of fibre spatial arrangement are now possible with the use of automatic image analysis. The method described here generally leads to descriptions of the microstructure which may be utilised directly in analytical techniques such as the finite element analysis method. Alternatively, the descriptions may be used to calculate further topological parameters for the quantitative description of complex microstructures, such as those based on the concepts of contiguity [46].

The first methods used for the description of fibre arrangement may be described as functional methods; the microstructure is described by some parameter or function. The spatial distribution of the parameter is measured on the image analyser and the data are often transferred for further computation. Increasing computer power now allows more exact methods, that is, the distribution itself is measured, not a function of it.

6.5.1 Functional methods

6.5.1.1 Variance analysis

The spatial distribution exhibited by continuous fibres viewed on the plane perpendicular to their longitudinal direction may be compared with spatial distributions encountered in ecology and geography. Methods for describing such distributions have been developed in those subjects; the methods may be applicable to the definition of microstructures.

A method of examining spatial distribution was first proposed in 1952 [47]. The method is described more fully in ecological texts [48,49]. A parameter defining the distribution, for example fractional area covered, is measured in a contiguous grid of cells, the size of the test cell being smaller than the anticipated cluster size or scale of pattern. Larger cells, square or rectangular, are often formed from two cells of the preceding size, with cells being adjacent and not overlapping. The variance of the parameter values for the different cell sizes is calculated and a graph is drawn. It is apparent that if the cell size is such as to contain a pattern exactly, either fibre clusters or empty space, the variance at that cell size would be a maximum [49]. Any peaks in the graph are therefore deduced to indicate scales of pattern at those cell sizes. Alternatively it has been suggested that larger cells may be long thin cells, transects, in two perpendicular directions, resulting in a measure of dimension instead of area being represented by peaks in the graph [50]. Transects have been used for both artificial and real distributions [51, 52]. The values of the parameter can be compared with values calculated assuming a known well defined distribution; for example, values could be calculated assuming a random distribution. The random distribution would then be described as the null hypothesis. Features of the distribution can be deduced from comparison with values from the null hypothesis even if no peaks occur. It is important to note that the derivation of theoretical values for the null hypothesis are often calculated assuming a Poisson process, that is, for a random distribution of points. The finite size of the reinforcement is neglected.

This method of analysis has been used for the definition of the microstructure of glass fibre-reinforced polyester resin [21] and for particulate filled metal matrix composite [53]. The results from both materials showed marked deviations from the theoretical results for the null hypothesis, calculated assuming that the filler was randomly distributed. Results from variance analysis have been compared with results from matrix intercept analysis and results from computer simulations by Li *et al.* [54]; the material used was aluminium alloy reinforced with continuous fibres. The variance analysis clearly indicated that the fibre distributions are non-random. The effect of neglecting the finite size of the reinforcement was observed. A similar method of analysis has been described for continuous silicon carbide fibre-reinforced glass produced by an extrusion process [55]; the variation of local volume fraction for the test cell is presented as a histogram. However, the value of variance was not calculated, nor were results from larger cell sizes calculated.

This method of analysis is sometimes described as quadrat analysis. The grid of cells, or quadrats, sampled is often contiguous as described above. Miles [56] describes the statistical basis for the choice of random quadrats. This method of sampling could be applied to microstructural analysis.

6.5.1.2 Structuring elements

An alternative method of microstructural definition describes the microstructure in terms of a given unit or structuring element. For fibre-reinforced composites containing clustered fibres, and thus regions containing no fibres, a suitable structuring element might be empty space [21]. Results from this analysis are comparable with the results from variance analysis at the smallest cell size [57]. The smallest cell size had been chosen to correspond to the smallest pattern element observed in the microstructure, namely three contiguous fibres. The variance values describe only the second order structure, that is, the position of a particle with respect to another particle. Thus correlation would be anticipated at the smallest cell size. Results from larger cell sizes can be related to larger scales of pattern. This method of analysis may be more useful than variance analysis for fibre composites containing pattern at various scales, such as can arise from tow structure.

Alternative structural units can be based on morphological definitions of the microstructure [7]. This method was successfully applied to define the complex microstructures of polycrystalline graphite. However, most fibre composites may be better described using a simple structuring element such as the region of empty space as described above.

6.5.2 Rigorous quantitative methods

The development of more powerful image analysers has opened up the possibility of more rigorous microstructural definition; the positions of the fibres themselves can be recorded. A means of description of the fibre arrangement must therefore be sought.

Davy and Guild [58] have presented a description of composite materials in terms of Voronoi cells. The Voronoi cell about a particle is the region around that particle closer to it than any other particle. The variability of the half interparticle distance, or radius of the Voronoi cells, was calculated for a random distribution, assuming that the particles were randomly distributed. The particles were assumed to be of constant radius and were not allowed to overlap. The important distinction between this null hypothesis derived by Davy and Guild and that described above for the variance analysis, is that the finite size of the reinforcement is taken into account. This theoretical approach is known as the random hard core model. The division of a plane into Voronoi cells is also known as the Dirichlet cell tessellation [59]. Pyrz [60, 61] has compared parameters of Voronoi cells derived from the Poisson process, assuming a distribution of points, and using a hard core model. The results were compared with measured material parameters. The importance of including the hard core model is clearly demonstrated, although the analysis does not include the derivation of the half interparticle distance. Yang and Colton [36] described Voronoi cell distance by approximating measured Voronoi cells to the nearest square, in a given orientation. This application of arbitrary axes to define the orientation of the squares is hard to justify.

The theoretical distribution of half interparticle distance for a random distribution, using the hard core model, has been compared with measurements of a single tow of carbon fibre pre-preg. Good agreement was found between the measurements and the theoretical distributions [62]. The fibre distributions arising from different processing parameters in vacuum-bag preparation of carbon fibre/epoxy plates have been described using this method [63]. Figure 6.5 shows comparison of the observed cumulative distribution functions of Voronoi cell radius and the corresponding theoretical cumulative distribution functions calculated assuming the null hypothesis. The null hypothesis used is the random hard core model, namely that the fibres are randomly distributed but not allowed to overlap. The features of the fibre arrangement can be assessed from the deviations of the graph from the (0, 0), (1, 1) line. This line corresponds to agreement between the observed and theoretical distribution functions. Significant deviations are observed.

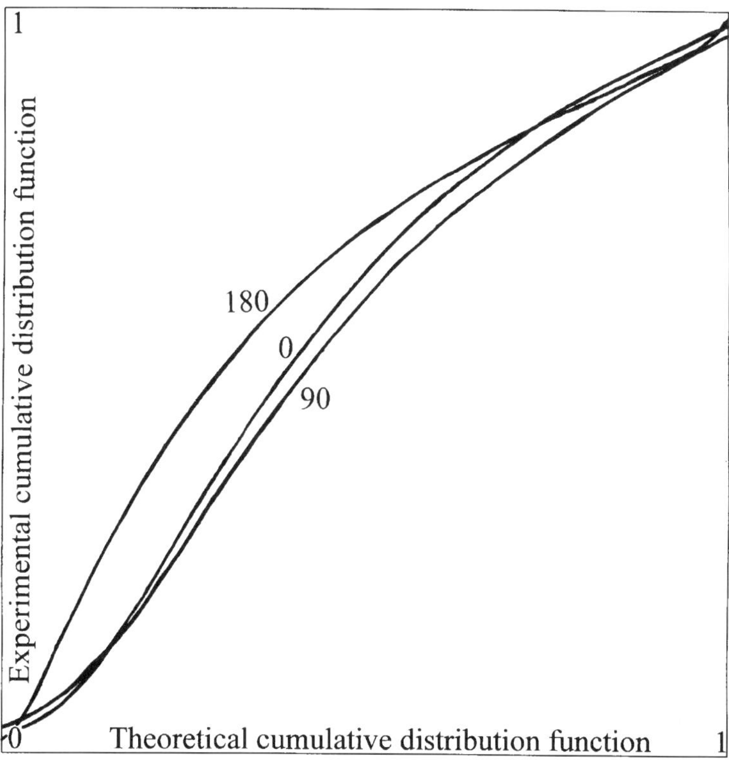

6.5 Comparison of theoretical and observed cumulative distribution functions of Voronoi cell radius from sections of unidirectional continuous carbon fibre/epoxy laminates manufactured by the vacuum-bag process with dwell times of 0, 90 and 180 min before the application of the vacuum.

The definition of clusters can also arise from the use of Voronoi cells. The geometric structure of a dot pattern was represented by the geometric properties of the Voronoi cells. The method produces an algorithm which assigns the dots into clusters [64]. Such definition of clustering may be applicable to fibre composites.

An alternative method of description of microstructure arises from definition of nearest neighbour distances. The results are highly dependent on the number of nearest neighbours considered [65]. The results for real composites were compared with results from simulation; the real composites were found to have higher variability of nearest neighbour distance than the simulations. This higher variability may be described as a tendency for the fibres to cluster. Everett and Chu [66] quantified clustering by the nearest neighbour distance distribution skewness. The material used was a ceramic matrix reinforced with silicon carbide fibres. This method of quantification was found to be the most sensitive measure for the 'chain' clustering observed.

A method based on the 'included angle' has been proposed as a measure of fibre spatial distribution [67, 68]. The measured angles are those between the vectors joining the centres of fibres to nearest neighbours. From simple geometry, it is clear that the measured angles for a perfect cubic array would be 45° and 90°, while those for a perfect hexagonal array would be 60°. Histograms of measured included angles therefore yield a description of the type of packing. However, it is important to note that such measurements are independent of both overall and local volume fraction. Thus clustering, which may be the most important parameter for fracture behaviour, cannot be detected using this technique.

Methods of rigorous quantitative definition of the microstructure of fibre composite materials are presently becoming more rigorously defined, but further experimental and theoretical work is required. The need for such definition is becoming increasingly apparent as the links between detailed microstructure and properties are being investigated.

6.6 Property–microstructure relationships

6.6.1 Elastic properties

The elastic properties of aligned continuous fibre composites have been theoretically investigated using numerous different approaches. Methods based on classical mechanics assume a uniform fibre array; a hexagonal or square array is generally assumed. Different predictions are obtained for different fibre arrays. These approaches are summarised by Mansfield [69], who extended the analysis to take into account some randomness in fibre distribution. The predicted elastic properties were found to be highly dependent on the occurrence of variations in fibre distribution resulting in localised areas containing no fibres. Experimental measurement of specific damping capacity has found that lower values of specific damping capacity, indicative of greater stiffness, are found for unidirectional beams containing more resin-rich areas [57].

More recently, the finite element method has been used to predict the transverse stiffness of unidirectional continuous fibre composites. The method has been applied assuming a regular array [70] or assuming that the fibres are randomly distributed [71]. In both cases, the predictions of transverse modulus were found to be dependent on the fibre array, either the type of packing [70] or the degree of randomness [71].

6.6.2 Fracture properties

The dependence of fracture properties of continuous fibre composites on fibre arrangement is demonstrated by consideration of the interactive stresses between the constituents [61]. Using classical analytical methods, it is shown that the interactions are dependent on fibre arrangement. The finite element method has been used to predict the dependence of fracture strength on fibre arrangement. Failure in longitudinal compression has been related to regions containing no fibres [72]. The compression strength of laminates fabricated from specialist fabrics designed for resin transfer moulding has been measured [73]. These laminates contain regions of 'pore space', that is, regions containing no fibres. The values of compression strength for a range of such laminates has been related to the measured regions of pore space; the compression strength decreases with increasing pore space.

Compression strength has been widely related to the 'waviness' of fibres, which may be measured as fibre misalignment; these effects have recently been reviewed by Adams and Bell [74]. Fibre misalignment or waviness is predicted to have significant effects on a further wide range of laminate properties, including fatigue endurance, shear strength and delamination resistance [75].

Failure in longitudinal tension may also be related to fibre arrangement. Such dependence has been predicted using a micromechanical model for failure [76]. Stress transfer around broken large diameter silicon carbide fibres has been observed experimentally [77]; interaction between neighbouring fibres was found to be negligible when the fibre separation exceeds about 8 fibre diameters. The effects were considered on a statistical basis [78]. The strength of the model array was dependent on the arrangement of fibres. Similar dependence has been found for the fracture of carbon fibre-reinforced epoxy [79]; failure processes were observed *in situ* and found to be dependent on neighbouring fibres. The importance of neighbouring fibres has also been demonstrated for the process of stress corrosion cracking [80].

The tensile strength and fracture energy of carbon fibre-reinforced aluminium are predicted to increase if the fibres are arranged in bundles [81]. The predictions are based on observed change in yield stress for narrow interfibre spacing, presumably arising from metallurgical change in the matrix. This effect would not therefore be expected for polymer matrix fibre composites. An effect of fibre distribution on longitudinal strength of polymer composites may arise from debond length. It is generally accepted that longer pull-out lengths are desirable as energy is absorbed both by the debonding process and the work done in friction during pull-out (e.g. Harris *et al.* [82]). Longer debond lengths have been generally observed in fibre bundles [57].

The fracture energy arising from fibre matrix debonding and pull-out has been considered by Karbhari and Wilkins [83]. The analysis indicates that energy absorption may be increased if the fibres are arranged in bundles. Values of critical strain energy release rate have been related to an inhomogeneity parameter for laminates prepared using different methods [84]. The values of fracture energy were predicted to increase with increasing inhomogeneity. This increase was found experimentally. However, for brittle-matrix composites this effect may be reversed since fibre bundles have been predicted to lead to catastrophic crack growth [85].

Stress concentrations in transverse loading have been investigated for metal matrix and polymer matrix unidirectional continuous fibre composites [70, 71]. The analysis of Wisnom [70] takes into account the residual stresses arising from processing. For silicon carbide fibres in aluminium alloy matrix, the stress concentrations were found to be unaffected by fibre arrangement or fibre spacing. A similar result was found by Guild *et al.* [71] for glass fibres in polymer matrix. However, the analysis of high strength carbon fibres, with a lower transverse modulus than the matrix material, found a strong dependence of stress concentrations on fibre spacing. For this material, the transverse strength is expected to be highly dependent on fibre arrangement.

6.7 Process–structure relationships

6.7.1 Vacuum-bag manufacture

Summerscales *et al.* [63] manufactured unidirectional carbon fibre composites with cold curing (25°C) epoxy resin in silicone rubber bag vacuum tables. Plates were produced with three different dwell times before the vacuum (0.8 bar) was applied: zero, 90 or 180 min. The resulting laminates had mean thicknesses of 1.4 mm, 1.3 mm and 2.1 mm, respectively. The thinnest plate (90 min dwell) was processed within the probable optimum dwell-time window described by Stringer [86] where the viscosity lies between 75 and 165 poise. This timing is a function of the increasing flow (decreasing viscosity) as the resin is warmed by the exotherm and the decreasing flow as crosslinking inhibits the movement of the resin. Analysis of the Voronoi distances measured from polished sections revealed that the clustering for the 90 min dwell plate was slightly lower than for the zero dwell sample whilst clustering was most marked in the 180 min dwell material (Figure 6.5).

6.7.2 Resin transfer moulding

The resin transfer moulding (RTM) process involves the long flow of resin through the pore space between and within the reinforcement fibres. The flow processes are generally modelled using Darcy's law [87] to predict the linear flow rate, q, using:

$$q = \frac{K}{\eta}\frac{\Delta P}{L} \tag{1}$$

where ΔP is the pressure drop, η is the fluid viscosity and L is the length of the bed of porous medium. The constant of proportionality, K, is normally known as the permeability of the medium. Darcy's law is empirical and cannot predict permeability from a knowledge of the microstructure.

It may be possible to predict the effect of changing the micro/mesostructure of the reinforcement using the expression developed by Kozeny [88] and Carman [89]. They suggest that the volumetric flow rate of the fluid, Q $(= qA)$, is governed by the equation:

$$Q = \varepsilon\frac{Am^2}{k\eta}\frac{\Delta P}{L} \tag{2}$$

where ε is the porosity (fractional free volume in the bed), A is the cross-sectional area normal to the flow direction, m is the mean hydraulic radius of the bed and k is the Kozeny constant. The concept of the hydraulic radius was introduced by Blake [90]. The mean hydraulic radius can be determined from the expression presented by Williams *et al.* [91]:

$$m = \frac{r}{2}\frac{\varepsilon}{(1-\varepsilon)} = \frac{rV_m}{2V_f} \tag{3}$$

where r is the fibre radius, V_m is the volume fraction of the matrix in void-free cured composite and V_f is the fibre volume fraction.

Experiments in France [92] have demonstrated that the linear flow rate through fabrics at the same fibre volume fraction, V_f (note that $\varepsilon = 1 - V_f$), and identical flow conditions results in markedly different linear flow rates when the fabric architecture is changed.

Summerscales [93] used simple idealised models, with a specific hydraulic radius following the original Blake definition, to predict the effect of changing the reinforcement architecture on the resulting flow. The models predicted significant increases in flow rate when changing from a uniform distribution of non-touching fibres to a regular close-packed array of clustered fibres. The prediction is in line with the previous French results.

Griffin and co-workers [94–96] manufactured laminates from twill weave fabrics with a varied proportion of flow-enhancing tows (FET) present. The permeability of the reinforcements increased with the proportion of FET. However, the inclusion of sparse FET (every eighth tow in one direction within a

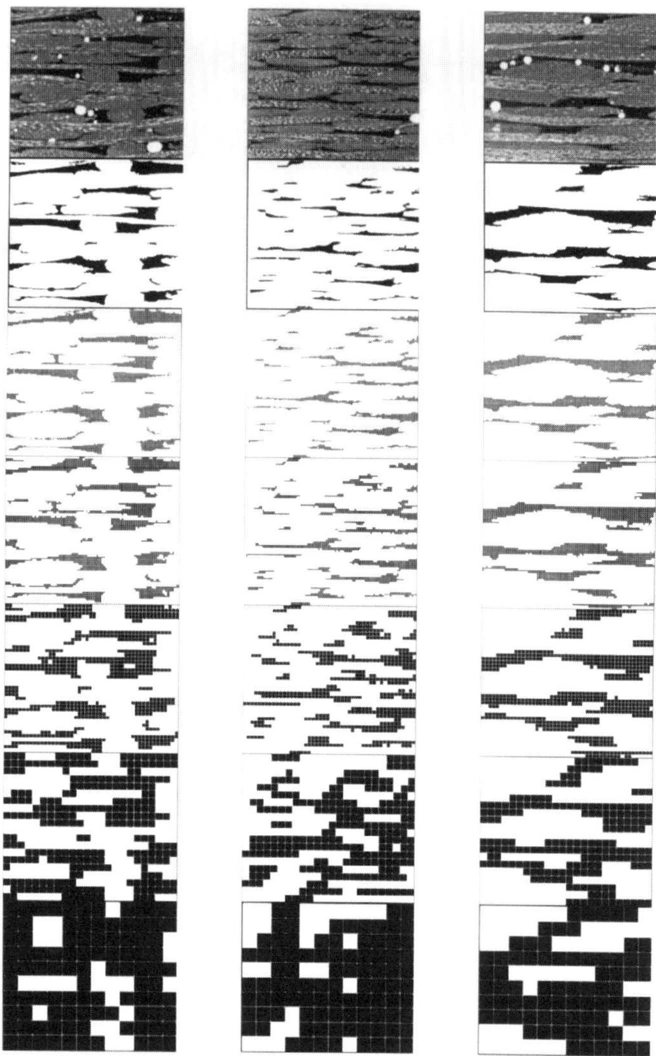

6.6 Representative images showing the progressive stages of fractal data generation for Injectex flow-enhancing satin (left column), satin (centre column) and twill (right column) fabrics. The first row is the digitised grey-level optical image of the microstructure. The second row is the equivalent binary image (each pixel is either black or white) and each subsequent row shows the boxes (of increasing size down the page) in which pore space is detected.

2×2 twill weave) was far more effective than subsequent increases in the proportion of FET. Quantitative microscopy of sections normal to the principal axis of the FET revealed that each FET creates significant pore spaces adjacent to their positions. The sharp increase in the permeability of the reinforcement stack with the introduction of the initial FET was correlated to a sharp increase in flow areas in the range 0.10–0.25 mm^2. Note that the improved permeability was accompanied by a decrease in the interlaminar shear strength (ILSS) and compression strength [73] as discussed in Section 6.6.2.

Pearce *et al.* [97] reported a similar study on three carbon fabrics (twill, normal 5-harness satin and FET 5-harness satin) with the same fibre type, same fibre surface treatment and equivalent fibre volume fractions ($50 \pm 10\%$). The permeabilities increased in the sequence normal-satin < twill < FET-satin for unidirectional laminates and in the sequence normal-satin < FET-satin < twill for cross-plied laminates. The normal-satin fabric had the highest number and highest proportion of small flow areas (<0.065 mm^2) and very few large flow areas (>0.25 mm^2). The twill fabric had the smallest number of flow areas but a significant number of large pore spaces. The FET-satin had a significant number of porespace areas in the range 0.08–0.30 mm^2. The pore spaces in this range are thought to explain the increase in permeability and decrease in ILSS relative to the values for the normal-satin fabric.

Pearce *et al.* [98] subsequently reported tension and compression properties for the above fabrics measured to Composite Research Advisory Group (CRAG) standards. The microstructure of the composites (Fig. 6.6) was quantified using fractal dimensions determined from a Richardson plot (a graph of the $\log_{10}$ detected area versus $\log_{10}$ box size). The ranking of the fractal dimensions of the three fabrics corresponded to the ranking of the secant moduli and failure stresses in both tension and compression.

Whilst the magnitude of the significant porespace areas is similar in the Griffin and Pearce studies above, it is not possible to make direct comparisons because different fluids were used in the experiments. Steenkammer and co-workers [99, 100] and Griffin *et al.* [95] have both reported significant differences in the permeabilities measured in notionally similar reinforcement stacks when the permeating fluid is changed. There is scope for an extensive study of the interrelationship between permeability and the microstructure parameters controlling the flow process.

6.8 Concluding remarks

The use of computer-based techniques for image analysis in the microscopical study of fibre-reinforced composites has been described. Such techniques have a significant advantage in that they provide access to spatial information on fibre distribution in addition to the volume fraction data. This is in contrast to destructive techniques which require that volume fraction is inferred from weight ratios and require accurate data on voidage and component densities. Quantitative microstructural definition of composite materials will be of increasing importance in the expanding field of property–microstructural relationships.

Acknowledgements

The authors would like to thank Neil Pearce (UoP SMMME) and Paul Russell (UoP Biological Sciences) for creating Fig. 6.6 and for giving us permission to use it.

References

1. D Hemsley, The importance of microstructure to the in-service survival of plastics products, *Plastics, Rubber Composites Processing Applic.* 1991 **16** 77–78.
2. T W Chou, *Microstructural Design of Fibre Composites*, Cambridge University Press, Cambridge, 1992.
3. G K Haritos, J W Hager, A K Amos, M J Salkind and A S D Wang, Mesomechanics: the microstructure–mechanics connection, *Int. J. Solids Structures* 1988 **24** 1081–1096.
4. R C Gonzalez and P Wintz, *Digital Image Processing*, 2nd edn, Addison-Wesley, Reading, MA, 1987.
5. J S Lim, *Two Dimensional Signal and Image Processing*, Prentice Hall, Englewood Cliffs, NJ, 1990.
6. T Y Young and K-S Fu, *Handbook of Pattern Recognition and Image Processing*, Academic Press, Orlando, FL, 1986.
7. D Jeulin, Recent developments of image analysis in materials sciences, *Memoires Etudes Sci. Revue Metallurgie* 1988 **85** 135–149.
8. J B Sturgeon, *Specimens and Test Methods for Carbon Fibre Reinforced Plastics*, Royal Aircraft Establishment Technical Report RAE TR 71-026, Farnborough, 1971.
9. P T Curtis, *CRAG Test Methods for the Measurement of the Engineering Properties of Fibre Reinforced Plastics*, Royal Aircraft Establishment Technical Report RAE TR 88-012, Farnborough, 1988.
10. P Green, Fibre volume fraction determination of carbon-epoxy composites using an acid digestion bomb, *J. Mater. Sci. Lett.* 1991 **10** 1162–1164.
11. A Rubio, P. Eguizabal, M A Mendizabal and J F Liceaga, Influence of processing parameters on glass mat thermoplastics (GMT) stamping, *Composites Manufac.* 1992 **3**(1) 47–52.
12. J A Hendron, K K Groble and W Wangard, The determination of the resin-to-glass ratio of glass-epoxy structures by beta-ray backscattering, *Mater. Evaluation* 1964 **22** 213–216.
13. G Hofer and P Gayer, Non-destructive determination of the glass and void content in glass-fibre reinforced plastics, *Materialpruefung* 1975 **17** 17–19. CEGB translation 7782.

14. B G Martin, An analysis of radiographic techniques for measuring resin content in graphite fibre reinforced epoxy resin composites, *Mater. Evaluation* 1977 **35** 59–64.
15. L C Sawyer and D T Grubb, *Polymer Microscopy*, Chapman and Hall, London, 1987.
16. D A Hemsley, *Applied Polymer Light Microscopy*, Elsevier, London, 1989.
17. B Bousfield, Micrographs and metallography: fingerprinting microstructures, *Microscopy and Analysis* 1992 **28** 7–9.
18. T Wilson, *Confocal Microscopy*, Academic Press, 1990.
19. J L Thomason and A Knoester, Application of confocal scanning optical microscopy to the study of fibre-reinforced polymer composites, *J. Mater. Sci. Lett.* 1990 **9** 258–262.
20. D W McNeil and B Bennett, Light microscopy and reinforced plastics, *Modern Plastics* 1964 143–145.
21. F J Guild and B W Silverman, The microstructure of glass fibre reinforced polyester resin composites, *J. Microscopy* 1978 **114** 131–141.
22. B L Metcalfe and R J Wilson, Enhancement of optical contrast on carbon/carbon fibre composites, *J. Mater. Sci. Lett.* 1985 **4** 235–236.
23. I Stapf and G Elssner, Contrasting of a composite material using the Leitz contrasting chamber, *Prakt Metallogr.* 1984 **21** 489–493.
24. A Ribbens, U Rosenfelder and E Berghof-Hasselbacher, Methods of contrasting fibre reinforced light metal alloys for automatic image analysis, *Prakt Metallogr.* 1991 **28** 621–632.
25. E G Bennett, J D Lord and B Roebuck, Interference films for microstructural characterization of metal matrix composites, *Composites* 1989 **20** 519–526.
26. G W Ehrenstein and V Alstadt, Methods of preparation to reveal the structural arrangements of glass fibre reinforced thermoplastics, *Prakt. Metallogr.* 1984 **21** 440–459.
27. E E Underwood, *Quantitative Stereology*, Addison-Wesley, 1970.
28. R T de Hoff and F N Rhines, *Quantitative Metallography*, McGraw-Hill, 1969.
29. S Srinivasan, J C Russ and R O Scattergood, Grain size measurements using the point-sampled intercept technique, *Scripta Metallurgica Materialia* 1991 **25** 931–934.
30. J E Hilliard and J W Cahn, An evaluation of procedures in quantitative metallography for volume-fraction analysis, *Trans. Metall. Soc. AIME* 1961 **221** 344–352.
31. D Shi and D Winslow, Accuracy of a volume fraction measurement using areal image analysis, *J Testing Evaluation* 1991 **19** 210–213.
32. H F Fischmeister, Applications of quantitative microscopy in materials engineering, *J. Microscopy* 1972 **95** 119–143.
33. W L Nicholson, Application of statistical methods in quantitative microscopy, *J. Microscopy* 1978 **113** 223–239.
34. R A Taylor and L C Godbey, How illumination and contrast affect the area measurement of small particles by digital imaging, *Optical Eng.* 1992 **31** 219–226.
35. M C Waterbury and L T Drzal, Determination of fiber volume fractions by optical numeric volume fraction analysis, *J. Reinforced Plastics Composites* 1989 **8** 627–636.

36. H Yang and J S Colton, Quantitative image processing analysis of composite materials, *Polymer Composites* 1994 **15** 46–54.
37. S H Saidpour, *The Effect of Fibre/Matrix Interfacial Interactions on the Mechanical Properties of Unidirectional E-glass Reinforced Vinyl Ester Composites*, PhD thesis, Loughborough University of Technology, 1991.
38. A R Clarke, G Archenhold and N C Davidson, 3D confocal microscopy of glass fibre reinforced composites, Chapter 3 in this book.
39. S W Yurgartis, Measurement of small angle misalignments in continuous fibre composites, *Composites Sci. Technol.* 1987 **30** 279–293.
40. D A G Marshall, *The Mechanical Properties of Carbon-Glass Composites Manufactured by a Vacuum Box Moulding Technique*, MSc thesis, Lancaster University, 1974.
41. S V Kulkarni, B W Rosen and H C Boehm, Evaluation of cost effectiveness of hybrid composite laminates, *J. Aircraft* 1977 **14** 1153–1154.
42. M G Bader and P W Manders, Failure strain enhancement in carbon glass fibre hybrid composites, *Second International Conference Composite Materials*, Toronto, 1978, paper 14.3, pp 1147–1165.
43. J Aveston and A Kelly, Tensile first cracking strain and strength of hybrid composites, *Phil. Trans Roy. Soc. London A*, 1980 **294** 519–534.
44. D Short and J Summerscales, Hybrids – a review. Part 1: Techniques design and construction, *Composites* 1979 **10** 215–221.
45. D Short and J Summerscales, The definition of microstructures in hybrid reinforced plastics, *Proceedings International Conference, SAMPE*, Montreux, 1984, Volume 2, paper 19.
46. Z Fan, A P Midownik and P Tsakiropoulos, Microstructural characterisation of two phase materials, *Mater. Sci. Technol.* 1993 **9** 1094–1100.
47. P Greig-Smith, The use of random and contiguous quadrats in the study of the structure of plant communities, *Annal. Botany* 1952 **NS16** 293–316.
48. P Greig-Smith, *Quantitative Plant Ecology*, Butterworth, 1964.
49. K A Kershaw, *Quantitative and Dynamic Ecology*, Edward Arnold, 1964.
50. P Greig-Smith, Data on pattern within plant communities 1: The analysis of pattern, *J. Ecology* 1961, **49**.
51. K A Kershaw, The use of cover and frequency in the detection of pattern in plant communities *J. Ecology* 1957 **38**(2) 291–299.
52. A R Hill, The distribution of drumlins in County Down, Ireland, *Annal. Assoc. Amer. Geog.* 1973 **63**.
53. A Larsson, L Eklund, S Karlsson, R Carlsson, S Bengtsson and R Warren, Determination of fibre distribution in metal matrix composites using image analysis, *Proceedings Conference Metal Matrix Composites*, Institute of Materials, London, 1989.
54. Q F Li, R Smith and D G McCartney, Quantitative evaluation of fiber distributions in continuously reinforced aluminium alloy using image analysis, *Mater. Characterisation* 1992 **28** 189–203.
55. N Klein and E Roeder, Investigation of SiC-long-fiber reinforced glass composites by automatic image analysis, *Prakt. Metallogr.* 1991 **28** 350–358.
56. R E Miles, The sampling, by quadrats, of planar aggregates, *J. Microscopy* 1978 **113**(3) 257–267.

57. F J Guild and B Ralph, Microstructure–property relationships of GRP, *J. Mater. Sci.* 1979 **14** 2555–2562.
58. P J Davy and F J Guild, The distribution of interparticle distance and its application in finite-element modelling of composite materials, *Proc. Roy. Soc. London A* 1988 **418** 95–112.
59. A Getis and B Boots, *Models of Spatial Processes*, Cambridge University Press, Cambridge, 1978.
60. R Pyrz, Stereological quantification of the microstructure morphology for composite materials, In *Optimal Design with Advanced Materials*, ed. P Pedersen, Elsevier, Amsterdam, 1993, pp 81–93.
61. R. Pyrz, Quantitative description of the microstructure of composites. Part 1: morphology of unidirectional composite systems, *Composites Sci. Technol.* 1994 **50** 197–208.
62. D Green and F J Guild, Quantitative microstructural analysis of a continuous fibre composite, *Composites* 1991 **22** 239–242.
63. J Summerscales, D Green and F J Guild, Effect of processing parameters on the microstructure of a fibre reinforced composite, *J. Microscopy* 1993 **169** 173–182.
64. N Ahuja and M Tuceryan, Extraction of early perceptual structure in dot patterns: integrating region, boundary, and component Gestalt, *Computer Vision, Graphics Image Processing* 1989 **48** 304–356.
65. A S Wimolkiatisak, J P Bell, D A Scola and J Chang, Assessment of fiber arrangement and contiguity in composite materials by image analysis, *Polymer Composites* 1990 **11** 274–279.
66. R K Everett and J H Chu, Modelling of non-uniform composite microstructures, *J. Composite Mater.* 1993 **27** 1128–1144.
67. S W Yurgartis and M N Purandare, Describing fibre spatial distribution in unidirectional composite materials, *J. Computer-Aided Microscopy* 1991 **3** 117–125.
68. S W Yurgartis, Techniques for the quantification of composite mesostructure, *Composite Sci. Technol.* 1995 **53** 145–154.
69. E H Mansfield, *The Influence of Fibre Distribution on the Moduli of Unidirectional Fibre Reinforced Composites*, Royal Aircraft Establishment Technical Report RAE TR 74-182, Farnborough, 1974.
70. M R Wisnom, Factors affecting the transverse tensile strength of unidirectional continuous silicon carbide fibre reinforced 6061 aluminium, *J. Composite Mater.* 1990 **24** 707–726.
71. F J Guild, P J Hogg and P J Davy, A predictive model for the mechanical behaviour of transverse fibre composites, *Proceedings Conference Fibre-Reinforced Composites (FRC90)*, IMechE, Liverpool, 1990.
72. F J Guild, P J Davy and P J Hogg, A model for unidirectional composites in longitudinal tension and compression, *Composites Sci. Technol.* 1989 **36** 7–26.
73. D M Basford, P R Griffin, S M Grove and J Summerscales, Relationship between mechanical performance and microstructure in composites fabricated with flow-enhancing fabrics, *Composites* 1995 **26** 675–679.
74. D O Adams and S J Bell, Compression strength reductions in composite laminates due to multiple-layer waviness, *Composites Sci. Technol.* 1995 **53** 207–212.

75. M R Piggott, The effect of fibre waviness on the mechanical properties of unidirectional fibre composites: a review, *Composites Sci. Technol.* 1995 **53** 201–205.
76. W-C Zhang and K E Evans, Micromechanical failure analysis for composites, *Proceedings ICCM-7*, China, 1989.
77. D A Clarke and M G Bader, The strength of single fibres and simple arrays of fibres in a model composite system. Part 1: Phenomenological aspects of failure, *Proceedings ICCM-7*, China, 1989.
78. L C Wolstenholme, The strength of single fibres and simple arrays of fibres in a model composite system. Part 2: Statistical aspects of failure, *Proceedings ICCM-7*, China, 1989.
79. N Sato, T Kurauchi and O Kamigaito, Fracture mechanism of unidirectional carbon-fibre reinforced epoxy resin composite, *J. Mater. Sci.* 1986 **21** 1005–1010.
80. F R Jones and J W Rock, A method for determining crack velocity stress intensity curves for stress corrosion cracking of GRP, *Proceedings 6th International Conference Fracture*, New Delhi, 1984.
81. X Zhenhai, M Zhiying and Z Yaohe, Effect of fibre distribution on infiltration processing and fracture behaviour of carbon fibre-reinforced aluminium composites, *Zeit. Metallkunde* 1991 **82** 766–768.
82. B Harris, J Morley and D C Phillips, Fracture mechanisms in glass-reinforced plastics, *J. Mater. Sci.* 1975 **10** 2050.
83. V M Karbhari and D J Wilkins, Constituent scale and property effects on fibre-matrix debonding and pull-out, *J. Mater. Sci.* 1991 **26** 5888–5964.
84. D S Cairns, L B Ilcewicz, T Walker and P J Minguet, Fracture scaling parameters of inhomogeneous microstructure in composite structures, *Composites Sci. Technol.* 1995 **53** 223–231.
85. C Bascoe and A Carpinteri, Discontinuous constitutive response of brittle-matrix fibrous composites, *J. Mech. Phys. Solids* 1995 **43** 261–274.
86. L G Stringer, Optimization of the wet lay-up/vacuum bag process for the fabrication of carbon fibre epoxy composites with high fibre volume fraction and low void content, *Composites* 1989 **20**(5) 441–452.
87. H P G Darcy, *Les Fontaines Publiques de la Ville de Dijon*, Dalmont, Paris, 1856.
88. J Kozeny, Uber die kapillare Leitung des Wassers in Boden, *Sitzungsberichte Akademie der Wisscenschaft Wien Math-naturw* 1927 **136(Kl.abt.IIa)** 271–306.
89. P C Carman, Fluid flow through a granular bed, *Trans. Inst. Chem. Eng.* 1937 **15** 150–166.
90. F C Blake, The resistance of packing to fluid flow, *Trans. Amer. Inst. Chem. Eng.* 1922 **14** 415–421.
91. J G Williams, C E M Morris and B C Ennis, Liquid flow through aligned fibre beds, *Polymer Eng. Sci.* 1974 **14**(6) 413–419.
92. J-M Thirion, H Girardy and U Waldvogel, New developments in resin transfer moulding of high-performance composite parts, *Composites (Paris)* 1988 **28**(3) 81–84.
93. J Summerscales, A model for the effect of fibre clustering on the flow rate in resin transfer moulding, *Composites Manuf.* 1993 **4**(1) 27–31.

94. P R Griffin, S M Grove, F J Guild, P Russell and J Summerscales, The effect of microstructure on flow promotion in RTM reinforcement fabrics, *J. Microscopy* 1995 **177**(3) 207–217.
95. P R Griffin, S M Grove, P Russell, D Short, J Summerscales, F J Guild and E Taylor, The effect of reinforcement architecture on the long-range flow in fibrous reinforcements, *Composites Manuf.* 1995 **6**(3/4) 221–235.
96. J Summerscales, P R Griffin, S M Grove and F J Guild, Quantitative microstructural examination of RTM fabrics designed for enhanced flow, *Composite Structures* 1995 **32** 519–529.
97. N R L Pearce, F J Guild and J Summerscales, An investigation into the effects of fabric architecture on the processing and properties of fibre-reinforced composites produced by resin transfer moulding, *Composites Part A: Appl. Sci. Manuf.* 1998 *A29*(1/2) 19–27.
98. N R L Pearce, F J Guild and J Summerscales, The use of automated image analysis for the investigation of the effect of fabric architecture on the processing and properties of fibre reinforced composites produced by RTM, *Preprints of 5th International Conference Automated Composites*, Institute of Materials, Glasgow, 4–5 September 1997, pp 319–326.
99. D A Steenkammer, D J Wilkins and V M Karbhari, Influence of test fluid on fabric permeability measurements and implications for processing of liquid moulded composites, *J. Mater. Sci. Lett.* 1993 **12**(13) 971–973.
100. D A Steenkammer, S H McKnight, D J Wilkins and V M Karbhari, Experimental characterization of permeability and fiber wetting for liquid moulding, *J. Mater. Sci.* 1995 **30**(12) 3207–3215.

7
Electron microscopy of polymer composites

MICHÈLE GUIGON

7.1 Introduction

The behavior and performance of a polymer composite cannot be understood solely on the basis of the specific properties of its principal components (the fibres and the matrix). The interface that exists between fibres and matrix is an essential element of the composite. Adequate adhesion between the fibres and matrix is a precondition for an optimized stress transfer in a fibre-reinforced polymer.

Nowadays, the scientific community is unanimous in recognizing that improvement in composites is possible by fully understanding and controlling the interfacial phenomena [1, 2]. The work that we have undertaken on interface and interphase characterization is part of an active research programme in a new materials field [3–6].

This chapter focuses on the characterization of interfacial phenomena in unidirectional PAN (polyacrylonitrile)-based carbon fibre-reinforced organic matrix composites. It deals with three themes:

- Effects of various treatments on carbon fibre surfaces
- Benefits of a functionalized sizing
- Contribution of a coating with specific mechanical properties.

High resolution transmission electron microscopy associated with mechanical tests has been used to characterize the interface in carbon fibre-reinforced polymeric matrix composites. The observations have shown that:

- surface treatment improves the fibre–matrix adhesion by modifying the structure and the chemical composition of the fibre surface,
- when sizing is used, stress transfer from the matrix to the fibres depends both on the fibre–resin adhesion and the quality of the sizing–matrix junction.

7.2 Experimental

7.2.1 Description of the materials studied

Three families of composites have been studied.

7.2.1.1 Surface treatment study

These composites were provided by the Office National d'Etudes et de Recherche Aérospatiale (ONERA) in order to study the effects of various surface treatments on carbon fibres [3–9].

The selected fibre was a high strength PAN-based carbon fibre (Courtaulds fibre). This fibre was delivered either without any surface treatment (XAU fibre) or after having undergone a commercial oxidizing process (XAS fibre), or an electrochemical amine surface treatment (ethylene diamine (EDA) and triethylene tetramine (TETA)). The thermoset resin used was Narmco 5208 epoxy.

7.2.1.2 Study of a functionalized sizing

New acrylic resins have been developed by Aérospatiale: Laboratoire d'Aquitaine to be polymerized by electron bombardment for use in aerospace applications [10–12]. The composites were supplied by Aérospatiale: Laboratoire d'Aquitaine. In this work, the matrix was an urethane–acrylic resin (C59) and the fibre was Hercules IM7 PAN-based carbon fibre subjected to an oxidizing surface treatment.

Two sizings have been chosen:

- A bifunctional isocyanate–acrylate sizing named ISOH, consisting of an isocyanate function that creates bonds with the hydroxyl sites present on the surface of the fibre. We distinguish the ISOH1 with a pretreatment at 50°C for 17.5 s and the ISOH2 with a pretreatment at 200°C over 210 s.
- The ISOP2 sizing consisting of the same functions as above at the extremities of a macromolecule prepared with a much longer length that leads to a soft polymer (pretreatment at 50°C for 17.5 s).

7.2.1.3 Effect of a specific coating

With a view to improving the impact strength properties of some composites, Aérospatiale: Laboratoire Central and INSA de Lyon have developed an elastomeric sizing of DGEBA–CTBN–IPD (diglycidyl ether of bisphenol A–carboxy terminated butadiene acrylonitrile–isophorone diamine) type [13]. The high strength Toray T300 fibres and a Structil DA508 epoxy matrix were chosen. The composites came from Aérospatiale: Laboratoire Central and from INSA de Lyon.

This composite has been studied in comparison with an industrial composite made of carbon fibres coated with the Toray number 5 sizing.

7.2.2 Experimental techniques of the study

Transmission electron microscopy, in a high resolution mode, is necessary to obtain useful information about the fine structure of the fibre–matrix interface. The interference fringes coming from the interference between the undeviated beam and the 002 beam diffracted by the carbon layers of the fibre have been the basis for the study of fibre–matrix interfaces. All the observations of structural details at the nanometer scale were made on thin transverse and longitudinal sections (40–60 nm thick) prepared by ultramicrotomy with a diamond knife.

7.2.2.1 The microscope

Fine structural details smaller than 1 nm were sought. In such cases samples must be used in which the variation of the phase ϕ is small compared with the average phase ϕ_o (weak phase object). For these conditions, the optical system (i.e. the transfer mechanisms) plays a significant part in the formation of the images and their interpretation.

Figure 7.1a illustrates the mechanism of image formation in a transmission electron microscope. The parallel illuminating electron beam is scattered by the specimen S. The scattered beam passes through the convergent objective lens L and is focused in the image focal plane A. In this plane, corresponding to the first Fourier transform of the object, the electron diffraction pattern is formed. The optical image (second Fourier transform of the object) is formed in the Gauss plane, G.

In order for the magnified image of the object to be a reliable Fourier synthesis, all the beams must imperatively (for a very thin object) interfere while keeping their initial phase shift. Unfortunately, the objective lens (transfer function) introduces additional phase shifts. They vary as a function of spherical aberration C_s of the lens, of the diffraction angle 2θ of the beams relative to the optical axis of the lens and of the defocus Δf.

Let ϕ_o and ϕ_i be the complex amplitudes of the object and image waves, respectively. The object–image relationship is expressed as $\phi_i = K\phi_o$, with K known as the impulsion response of the objective lens. The relation of convolution is equivalent to the relation of transfer $\boldsymbol{\phi}_\mathbf{i} = \boldsymbol{K}\boldsymbol{\phi}_\mathbf{o}$ where $\boldsymbol{\phi}_\mathbf{i}$, $\boldsymbol{\phi}_\mathbf{o}$ are the Fourier transforms of ϕ_i, ϕ_o and $\boldsymbol{K}$ is the transfer function.

It has been shown [14–16] that the transfer function is:

$$\boldsymbol{K}(u) = D(\lambda fu) \exp -i\chi(u) \quad \text{with} \quad \chi(u) = \pi\left(\frac{1}{2}C_s\lambda^3u^4 - \lambda\Delta fu^2\right) \qquad (1)$$

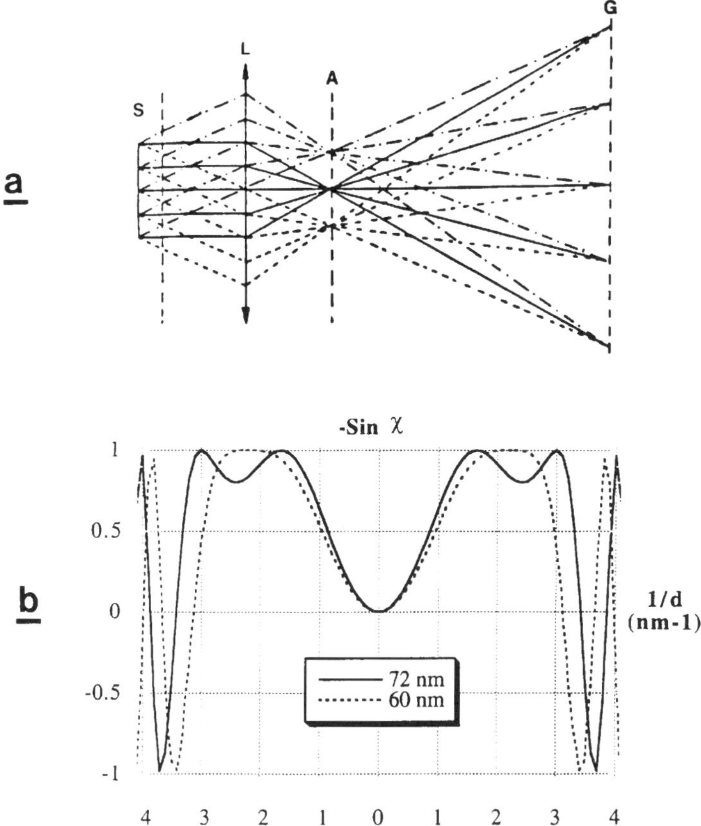

7.1 (a) Image formation in a transmission electron microscope. (b) Transfer function with $\Delta f = 60\,\text{nm}$ and $\Delta f = 72\,\text{nm}$.

where D is the characteristic function of the objective aperture, $\chi(u)$ is the aberration function, u is the spatial frequency ($u = 1/d = 2\theta/\lambda$), f is the focal image distance, Δf is the defocusing ($\Delta f > 0$ for underfocusing), λ is the wavelength and C_s is the spherical aberration.

Thus, we can state that the wavelength (λ) and the spherical aberration (C_s) play an important part in a microscope. The objective lens acts like a filter for spatial frequencies. It selects and modifies the phase of planar waves that can interfere in the image plane. An axial mode was used with partially coherent illumination. Under these conditions the transfer function of the phase contrast is, to a first approximation, proportional to $-2\sin\chi(u)$ [16].

It has been shown [14] that the phase shift introduced by the objective lens is constant around the optical axis for a given value $\Delta f_o = (C_s\lambda)^{1/2}$ and $\sin\chi = -1$, the so-called Scherzer's optimum (Scherzer's plateau).

For $\lambda = 3.35 \times 10^{-3}$ nm (acceleration voltage, $E = 120$ kV) and $C_s = 1.1$ mm, Δf_o is equal to 60 nm (Fig. 7.1b). In this zone, the image contrast is a maximum. However, several workers in this field [14,16–21] have shown that the image is not altered up to a value $\Delta f = 1.2\ (C_s\lambda)^{1/2}$ and $\sin\chi = -0.8$ (in our case $\Delta f = 72$ nm, Fig. 7.1b).

In our work, the interference of the undeviated transmitted beam and the 002 beam diffracted by the carbon layers of the fibre was used. For the high strength fibres studied here, the interplanar distance d_{002} is of the order of 0.36 nm which corresponds to a peak centred on 2.8 nm^{-1} with a half maximum intensity width of 0.5 nm^{-1} for a thickness of five carbon layers [4].

Therefore, from these data, it became necessary to find a value of the Δf defocalization in order to obtain an optimal value of the transfer function in the frequency band (2.3–3.3 nm^{-1}). The value satisfying $\Delta f = 72$ nm (under focus) for which $\Delta f = 1.2\ (C_s\lambda)^{1/2}$ and $\sin\chi = -0.8$ is suitable for our observations (see Fig. 7.1b).

In order to eliminate the noise caused by the scattered beams from the Scherzer's plateau (i.e. in the oscillations of the function sin χ after its first minimum) it is necessary to use an appropriate objective aperture. In our case we used an aperture 7 nm^{-1} in diameter. Therefore, the maximal resolving power defined by the first minimum of the transfer function in Fig. 7.1b is $d_{min} = 0.29$ nm ($u = 1/d_{min} = 3.5$ nm^{-1}).

A system of 002 lattice fringes is thus obtained which reliably represents the projection of aromatic layers on the observation plane for all the carbon fibres (one-dimensional Fourier transform).

7.2.2.2 *Definitions*

The electron microscope gives a bidimensional image of a three-dimensional object. Therefore, in our case, the contact ratio of the fibre with the matrix is related to the length of the interface and is obviously correlated to the modified external structure of the fibre. Consider Fig. 7.2, where a_i is the length of a fibre–matrix junction and p the perimeter of the fibre. The 'contact index' is defined [4] as:

$$C = \sum_i a_i/p \qquad (2)$$

It is important to observe that the ultramicrotomy process is destructive, therefore only the undamaged regions coming from the strongest fibre–matrix junctions are studied. It is thought that they are mainly responsible for the adhesion in composite materials. Thus, the contact index was measured on several slices of several similar fibres from the same batch. The statistical study gives a measurement precision of about 10%.

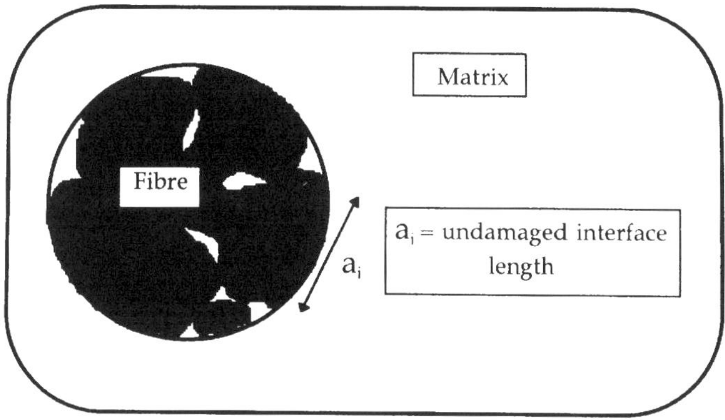

Transverse section of a composite

$$\text{contact index} = \sum_i a_i \,/\, \text{perimeter}$$

7.2 Definition of the contact index.

The mechanical integrity of the interfaces was evaluated by interlaminar shear strength tests (ILSS) carried out on composite materials [7,8,10,13].

7.3 Experimental results

7.3.1 Study of carbon fibres

A carbon fibre is formed by a set of turbostratic stacks of aromatic layers lying approximately parallel to the fibre axis known as elemental microtextural units (EMU) as represented in Fig. 7.3 [9]. The EMU are made of several perfect units, which are similar for all carbon fibres, called basic microstructural units (BSU). The BSU always have the same diameter of about 1 nm. The differences between carbon fibres are simply the number of BSU joined together and differ from one type of fibre to another. This observation leads us to distinguish BSU without specific characteristics from EMU which represents an intrinsic element for a given fibre. The way in which the EMU are arranged in space (microtexture) is responsible for the fibre properties, especially for mechanical properties [9]. In longitudinal sections of a high strength fibre, the EMU are linked together in wrinkled sheets oriented roughly parallel to the fibre axis. Longitudinal bending forms elongated voids between the carbon sheets. In transverse sections, carbon

(a)

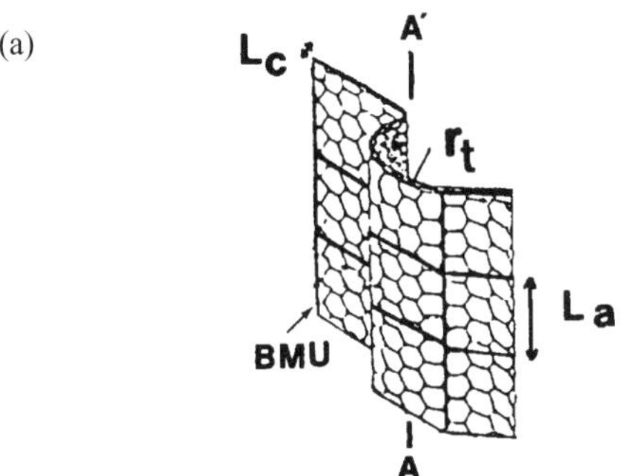

(b)

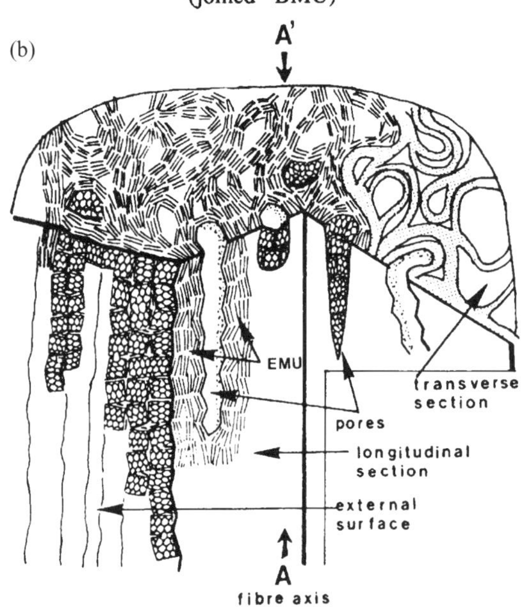

7.3 (a) Model of elemental microtextural unit (EMU). (b) Model of microtexture of a carbon fibre.

layers exhibit an intricate structure resulting from their strong radius of curvature (Fig. 7.3). It is thought that the transverse radius of curvature is due to the presence of residual defects (disclinations, tetrahedral bonds and crosslinking heteroatoms such as nitrogen atoms) [9, 22–24]. If we consider the model drawn in Fig. 7.3, the carbon layers of EMU show either their edge or their surface outcropping the external surface of the fibre.

From detailed observations of transverse cross-sections, PAN-based carbon fibres exhibit three different structures at their surface (Fig. 7.4).

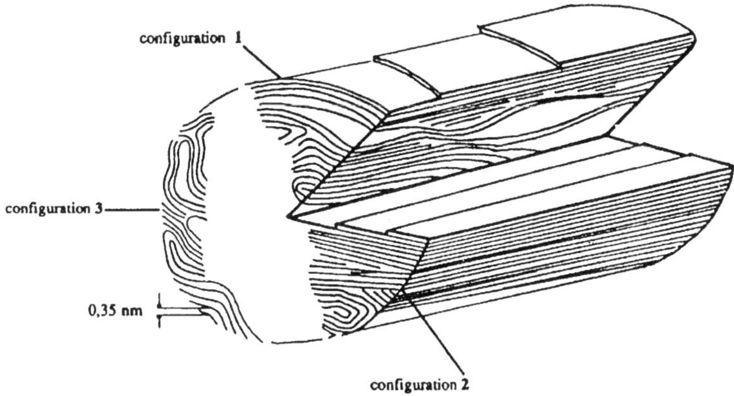

7.4 Arrangement of the carbon layers at the surface of high strength PAN-based carbon fibres.

Configuration 1: the carbon layers, which have a large radius of curvature, are more or less parallel to the fibre surface. In this case, the surface consists of basal planes which are known to have poor reactivity [25, 26]. This configuration is not conducive to a strong fibre–matrix adhesion.

Configuration 2: the carbon layers are mainly perpendicular to the surface. The unsaturated atoms present at their edges are very reactive. They tend to chemisorb functional groups that can bond with the resin molecules. This configuration is favourable to a strong fibre–matrix adhesion.

Configuration 3: the carbon layers exhibit a small radius of curvature which leads to a succession of basal and prismatic planes at the surface. This configuration is an intermediate arrangement between the two previous ones.

Because of the difference between the reactivity of saturated bonds (basal planes) and unsaturated bonds (prismatic planes), the fibre–matrix adhesion will be more or less strong depending on the predominant outer structure of the fibre. The surface treatment carried out on carbon fibres is designed to attack carbon atoms which are slightly linked in defect areas. This process results in an increasing number of reactive sites, the development of a microrelief and the grafting of functional groups.

7.3.2 Study of the surface of the untreated fibre

The Courtaulds XAU fibre is composed mainly of regions corresponding to configuration 1 which is represented by Fig. 7.5 (T300 and Hercules fibres were not made available without oxidization treatment). The fibre consists of a thin 'skin' of a few well-organized basal planes which cover a structure that is clearly more disordered. This outer structure does not promote a large number of bonds between the carbon fibre and the resin molecules.

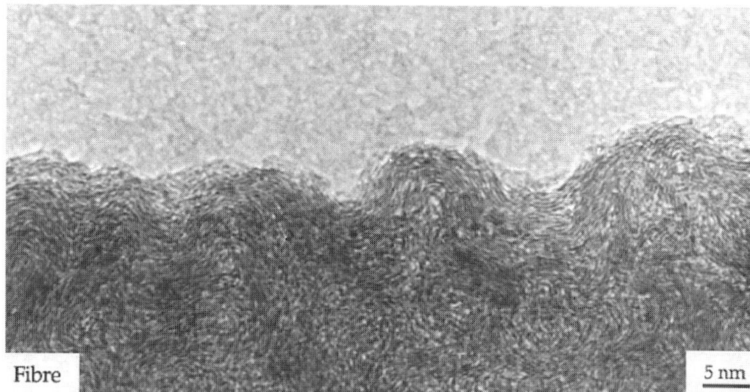

7.5 Outer structure of the XAU fibre in transverse section (configuration 1).

7.3.3 Study of the surface of the treated fibres

The Courtaulds XAS fibre consists of regions that correspond mainly to configuration 2, as represented by Fig. 7.6, and sometimes to configuration 3, as represented by Fig. 7.7. The Toray T300 fibre (oxidized fibre) shows the structure described for configuration 2. The carbon layers are disordered, 'exfoliated' and oriented towards the surface. The Hercules IM7 fibre (oxidized fibre) has an outer surface which corresponds to configuration 3.

Some XAU fibres were treated at ONERA to graft amine molecules. The amine molecules used in this study were ethylene diamine (EDA) and triethylene tetramine. The amines were designed to react as a hardener with the epoxy functions of the resin. The amine-treated fibre surface is intermediate between the

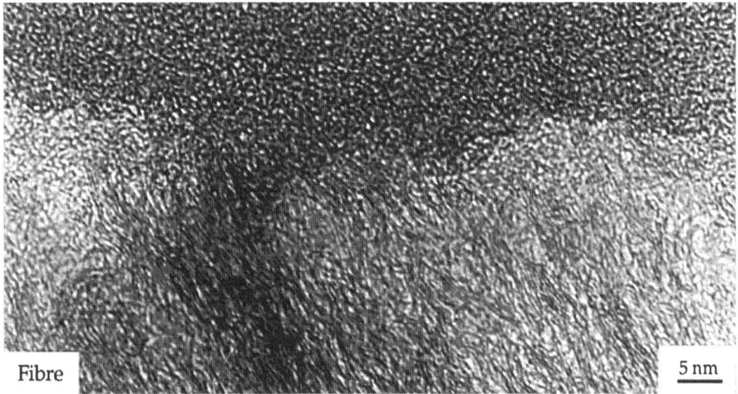

7.6 Outer structure of the XAS fibre in transverse section (configuration 2).

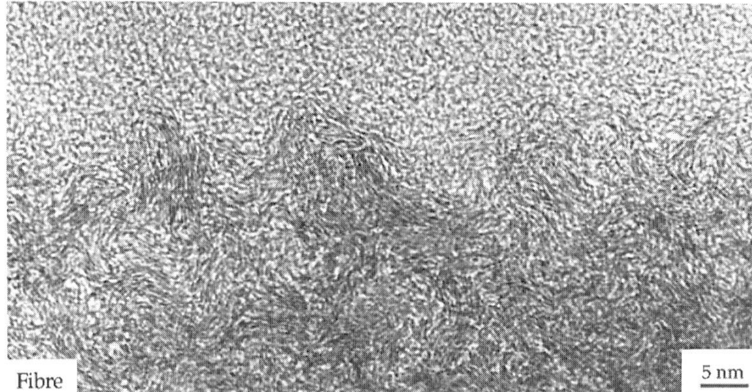

7.7 Outer structure of the XAS fibre in transverse section (configuration 3).

oxidized and untreated fibre surface. The predominant structure depends on the intensity of the treatment.

Changes coming from the external structure as well as the chemical composition of fibres are therefore recorded after surface treatments for the Courtaulds fibres [5, 6, 8]. The structural modifications cause the appearance of a less organized underlying structure. They result from the attack of basal planes on vacancies, disclinations, etc. The erosion starts on the carbon layers under tension. This phenomenon allows the formation of new prismatic surfaces (Fig. 7.8) and is represented schematically in Fig. 7.9.

The oxidative treatments (XAS fibres, Fig. 7.6) seem to have removed all the contaminants and the poorly bonded carbon layers. The more disordered structure

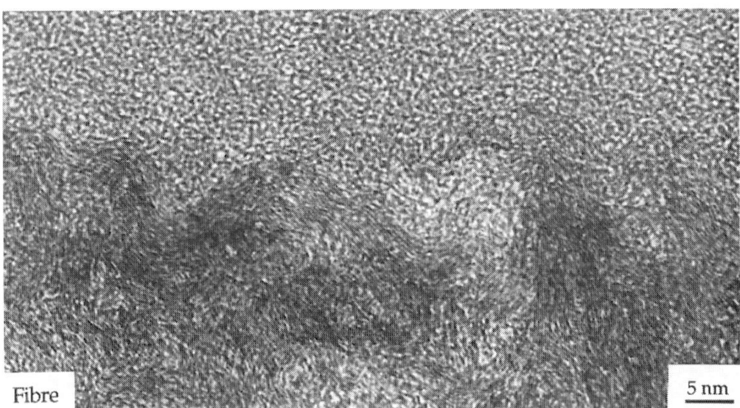

7.8 Outer structure of the XAU/EDA fibre in transverse section (configuration 3).

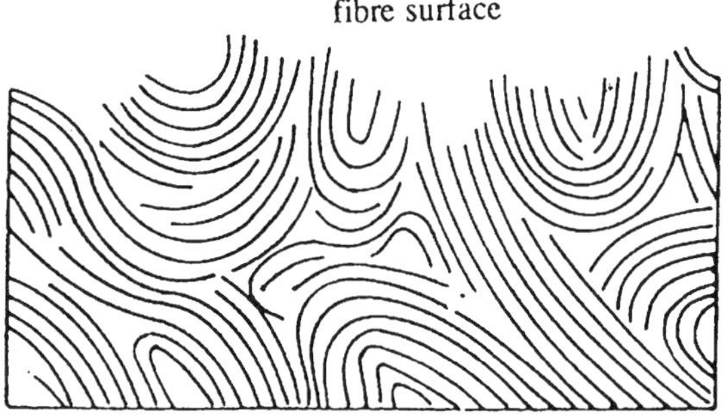

7.9 Erosion process of the XAU fibre surface.

which appears associated with the presence of edges of carbon layers at the surface is highly reactive and therefore favourable to chemical grafting (CO, COOH, COH). As far as this process of cleansing the surface with the help of an oxidizing treatment is concerned, it is to be noted that we have observed the same phenomenon, although less intense, with amine treatments. In this case, in addition to grafting of amine molecules, the surface treatment can also oxidize the fibre.

The nature and the number of grafted molecules or functions depend on the type of treatment applied (oxidization or amination) and the density of available sites on the fibres.

7.3.4 Scanning tunnelling microscopy

The surfaces of XAU and XAS fibres have been characterized using a scanning tunnelling microscope (STM). No special preparation was needed prior to placing the carbon fibres in the microscope.

A scanning tunnelling microscope works on the basis of the quantum mechanical tunnelling of electrons under the influence of a small bias voltage (a few tens of mV) between an extremely sharp metallic tip and a conducting surface separated by a gap of less than a nanometer [27–29]. As the tip is moved in three dimensions with an *x*-*y*-*z* piezoelectric translator, the surface can be imaged with very high resolution. The details were observed in a constant current mode (a few nA).

The results are described in Fig. 7.10 and 7.11 and are in agreement with those obtained with transmission electron microscopy. We observe a smooth surface for

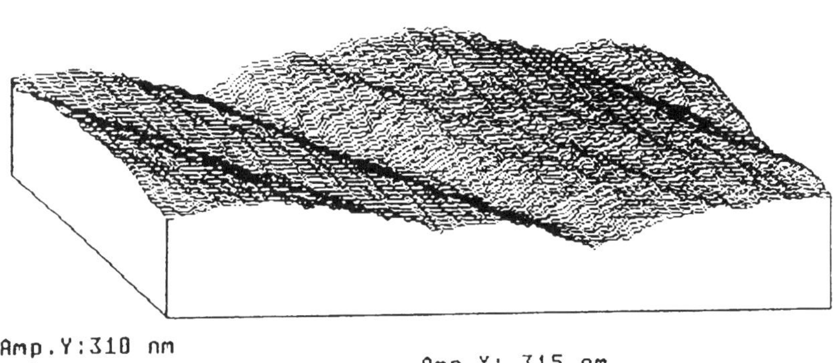

7.10 XAU fibre observed by scanning tunnelling microscopy.

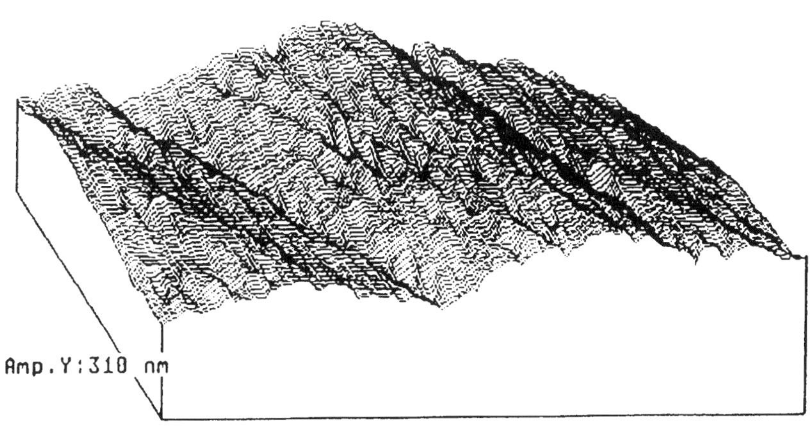

7.11 XAS fibre observed by scanning tunnelling microscopy.

the XAU fibre (Fig. 7.10), whereas a microrelief appears on the XAS fibre surface (Fig. 7.11). Indeed, a more defined microrelief produces a more reactive fibre.

7.3.5 Observation of interfacial phenomena

For the composites prepared with untreated fibres (XAU), the contact index is very small (25%). The most frequently found and most typical interface is illustrated in Fig. 7.12. The interface is undulated and the carbon layers (basal planes) are parallel to the fibre–matrix junctions. Such an interfacial structure, characteristic of the XAU fibre (configuration 1, Fig. 7.5), is only favourable to physical bonding and thus leads to a weak fibre–matrix adhesion [5].

The composites reinforced with XAS fibres present an excellent contact index (85%). The fibre–matrix junctions are achieved, mostly, with carbon-layer edges (configuration 2, Fig. 7.6) and the interfaces become 'hairy'. However a few undulated carbon layers are still visible in some areas of configuration 3 (Fig. 7.7). The surface treatment has attacked the surface of carbon layers making them less organized and therefore more reactive (Fig. 7.9). In this case, the probability of chemical links between functional groups of the fibre and the resin appears to be very high [5].

For the amine-treated fibres, the surface treatments lead to morphologies that are intermediate between the two described above dependent on the nature and the intensity of the treatment. The contact index varies between 50% and 70%. It is seen that this interface is achieved by alternating edges and surfaces of carbon layers (configuration 3, Fig. 7.8). In this case, in addition to grafting of amine molecules, the surface treatment can also oxidize the fibre [6].

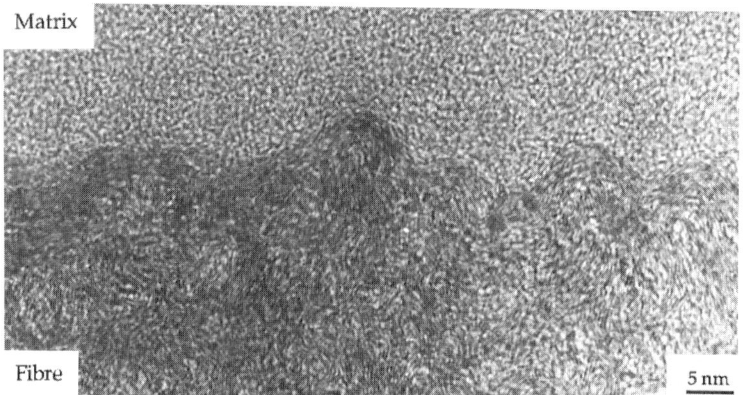

7.12 Typical interface in a transverse section of the XAU/5208 composite.

The Hercules IM7 carbon fibre-reinforced composites exhibit both edges and basal planes at the fibre–resin interface. According to the sizing characteristics (pretreatment, chain length), the contact index varies from 50% to 70%. When they exist, the fibre-sizing junctions are well established with the ISOH1 sizing (Fig. 7.13) and the ISOH2 sizing (Fig. 7.14).

However, new phenomena induced by the deposition process of the sizing on the fibres, are observed in the composites. For the ISOH2 sizing (pretreatment of the sizing at 200°C for 210 s), an interphase (10 nm thick), as represented in Fig. 7.14, exists between the fibre and the acrylic matrix [30, 31]. In some regions, a porosity has developed (Fig. 7.15). However, the contact index is high (70%).

Such phenomena are not observed in the case of the ISOH1 sizing (pretreatment 50°C, 17.5 s, Fig. 7.13) that leads to a contact index value of

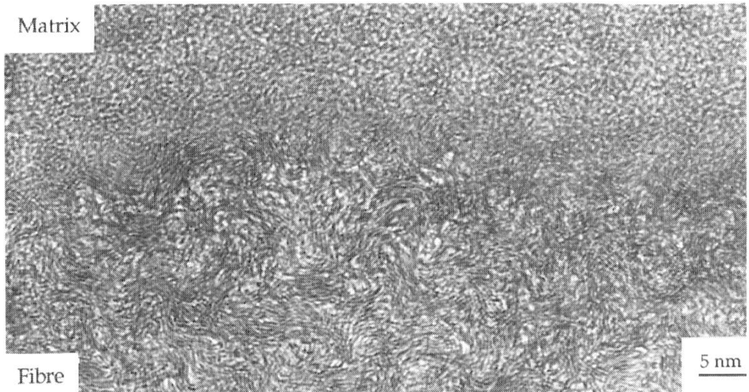

7.13 Typical interface in a transverse section of the IM7/ISOH1/C59 composite.

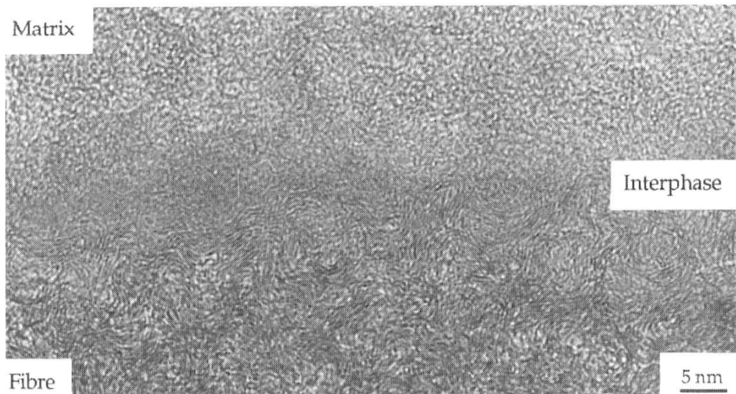

7.14 Typical interface in a transverse section of the IM7/ISOH2/C59 composite.

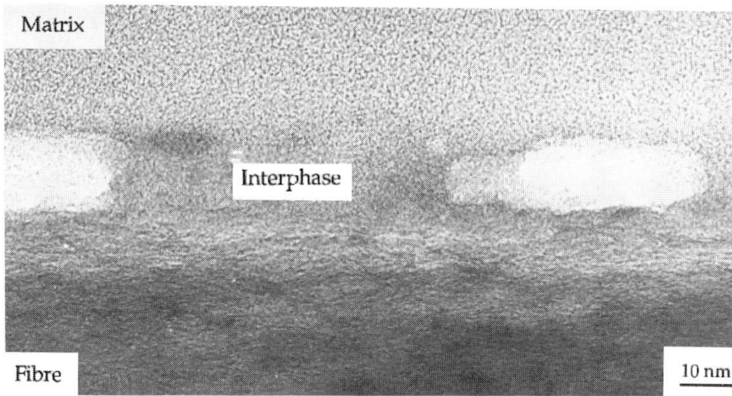

7.15 Interface in a longitudinal section of the IM7/ISOH2/C59 composite showing the interphase and the porosity.

50%. For the ISOP2 sizing (long chains, pretreatment 50°C, 17.5 s), the low value of the contact index (50%) is associated with a decohesion between the sizing and the bulk matrix (Fig. 7.16).

The outer structure of the fibre may allow interlocking bonds as well as chemical ones. The chemical bonding results in reactions between phenolic sites of the fibre and isocyanate sites of the sizing [12]. The parameters of sizing deposition (time, temperature) are of great importance for achieving specific fibre-sizing reactions. Temperature is a significant factor for effective chemical bonding, as previously shown for carbon-epoxy resins [32]. A comparison of ISOH1 and ISOH2 sizings suggests that chemical bonding is less effective for the former. Indeed, the higher the temperature, the more important the chemical

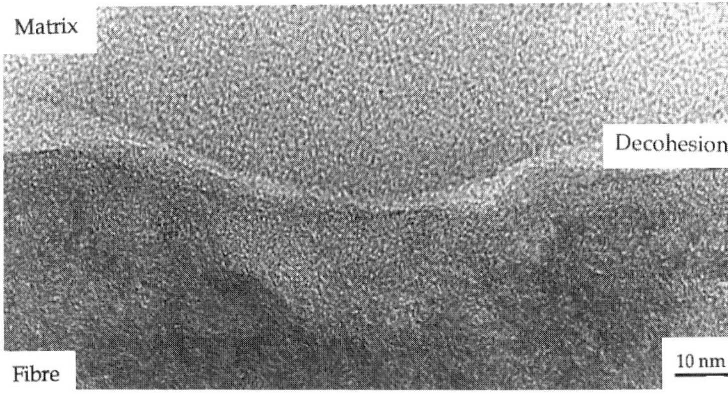

7.16 Decohesion between sizing and matrix as in IM7/ISOP2/C59 and T300/Elast./DA508 composites.

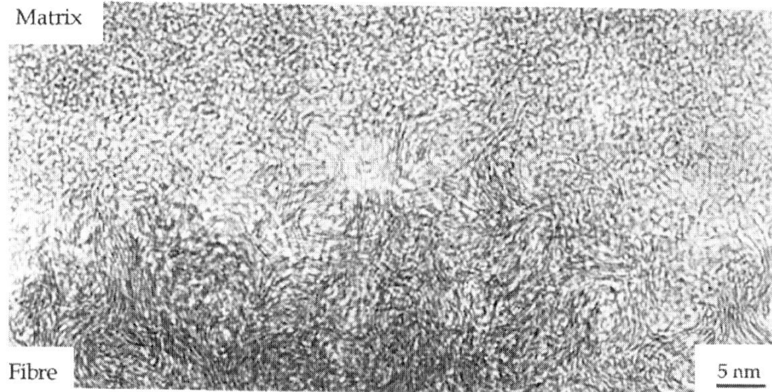

7.17 Interface in a transverse section of the T300/5/DA508 composite.

bonds are. However, for the ISOH2 sizing, the interphase might result from crosslinking of the macromolecules and the porosity might be correlated both with products and shrinkage by chemical reactions. In the ISOP2 composite, some well-established fibre–sizing interfaces are associated with a sizing–matrix incompatibility probably resulting from the great chain length of the sizing macromolecules and the subsequent difficulty of achieving chemical bonds between acrylate functions of the sizing and the matrix.

Special attention must be paid to the Toray T300 fibre. The interfaces observed are characteristic of configuration 2 [6]. With the Toray sizing number 5, the contact index is very high (85%) and the fibre–matrix junctions are perfect with no interphase and no porosity (Fig. 7.17).

It is thought that chemical bondings are numerous and strong (presence of CO, COOH, COH groups at the fibre surface) and there is an excellent compatibility between sizing and matrix. However, with the elastomeric additive, the contact index is very low (25%) and results from the incompatibility between the additive and the matrix (Fig. 7.16).

7.3.6 Mechanical characteristics of the composites

The mechanical evaluation of the resistance of the interfaces has been carried out by interlaminar shear strength tests. The precision of measurements is about 10%. Data are shown in Table 7.1.

The fibre–matrix junctions are weaker with untreated fibres. Oxidization improves the interlaminar strength to a large extent. The amine treatments are intermediate.

As has already been shown [5, 6], the oxidizing treatment does not increase the oxygen content but modifies the chemical nature of the surface. This process replaces a part of the initial oxygen originating from contamination with grafted

Table 7.1 Interlaminar shear strength values for the composites studied (each value is known with a precision of about 10%)

Composites	Contact index (%)	ILSS (MPa)
XAU/5208	25	69
XAS/5208	85	115
XAU/EDA/5208	50	87
XAU/TETA/5208	70	103
IM7/ISOH1/C59	50	52
IM7/ISOH2/C59	70	60–65
IM7/ISOP2/C59	50	NO
T300/5/DA508	85	117
T300/Elast./DA508	25	38

oxygen that is chemically active and therefore favourable to adhesion. In addition to an oxidizing process, the amine treatments also permit the grafting of nitrogen functions which could react as hardeners with the epoxy rings. It would seem, therefore, that the shear strength is due to physical adhesion (interlocking in the microroughness) and chemical adhesion (grafted functions reacting with the matrix) promoted by the surface treatment of the fibre.

The low value or the absence of interlaminar shear strength is associated with sizing–matrix incompatibility. It is interesting to note in this case that the interlaminar strength of the composite is mainly governed by the mechanical properties of the interphase which is a soft polymer (additive or ISOP2).

7.3.7 Relationships between interfacial characteristics and mechanical properties

The interlaminar shear strength values are plotted against the contact index in Fig. 7.18. Curve 1 (XAU and XAS fibres) shows quantitatively the benefits of a surface treatment. Physical bonding is probably predominant in the XAU composite (69 MPa) whereas chemical bonding (oxygen and nitrogen grafted on active sites) is the major component for the treated-XAU and XAS composites (up to 115 MPa).

Curve 2 (IM7 sized fibre) shows that the compatibility between the sizing and the resin used is of utmost importance and confirms that the contact index is very sensitive to chemical bonding between the fibre and the matrix. The shear strength thus obtained strongly suggests that the chemical bondings are more numerous in ISOH2 than in ISOH1. We have seen that ISOH2 sizing was slightly crosslinked by thermal treatment and was thus not able to ensure the maximum bonding with the resin. In this case, there is a conflict between the benefit of chemical links and the porosity due to the shrinkage of the cured ISOH2. When

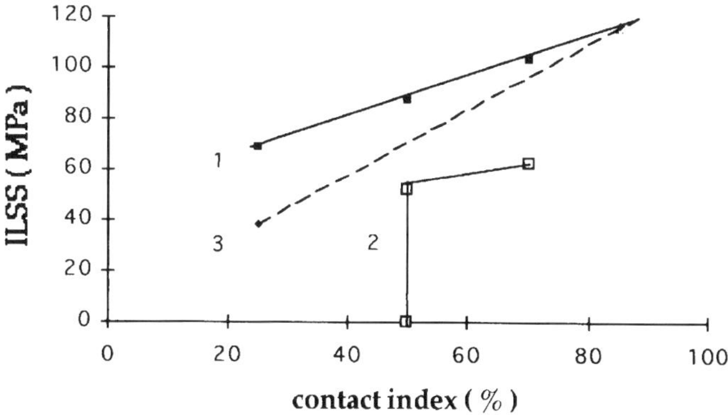

7.18 ILSS versus contact index.

the compatibility between the sizing and the matrix is low, the composites present no shear strength (ISOP2).

Curve 3 (T300 fibre) confirms the necessity of a good sizing–matrix compatibility and a strong fibre–matrix adhesion for the production of a high performance composite (from 38 MPa in the case of the elastomeric additive to 117 MPa for the commercial optimized sizing).

7.4 Conclusion

The nature of the interfacial bonding determines the final properties of the composites. The quality of this bonding must be as high as possible so as to achieve strong adhesion. Such bonding is obtained as a result of structural (cleansing of the surface, creation of microrugosites, creation of active sites) and chemical (grafting of oxygenated or amine functional groups) modifications of the fibre surface induced by a surface treatment.

In order to obtain an optimized stress transfer from the matrix to the fibres, the sizing must simultaneously meet three conditions:

- good adhesion between the fibre and the sizing
- good adhesion between the sizing and the matrix
- correct intrinsic physicochemical and mechanical characteristics in order not to create an interphase with poor mechanical properties.

The least satisfied condition will form the weakest link in the composite. It appears that the contribution of the physical fibre–polymer bonding is important. In addition, the chemical bonding would ensure a gain in the fibre–matrix adhesion and possibly in the sizing–matrix compatibility.

Acknowledgements

This work has been carried out in collaboration with ONERA, Aérospatiale, INSA de Lyon and CEMES-LOE de Toulouse (scanning tunnelling microscope). The author would like to thank ONERA, Aérospatiale, INSA de Lyon and CEMES-LOE for their collaboration, Didier Lanciaux for reading the proof and expresses special thanks to DRET for their financial support.

References

1. L T Drzal, The interphase in epoxy composites, *Adv. Polym. Sci.* 1986 **75** 1–32.
2. M R Piggott, Whither interfaces in fibre composites, *Composites Sci. Technol.* 1991 **41** 1–2.
3. M Guigon, Interface and interphase in carbon-epoxy composite materials: study by transmission electron microscopy, *Microsc. Microanal. Microstruct.* 1991 **2** 15–25.
4. M Guigon, Characterization of the interface in a carbon-epoxy composite using transmission electron microscopy, *J Mater. Sci.* 1992 **27** 4591–4597.
5. E Klinklin and M Guigon, Characterization of the interface in carbon-fibre reinforced composites by transmission electron microscopy, *Colloids Surface A: Physicochem. Eng. Aspects* 1993 **74** 243–250.
6. M Guigon and E Klinklin, The interface and interphase in carbon fibre reinforced composites, *Composites* 1994 **7** 534–539.
7. B Barbier, M Villatte, M Sanchez and G Désarmot, Greffage de fonctions aminées sur des fibres de carbone par un traitement électrochimique en milieu organique: mesure de propriétés mécaniques sue composites carbone-résine, *Preprints 6ièmes Journées Nationales sur les Composites*, AMAC, Paris, 1988, pp 115–130.
8. G Dandine, M Villatte, J Cinquin, D Beziers and G Désarmot, Amino groups grafting on the surface of carbon fibres: influence on the mechanical properties of carbon-epoxy composites, *Preprints 7ièmes Journées Nationales sur les Composites*, AMAC, Paris, 1990, pp 11–18.
9. M Guigon, Microtexture and mechanical properties of carbon fibres: Relationships with the fibre–matrix adhesion in a carbon-epoxy composite, *Polym. Eng. Sci.* 1991 **31** 1264–1270.
10. E Chataignier and D Béziers, Amélioration de l'interface fibre-matrice dans les composites carbone polymérisés par irradiation, *Rapport DRET 87-311*, Bordeaux, 1988.
11. E Chataignier and D Béziers, Amélioration de l'interface fibre-matrice dans les composites carbone-acrylique polymérisés par irradiation, *Rapport DRET 88-399*, Bordeaux, 1990.
12. Y Thomas, Synthèse d'isocyanates uréthannes w-insaturés: amélioration de l'interface fibre-matrice dans les composites carbone-acrylique polymérisés par bombardement électronique, *Thèse de l'Université de Montpellier II*, France, 1991.
13. J F Gérard, Characterization and role of an elastomeric interphase on carbon fibres reinforcing an epoxy matrix, *Polym. Eng. Sci.* 1988 **28** 568–577.

14. O Scherzer, The theoretical limit of the electron microscope, *J. Appl. Phys.* 1949 **20** 20–29.
15. A Lannes and J Ph Perez, *Optique de Fourier en microscopie électronique*, Masson, Paris, 1983.
16. L Reimer, *Transmission Electron Microscopy*, Springer-Verlag, Berlin, 1989.
17. N Uyeda, T Kobayashi, E Suito, Y Harada and M Watanabe, Molecular image resolution in electron microscopy, *J. Appl. Phys.* 1972 **43** 5181–5189.
18. D J Johnson and D Crawford, Defocusing phase contrast effects in electron microscopy, *J. Microscopy* 1973 **98** 313–324.
19. J M Cowley, *Physical Aspect of Electron Microscopy and Microbeam Analysis*, eds B Siegel and D R Beaman, John Wiley, 1975, pp 3–15.
20. G R Millward and D A Jefferson, Lattice resolution of carbons by electron microscopy, *Chemistry and Physics of Carbons*, Marcel Dekker, New York, 1978, pp 1–82.
21. G R Millward and J M Thomas, Thoughts on the feasibility of directly imaging the ultramicrostructural characteristics of imperfectly ordered graphitic carbons by high resolution electron microscopy, *Carbon* 1979 **17** 1–5.
22. M Guigon, A Oberlin and G Désarmot, Microstructure and structure of some high tensile strength, PAN-based carbon fibres, *Fibre Sci. Technol.* 1984 **20** 55–72 and 177–198.
23. M Guigon and A Oberlin, Heat treatment of high tensile strength PAN-based carbon fibres: microtexture, structure and mechanical properties, *Composites Sci. Technol.* 1986 **27** 1–23.
24. A Oberlin and M Guigon, The carbon fibres structure, *Fibre Reinforcements for Composites Materials*, Elsevier, Amsterdam, 1988, pp 149–210.
25. J M Thomas, Microscopic studies of graphite oxidation, *Chem. Phys. Carbon* 1965 **1** 121–202.
26. P Ehrburger, F Louys and J Lahaye, The concept of active sites applied to the study of carbon reactivity, *Carbon* 1989 **27** 389–393.
27. G Binning, H Rohrer, Ch Gerber and E Weibel, *Phys. Rev. Lett.* 1982 **49** 57–60.
28. W P Hoffman, W C Hurley, T W Owens and H T Phan, The advantage of the scanning tunnelling microscope in documenting changes in carbon fibre surface morphology brought about by various surface treatments, *J. Mater. Sci.* 1991 **26** 4545–4553.
29. W P Hoffman, W C Hurley, P M Liu and T W Owens, The surface topography of non-sheared treated pitch and PAN carbon fibres as viewed by the STM, *J. Mater. Res.* 1991 **6** 1–10.
30. L T Dzral, M J Rich and P F Lloyd, Adhesion of graphite fibres to epoxy matrix 1: the role of fibre surface treatment, *Adhesion* 1983 **16**(1) 1–30.
31. L T Dzral, M J Rich, M F Koenig and P F Lloyd, Adhesion of graphite fibres to epoxy matrix 2: the effect of fibre finish, *Adhesion* 1983 **16**(2) 133–152.
32. K Horie, H Murai and I Mita, Bonding of resin to graphite fibres, *Fibre Sci. Technol.* 1976 **9** 253–264.

8 Micromechanics of reinforcement using laser Raman spectroscopy

COSTAS GALIOTIS

8.1 Introduction

The determination of the state of stress in structural materials during service has always been one of the key issues that the structural engineer/designer has to address. The lack of experimental techniques in this area often leads the engineer to resort to analytical and/or numerical methods in order to assess the overall stress distribution within a structural component. As a result, stringent design rules have to be applied to ensure safety in a structural assembly. This is particularly well demonstrated in the case of advanced fibre composite materials; lack of knowledge of the complex state of stress generated by the anisotropy and quite often inhomogeneity in these materials leads to over-design and hence high component costs. Thus, in polymer-based composites, the savings gained in moving parts as a result of their light weight and corresponding high specific properties can quite often be offset by the volume of material required to address the safety design limits.

Another important issue is the detection of the propagation of damage in service. In metals visual inspection, combined occasionally with non-destructive diagnostic techniques, can provide information about the integrity of the material at various stages of its 'lifecycle'. In composites which incorporate brittle polymer matrices reinforced with brittle fibres, toughness is normally attained by the complexity of the propagation of damage at the microscopic scale [1]. Careful control of the strength of the fibre/matrix interface can also enhance tensile strength and toughness by increasing the overall crack propagation and by diluting the effect of stress concentrations emanating from isolated fibre breaks [2,3]. In all cases, the exact knowledge of the local stress field generated by an applied external load is of paramount importance in determining the efficiency of stress transfer and for monitoring the propagation of damage at various load increments and/or time intervals.

The technique of laser Raman spectroscopy (LRS) has now been established as the only experimental technique to date that can provide information on fibre stress at the microscopic level [4,5]. The principle of this technique is based on the anharmonicity of the interatomic bonds, which requires that a change in the interatomic separation, as a result of an applied stress, should result in a corresponding change in the interatomic force constants and, hence, atomic vibrational frequencies (wavenumbers). This effect is particularly apparent in crystalline materials for which an applied macroscopic stress can be transferred directly to the atomic bonds. Most reinforcing fibres are crystalline materials and, therefore, have been found to exhibit this effect. By loading individual fibres either in uniaxial tension or compression, the magnitude of the Raman wavenumber shift from the value of the free-standing fibre can be measured. In a composite material the inverse methodology is applied; the magnitude of the wavenumber shift is measured along the reinforcing fibres and converted to axial stress or strain via a fibre-specific calibration curve. All fibres situated near the surface of a polymer composite can be interrogated remotely (and non-destructively) provided that the matrix is reasonably transparent. Fibres that are located in the bulk of a composite can be interrogated by means of a waveguide, such as a fibre optic cable [6], but in this case a certain amount of stress perturbation is expected to occur.

The application of laser Raman spectroscopy to stress measurements in fibre composites was first attempted here at QMW in the early 1980s and has expanded dramatically ever since. Today apart from QMW, there are a number of centres worldwide which employ this technique to assess the state of stress in a variety of composite geometries. These are UMIST in the UK, Cornell, Drexel, RPI, UMASS and Texas University in the USA and DSM in the Netherlands. The results presented in this article refer exclusively to work performed at QMW over the last few years. The reader is advised to refer to Schadler and Galiotis [7] for a much broader overview of this technique applied to fibres and fibre composites.

The technique of laser Raman microscopy can be employed to determine the stress or strain in composite materials at a spatial resolution of maximum 1 μm. The principles governing this technique and the various experimental arrangements required are presented here. An overview of the application of this technique to study of the stress transfer mechanisms in single, as well as multifibre, composites has been presented [7].

8.2 Experimentation

8.2.1 Conventional measurements

The experimental requirements of a modern laser Raman setup are shown in Fig. 8.1. The monochromatic light normally in the visible range produced by a gas laser (Ar^+ or He-Ne) is directed to a Raman modified microscope via a system of mirrors and focused down to the specimen, which is housed on the microscope platform. The 180° backscattered radiation is sent to the spectrometer by a beam splitter through a variable size pinhole to attain confocality. The main task of the

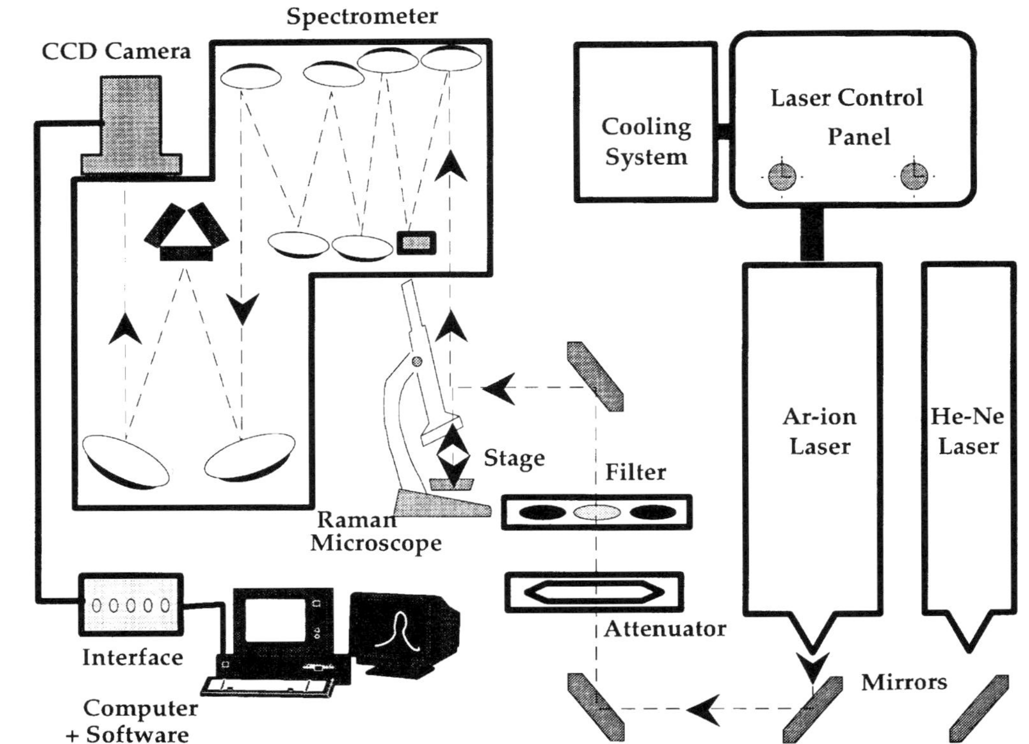

8.1 Schematic of the 'conventional' laser Raman experimental setup comprising two lasers, a Raman microscope, a spectrometer, a charge coupled device (CCD), associated optical components and a computer for data acquisition and analysis.

spectrometer (monochromator) is to block out the Rayleigh or elastic scattering through a system of mirrors and diffraction gratings and to detect the intensity of the inelastic (Raman) scattering, which is focused on the silicon chip of a CCD (charge coupled device) detector. The CCD converts the optical signal into an electrical output, which is subsequently stored in a PC. This experimental arrangement is ideal for small specimens, which can be translated in the laser beam with a suitable micromanipulator.

8.2.2 Remote laser Raman (ReRaM) measurements

To avoid the space restrictions imposed by the conventional setup, the spectroscopic assembly (laser plus spectrometer) can be decoupled from the testing area by employing a flexible fibre optic microprobe (Fig. 8.2) [8]. The main difference between the two setups lies in the design of the Raman microprobe; the incorporation of flexible fibre optic cables for laser light delivery and collection allows the positioning of the microprobe at any angle with respect to a system of reference such as the work bench. Furthermore, the length of the fibre optic can be as long as 300 m and, therefore, specimens at large distances from the laser source/monochromator can be interrogated (Fig. 8.2). The details of the design of the Remote Raman Microprobe (ReRaM) itself are shown in Fig. 8.3. Tailor-made optics at both input and output positions of each fibre optic, ensure laser collimation, maximum efficiency and enhancement of the Raman scattering [8]. Finally, the use of an incorporated CCTV camera allows optical observations of the specimen during Raman data acquisition. This remote setup is particularly useful for composite materials, which are subjected to mechanical loading, but it can also be used in a whole variety of other technological applications, such as, to conduct remote Raman measurements in chemically hostile environments/elevated temperatures, to monitor curing of polymer resins and the crystallisation of polymers during solidification, to provide non-destructive 'health' monitoring of sections of large structures (e.g. aircraft, ships, bridges, etc.) and to assess the quality of oil supplies or other chemical media.

8.2.3 Producing calibration curves for high performance fibres

The relationship between Raman wavenumber and tensile stress or strain is obtained by stressing a single fibre in air on a suitable mechanical tester operating at a low strain-rate while Raman spectra are taken at any position along the fibre length. The difference between the Raman wavenumber values obtained at each level of strain and that of the free-standing fibre represents the Raman wavenumber shift. In Fig. 8.4 the Raman wavenumber shift is plotted as function of applied stress for the $1615\,\text{cm}^{-1}$ Raman band of a Kevlar 49 fibre and, as can be seen, a linear relationship is obtained. The negative slope of

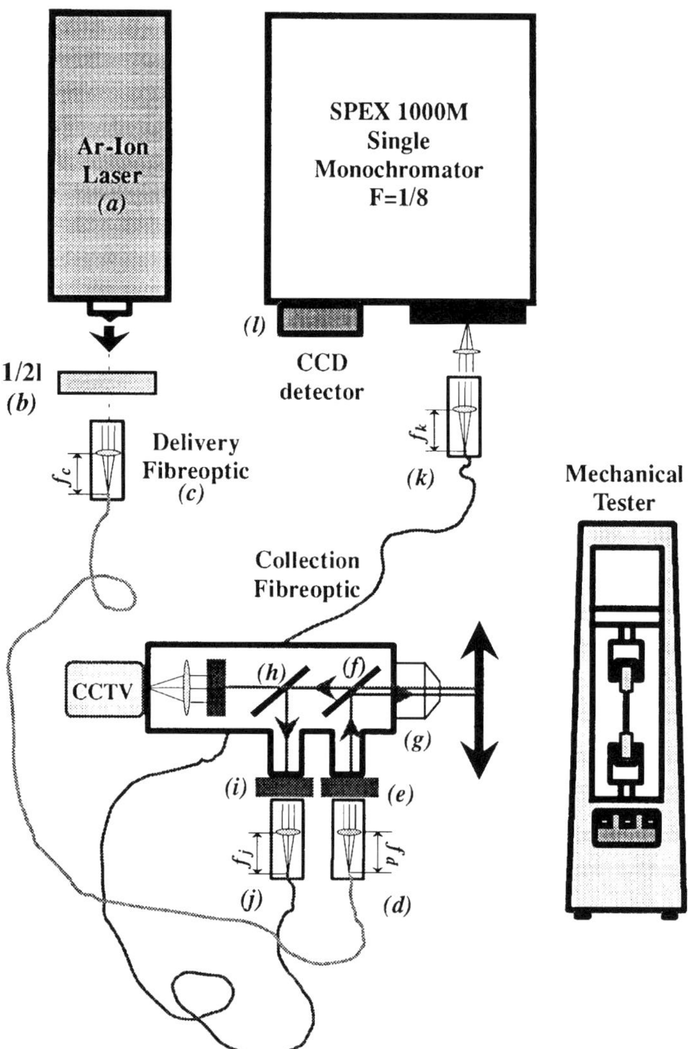

8.2 Schematic of the remote laser Raman experimental setup comprising one laser, two fibre optic cables for laser delivery and collection, a remote Raman microscope (ReRaM), a charge coupled device (CCD), associated optical components and a computer for data acquisition and analysis. (a) Argon-ion laser, (b) quarter wavelength plate, (c) laser collimation unit (input), (d) to (j) see caption Fig. 8.3, (k) laser collimation unit (output), (l) CCD detector.

$-3.4\ \mathrm{cm}^{-1}\ \mathrm{GPa}^{-1}$ of the least-squares-fitted straight line represents the sensitivity of the stress sensor. This value allows the conversion of the Raman frequencies obtained from fibres embedded in polymer-based composites into values of stress. In order to calibrate the Raman wavenumber as a function of tensile or compressive strain, single filaments are bonded to the surface of an

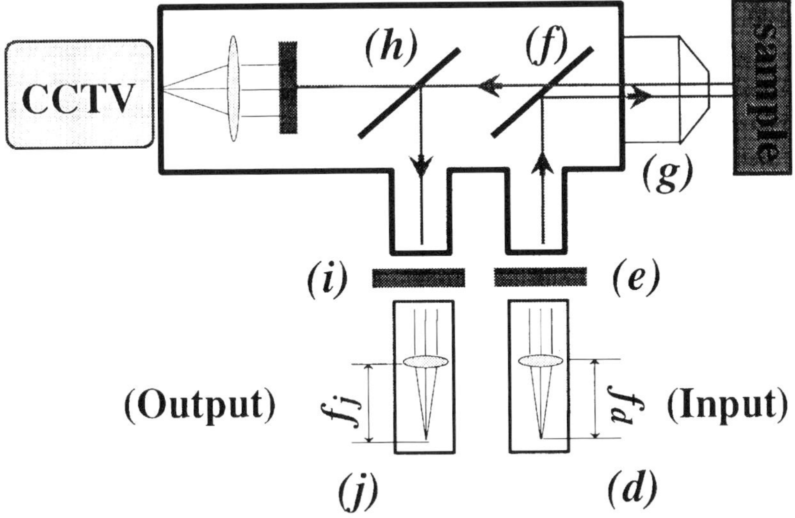

8.3 Schematic of the remote laser Raman microscope (ReRaM) comprising tailor-made optics at both input and output positions, a CCTV video camera and associated optical components to ensure laser collimation, maximum efficiency, confocality and enhancement of Raman scattering. The various components shown are: (d) laser collimation unit (input), (e) bandpass filter, (f) holographic beam splitter, (g) microscope objective, (h) mirror, (i) notch filter, (j) laser collimation unit (output).

elastic polymer cantilever beam which can be flexed up or down to subject the fibres to compression or tension, respectively [9]. Provided that no slippage takes place between the fibre and the bar, the strain (tensile or compressive) varies linearly along the length of the beam, and is only a function of the position on the bar as determined by the elastic beam theory. A typical graph of Raman wavenumber versus strain for a Kevlar 49 is presented in Fig. 8.5. It is interesting to note that the Raman wavenumber scales non-linearly with strain, whereas for aramid, as well as carbon fibres, the corresponding relationship with stress is always linear [9]. The spline polynomial functions represent the sensitivity of the strain sensor for converting Raman frequencies into strain in composites. Important mechanical parameters such as the critical compressive strain to failure and the molecular deformation of the fibre in the post-failure (post-buckling) region can also be determined from Fig. 8.5. Finally, for high modulus carbon fibre the relationship between Raman wavenumber and tensile stress (Fig. 8.6a) or strain (Fig. 8.6b) can be considered linear to a first approximation. The nominal values of stress or strain sensitivity of the Raman wavenumber for a number of commercial fibres are given in Table 8.1.

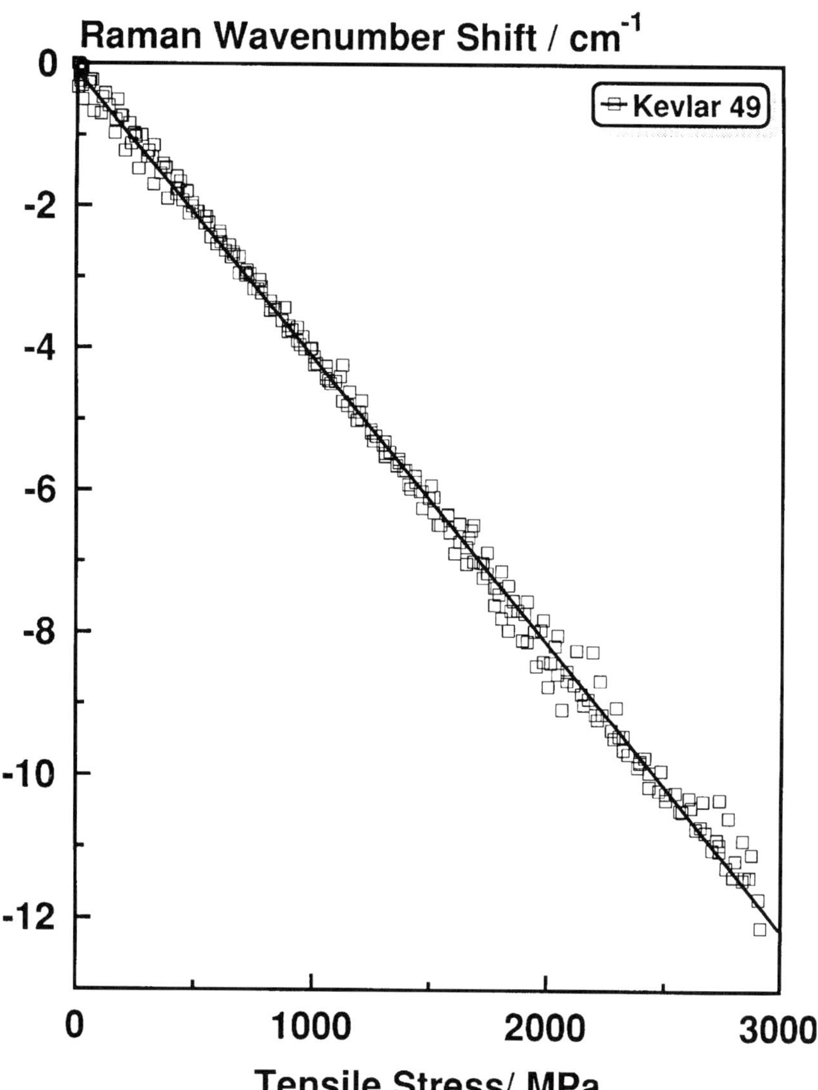

8.4 Raman wavenumber shift as a function of applied stress for a single Kevlar® 49 fibre. The sensitivity of the Raman stress sensor is given by the slope of the least squares fitted line which is $-3.4\,\text{cm}^{-1}$ GPa^{-1}.

8.3 Micromechanics of reinforcement in composites

The architecture of the fibre/matrix interface in composites is presented in Fig. 8.7. At the macroscopic level the bond in the representative volume element (RVE) is considered for design purposes to be perfect [10]. However, at the

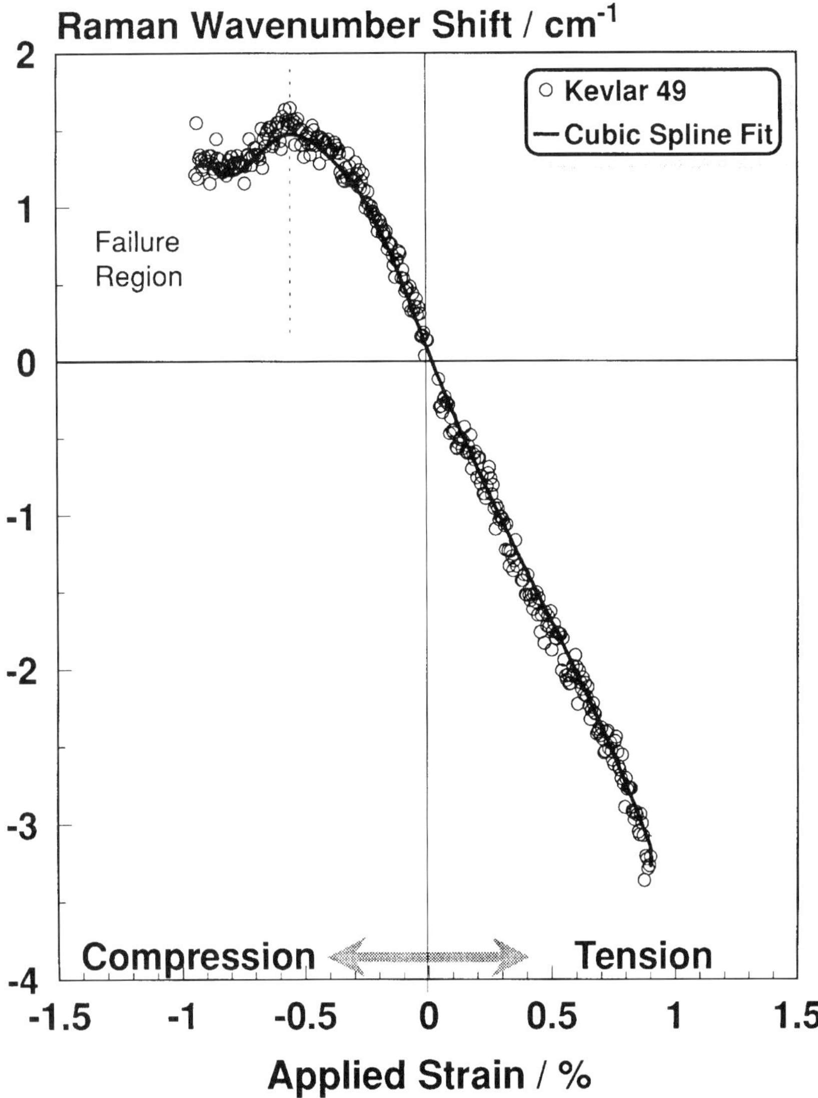

8.5 Raman wavenumber shift as a function of strain for a Kevlar® 49 fibre within the range of −1.0% to 1.0%. Each data point represents an average of four measurements. The solid line represents a cubic spline fit to the experimental data.

microscopic level the picture is extremely complex due to the existence of an interphase of variable thickness comprising fibre surface chemistry and topography, sizing, wetting and other coating agents, as well as diffused matrix material [11]. The properties of this intermediate phase are also affected by the local thermal and mechanical environment. Finally, the integrity of this interphase

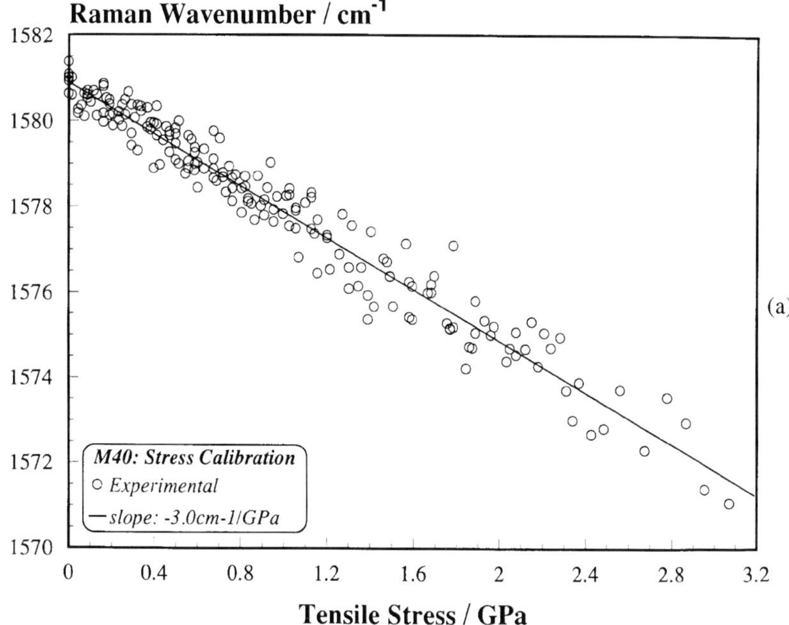

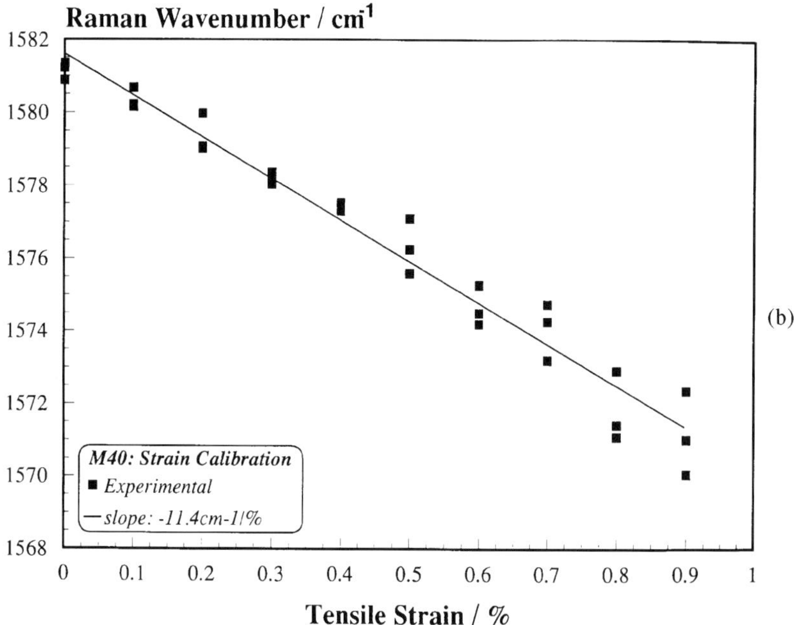

8.6 Raman wavenumber shift as a function of tensile (a) stress and (b) strain for the sized M40-3k-40B® fibre. The sensitivity of the Raman stress and strain sensors is given by the slope of the least squares fitted line which are $-3.0\,cm^{-1}\,GPa^{-1}$ and $-11.4\,cm^{-1}\,\%^{-1}$ respectively.

Table 8.1 Raman wavenumber stress/strain sensitivity of different commercial fibres

Fibre	Raman vibrational band (cm^{-1})	Strain sensitivity in tension (cm^{-1} $\%^{-1}$)	Stress sensitivity (cm^{-1} GPa^{-1})
Kevlar-49®[a]	1613	−4.3	±3.4
M40-3k-40B®[b]	1580	−11.4	±3.0
XAS®[c]	1600	−8.0	±2.9

[a]Trademark of E.I.DuPont de Nemours (USA).
[b]Trademark of Toray Industries, Soficar (France).
[c]Trademark of Courtaulds Grafil plc (UK).

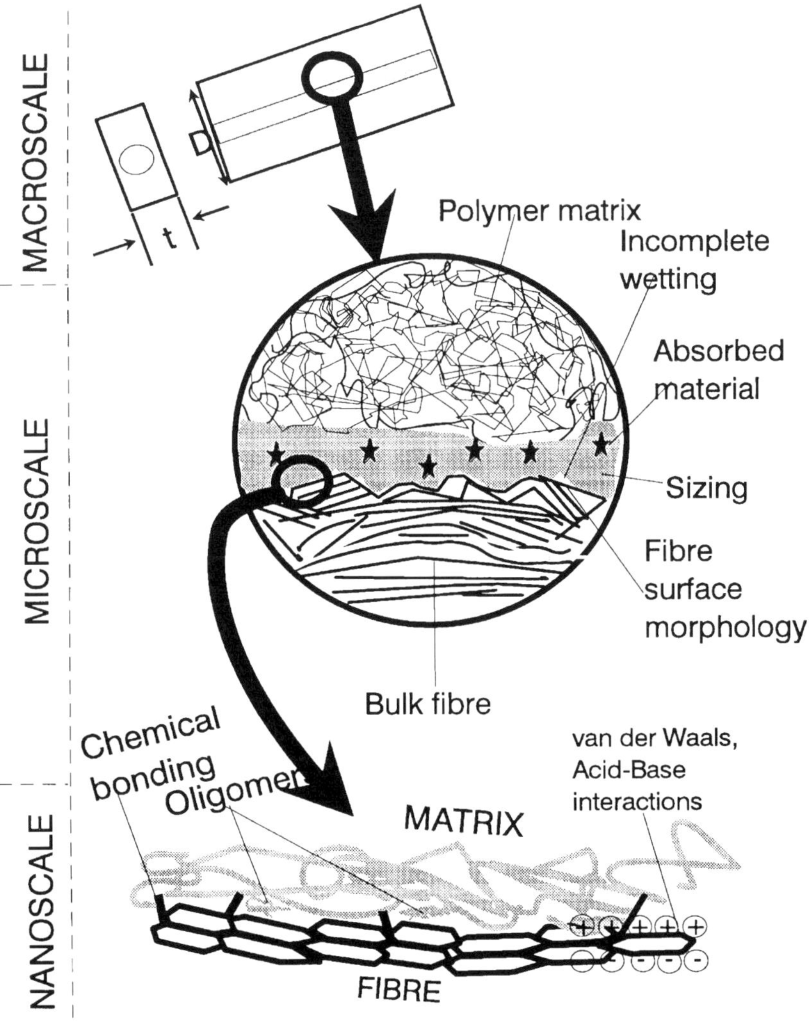

8.7 Architecture of the fibre/matrix interface.

and hence of the RVE/composite depends upon events taking place at the molecular/nanoscale level, as a result of adequate contact between the two dissimilar surfaces. These include the physicochemical interactions mentioned earlier and their effectiveness in the presence of oligomers, impurities, solvent molecules and other contaminants. Furthermore, molecular events, such as increases in fibre surface functionality, are precursors to increasing fibre–matrix adhesion through increases in fibre–matrix dispersion type interactions, surface energetics and wettability. The macroscopic, microscopic and molecular phenomena are strongly interrelated and therefore the overall physicomechanical performance can be successfully tailored by suitable manipulation of the various critical variables at the molecular and/or microscopic levels.

One of the most important functions of the interface is the efficient transfer of stress from the matrix to the reinforcing fibres. The degree of efficiency of stress transfer will depend primarily upon the fibre and matrix chemistry and the existence of residual thermal stresses at the interface. The latter is related to the geometry of the test coupon, the associated curing process and the fibre volume fraction.

In Fig. 8.8 the most important test geometries for the determination of interfacial parameters in composites are shown. In general, single fibre test coupons can be quite useful in the detection of the true interfacial phenomena without the influence of fibre–fibre interactions but their applicability is limited as they cannot be considered truly composite tests. Tests on full composites can only provide indirect evidence on the strength of the fibre/matrix interface through measurements of the off-axis properties such as the in-plane shear strength. As shown schematically in Fig. 8.8, laser Raman microscopy yields the axial stress distribution in the embedded fibres, through which the stress transfer efficiency is easily determined. Hence, the desirable link between single-fibre and multi-fibre test geometries, is established. Results obtained to date on single and multi-fibre test methods are briefly reviewed below.

8.3.1 Stress transfer in single fibre model composites

The stress transfer mechanism in fibre-reinforced composites is activated in the vicinity of discontinuities such as fibre ends and/or fibre breaks. Fibre ends are normally found in discontinuous fibre composites while fibre breaks are formed as a result of fibre fracture during fabrication or the influence of an externally applied stress field. Until the advent of the laser Raman technique [7], the stress transfer profiles activated at a fibre discontinuity could only be derived by means of analytical [12] or photoelastic modelling [13].

The stress transfer profiles obtained using the Raman technique can be converted to interfacial shear stress (ISS) profiles along the length of the fibre by means of a straightforward balance of shear-to-axial forces argument [14]. This

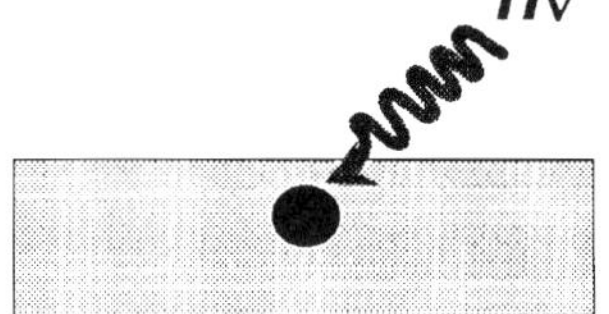

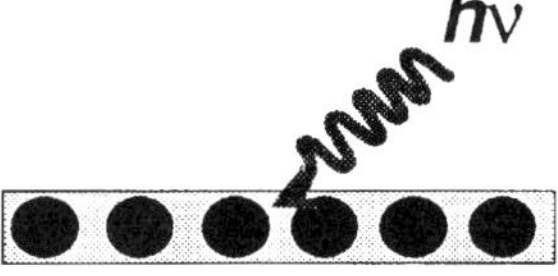

2D Microcomposites

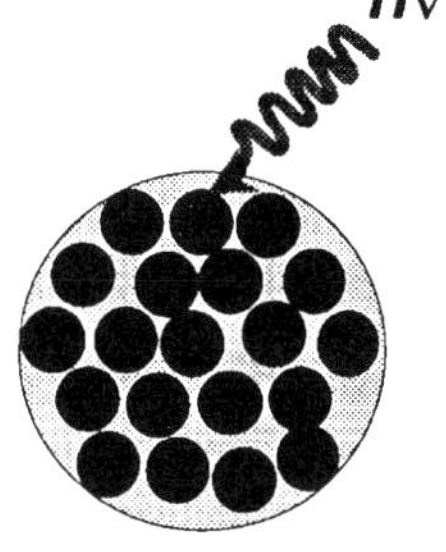

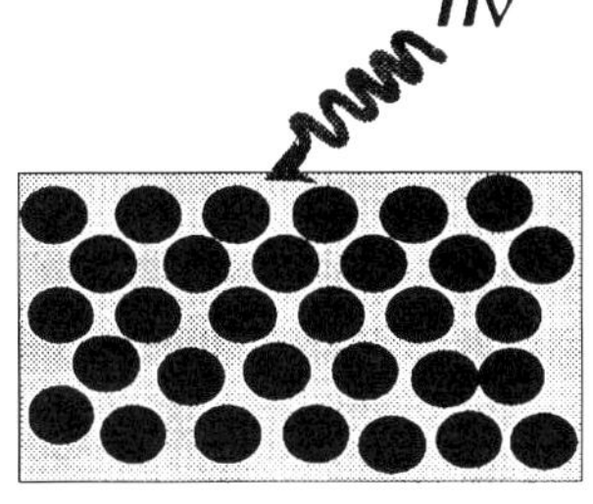

Standard Tensile
Coupons

8.8 Test geometries employed in this programme of work.

leads to a simple analytical expression between the ISS, $\tau_{(rx)}$, and the gradient $d\sigma_f/dx$ of the stress transfer profiles:

$$\tau_{(rx)} = -\frac{r}{2}\left(\frac{d\sigma_f}{dx}\right) \tag{1}$$

where σ_f is the axial stress in the fibre, r the fibre radius and x the distance along the fibre length. The ISS profiles, $\tau_{(rx)}$, are derived by (a) fitting a cubic spline to the raw data, (b) calculating the derivatives $d\sigma_f/dx$ from the spline equations and

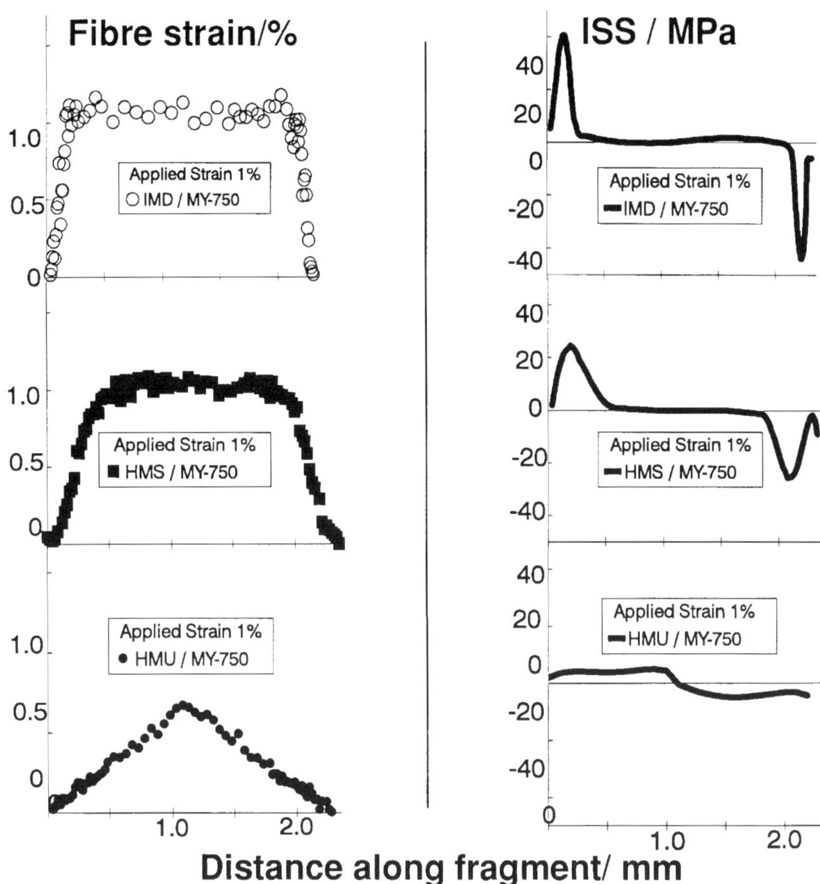

8.9 Fibre strain (left) and interfacial shear stress (right) distributions of representative fragments of similar length for the three different carbon fibre/epoxy systems listed in Table 8.1.

Table 8.2 Constituent properties of the carbon fibre[a]/MY750®,[b] composite system employed to examine the effect of fibre treatment upon stress transfer characteristics

Fibre	Tensile modulus (GPa)	Tensile strength (GPa)	Diameter (μm)	Surface treatment	Fibre sizing
HMU	390	3.2	7	None	None
HMS	390	3.2	7	Standard	None
IMD	305	5.5	5	Standard	None

[a]All fibres were supplied by Courtaulds Grafil plc (UK).
[b]Trademark of Ciba-Geigy plc (UK).

finally, (c) employing equation (1). For fibres exhibiting elastic stress–strain characteristics the $\tau_{(rx)}$ distribution can also be obtained from the axial strain distribution by substituting the σ in equation (1) by the corresponding product $E\varepsilon$, where E is Young's modulus and ε the axial fibre strain.

8.3.1.1 Effect of fibre treatment

A lot of work has been done recently using LRS to monitor fragmentation processes in carbon fibre/epoxy resin systems [14,15]. The fibre strain distributions of representative fragments of similar lengths for each fibre/resin system are shown in Fig. 8.9. The constituent properties of the materials used are given in Table 8.2. In all cases the maximum strain supported by the fibre is 1%. As can be seen, the strain profile for a non-surface treated high-modulus carbon fibre (HMU) is virtually linear indicating a frictional type of reinforcement. For the surface treated high- (HMS) and intermediate-modulus (IMD) fibre systems the strain take up is more or less in accordance with the elastic stress transfer models [12]. These fibre strain distributions are converted to ISS distributions via equation (1) and the resulting curves are also shown in Fig. 8.9. The following observations can be made at this point: (a) the interfacial shear stress is nearly constant along the HMU/epoxy fragment, (b) the surface treated fibre/resin systems, HMS and IMD/epoxy, exhibit distributions which reach a maximum value near the fibre end and decay to zero towards the middle of each fragment and finally, (c) the higher the maximum ISS value for the HMS and IMD/MY-750

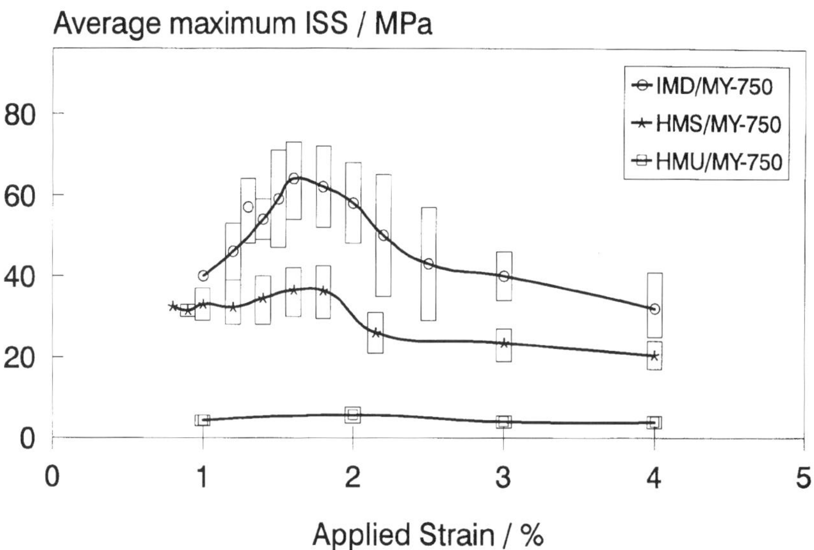

8.10 Average maximum interfacial shear stress (ISS) as a function of applied strain for the three different carbon fibre/epoxy systems listed in Table 8.1.

systems, the shorter the distance from the fibre end where this maximum appears. The interfacial shear strength (IFSS) of a fibre/resin system is normally defined as the maximum value of ISS developed throughout the fragmentation test. It should be stressed, however, that there is a statistical distribution of the ISS maxima at each level of applied strain, as a large number of fragments are sampled along the gauge length. It is, therefore, more appropriate to derive an average maximum ISS value at each level of applied strain, along with the standard deviation of the mean. Such a plot of average maximum ISS as a function of applied strain for all three systems examined in this project is shown in Fig. 8.10. As can be seen, the average maximum ISS increases with applied strain for both systems and reaches an upper limit of 36 ± 6 MPa and 66 ± 15 MPa for the HMS and IMD/epoxy systems, respectively. These values are good estimates of the IFSS of the two systems. The average maximum ISS for the untreated HMU/MY-750 system appears to be insensitive to applied strain and is approximately six times lower than that of the treated HMS/MY-750 system.

8.3.1.2 Effect of fibre sizing

To demonstrate the effect of fibre sizing upon the stress transfer efficiency, single fibre model systems incorporating sized and unsized carbon fibres have been studied. The model composites consist of the sized and unsized M40 carbon fibre (Table 8.1) supplied by Soficar embedded in the Ciba-Geigy MY750 epoxy. In Fig. 8.11 and 8.12, the axial fibre stress profiles of representative fragments at various increments of applied composite strain for the sized and unsized fibre system, respectively, are presented. In the case of the sized fibre system (Fig. 8.11), the propagation of interfacial damage with applied strain leads to the progressive reduction of the effective length for stress transfer and to the growth of zones of zero stress transfer on either side of the fragment [16]. These zones emanate from the fibre breaks and grow towards the middle of the fragment resulting in the characteristic S-shaped profiles of Fig. 8.11b and 8.11c. In contrast, in the case of the unsized system (Fig. 8.12), there are no areas in the fibre where the fibre axial stress is zero. This indicates that the stress is transferred efficiently along the whole fragment and, therefore, the interfacial damage zone is adequately bridged. As a result of this, the effective length required for stress transfer is reduced only marginally and therefore a new fibre fracture event occurs at higher strains (Fig. 8.12c).

The average maximum IFSS of the two systems is plotted as a function of applied strain in Fig. 8.13. The values of IFSS for each system do not seem to vary much with applied strain. Overall, the maximum IFSS values for the sized fibre system is higher than those of the unsized system at a confidence level of 95%, as determined by the Student t-test [16]. At this point of saturation, the IFSS of the sized and unsized systems are about 42 MPa and 35 MPa (Table 8.1), respectively. The average value of about 42 MPa measured for the sized M40/

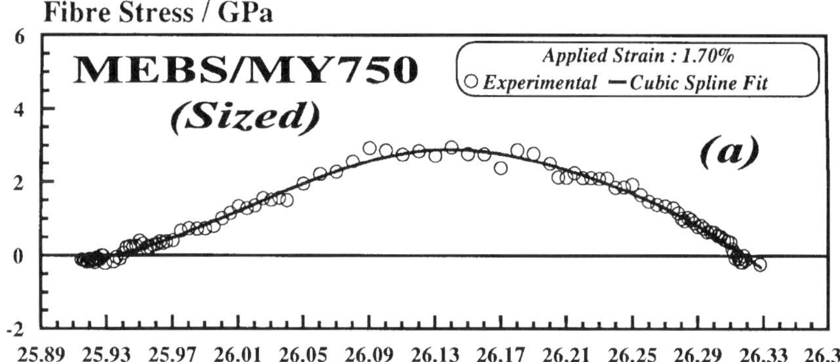

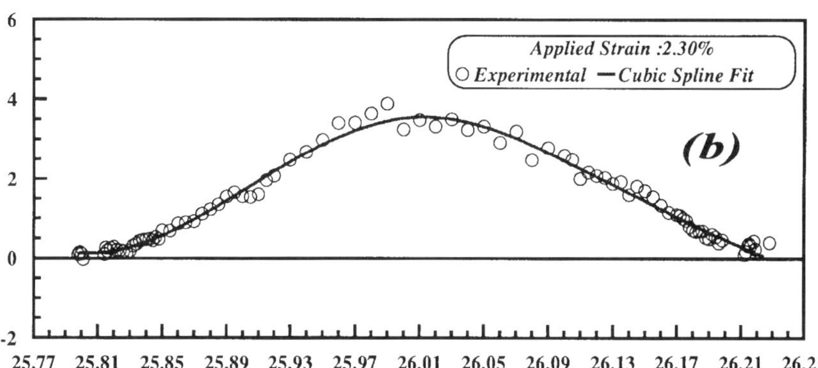

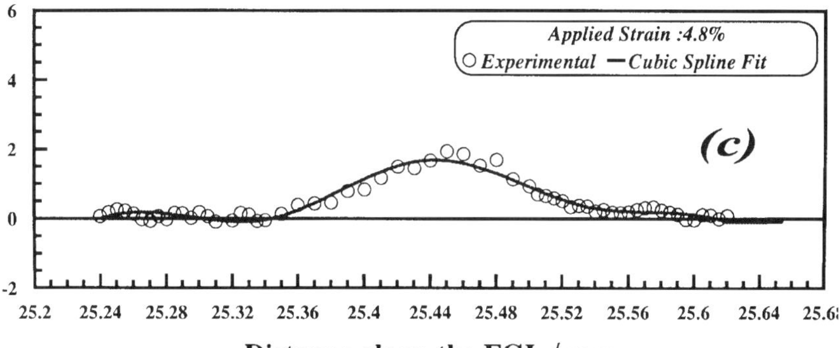

8.11 Fibre stress as a function of position for a representative fragment of a sized M40-3k-40B® fibre at (a) 1.7%, (b) 2.3% and (c) 3.8% applied composite strain.

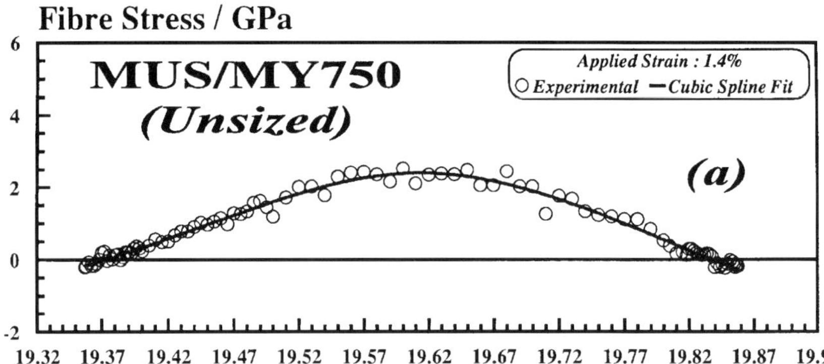

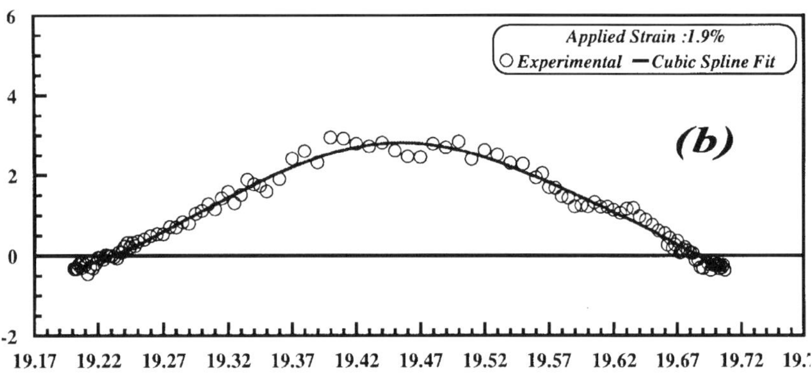

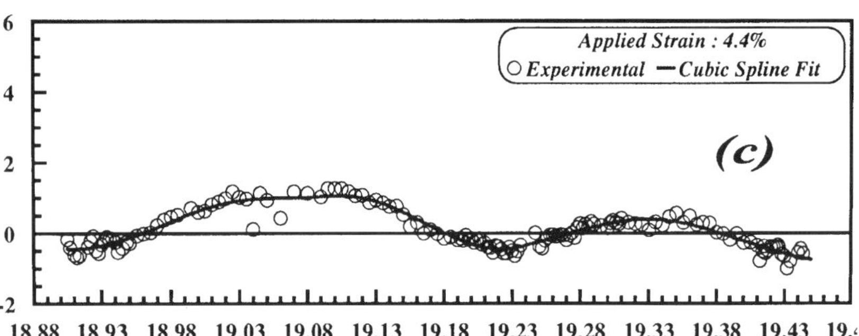

8.12 Fibre stress as a function of position for a representative fragment of an unsized M40-3k-40B® fibre at (a) 1.6%, (b) 1.9% and (c) 3.4% applied composite strain.

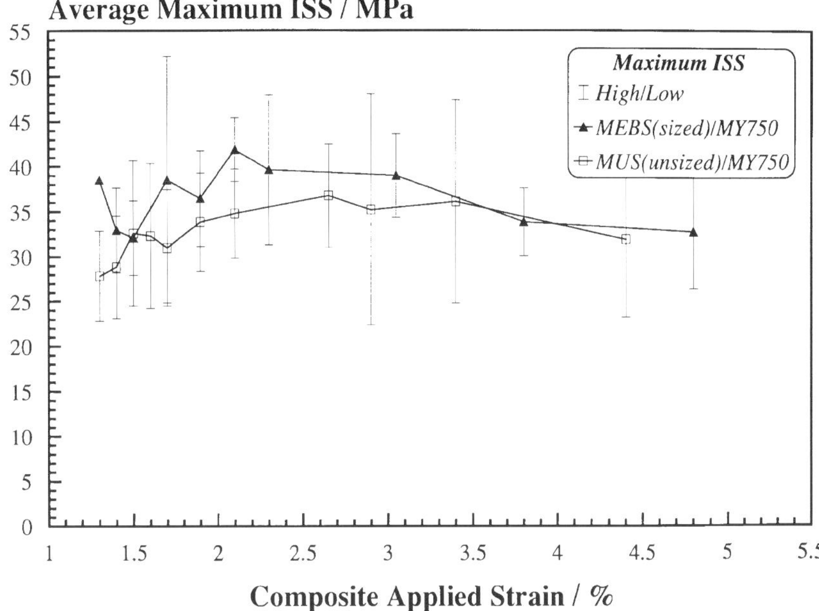

8.13 Average maximum interfacial shear stress (ISS) as a function of applied strain for the sized (MEBS) M40/MY-750 and the unsized (MUS) M40/MY-750 systems.

MY750 system compares well with the shear yield strength of the resin which is estimated to be of the order of 40 MPa [17]. SEM evidence points to two distinct modes for interfacial failure for the two systems; whereas in the unsized fibre system clear debonding can be seen at high strains, in the case of the sized system, the plane of interfacial damage appears shifted towards the matrix material in a mixed-mode fashion.

In conclusion, it can be said that the sized fibre system can sustain higher shear stresses at the interface (Fig. 8.13), which trigger a mixed mode type of failure. This is clearly undesirable at least at very high levels of applied strain, since there is no adequate bridging between matrix and fibre and, therefore, the effective length of the fibre (Fig. 8.11) is severely reduced. On the other hand, the relatively weaker interface observed in the unsized system fails by fibre/matrix debonding. This does not impair the ability of the interface to transfer stress since the two surfaces are adequately bridged due to the presence of a compressive radial stress field.

8.3.2 Stress transfer in composites

When an individual fibre breaks, the shear field perturbation which is generated at the point of fibre failure will cause a stress redistribution in the neighbouring fibres [2]. In the fully elastic case, the shear forces are maximum at the plane of

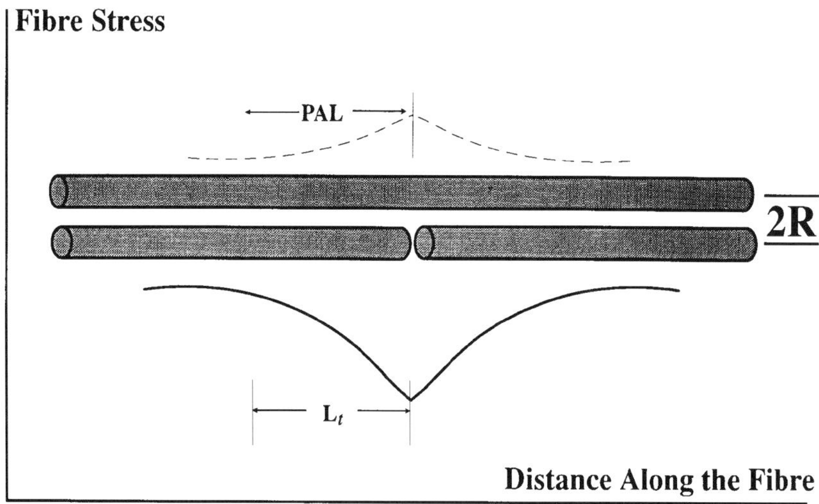

8.14 Schematic representation of the stress distributions in a fractured fibre and in its nearest neighbour. The fracture 'well' and the zone of 'positively affected length' are clearly shown.

fibre fracture ($\xi = 0$) and decay to zero at some distance away from the fibre break. As shown schematically in Fig. 8.14, the distance over which the axial stress reached its maximum value (positively affected length, PAL) is identical in magnitude to the transfer or ineffective length of the fractured fibre.

The stress concentration factor, K_q, for a fibre adjacent to q fractured fibres is defined as:

$$K_q = \frac{\sigma_n^{\xi=0}}{\sigma_{\text{applied}}} \tag{2}$$

where $\sigma_n^{\xi=0}$ is the stress of the intact fibre at the plane of first fracture ($\xi = 0$) and σ_{applied} is the far field stress in the fibre. Similar to the procedure for the derivation of the ISS mentioned earlier, the fibre stress profiles are fitted with cubic spline polynomials and the maximum stress is defined at the point for which the derivative $d\sigma_f/dx$ becomes zero. In almost all cases presented here, the maximum stress concentration is obtained at a displacement $\xi = 0$ from the plane of fracture but it is interesting to note that a kind of 'hill' [18] of stress concentration is observed rather than the 'spike' predicted analytically [2]. As it will be argued later, the existence of a 'hill' of stress concentration is indicative of the presence of interfacial failure in tandem with the fibre fracture. Finally, it is worth mentioning that the balance of forces argument of equation (1) is of general validity and can be also applied to the stress field acting on fibres adjacent to a fibre break. Thus, the interfacial shear stress and its decay in fibres located at a radial distance R from a given fibre break, can also be derived [18].

8.3.2.1 *Two-dimensional microcomposites*

The two-dimensional (2D) microcomposite tapes consist of regular arrays of typically three to maximum seven individual carbon fibres lying on the same plane of uniform interfibre distances. The advantages for investigating these model geometries are twofold. First, for a given fibre/matrix bond strength, the fracture behaviour can be monitored as a function of interfibre distance, which can be controlled and varied using special devices [19]. Secondly, the stress concentration in a fibre adjacent to a fibre break, can be measured without the presence of fibres lying on planes underneath the plane of observation. In contrast, the interfibre distance in full commercial composites cannot be adequately controlled and local variations near fibre breaks can be significant [4]. Also the presence of fibre breaks lying on planes underneath the plane of observation can affect the overall stress field and, hence, the exact value of the stress concentration factor [4].

In Fig. 8.15, a representative three-fibre microcomposite at an effective 'zero' interfibre distance is shown schematically. The fibre stress distribution as a function of position along the length of a fractured fibre (fibre 2, Fig. 8.15) and the corresponding interfacial shear stress profile at an applied composite strain of 0.75% are shown in Fig. 8.16. As can be seen, the fibre stress at the fracture point builds from a compressive stress of about -0.70 GPa to the far field value of 2.5 GPa at a 'transfer length' distance of about 200 μm from the fractured point (Fig. 8.16a). The maximum ISS that this particular combination of fibre/matrix can sustain at 0.75% of applied strain was about 40 MPa (Fig. 8.16b). It is worth noting, however, that the ISS maxima obtained on either end of the fibre fracture were not located on the plane of fracture but at a finite distance of about 40 μm away from it. As reported elsewhere [20], this is indicative of the onset of interfacial failure at the vicinity of a discontinuity such as a fibre break.

The corresponding axial stress and ISS profiles for the two fibres (fibres 1 and 3, Fig. 8.15) adjacent to the fractured fibre, are given in Fig. 8.17 and 8.18, respectively. In both cases, the axial stress builds from a far field value of 2.5 GPa to a maximum value of about 3.3 GPa at the plane of fracture ($\xi = 0$). For both adjacent fibres, the PAL is identical in magnitude to the transfer or ineffective length of the fractured fibre. A value of stress concentration factor, K_1 ($q = 1$), of 1.36 is estimated for the two nearest neighbours. The maximum value of ISS for the two nearest neighbours decreases dramatically to a value of about 8 MPa in spite of the close proximity of the three fibres (Fig. 8.15). Finally, the area of interfacial damage of about 80 μm observed in the fractured fibre results in the smooth decrease in the ISS distribution in the two nearest neighbours (Fig. 8.17b and 8.18b) on either side of the fracture plane ($\xi = 0$).

8.3.2.2 *Full unidirectional coupons*

The work on full composites is performed on ASTM-standard unidirectional tensile coupons (Fig. 8.19) [4] and fibre tows [21], which are loaded

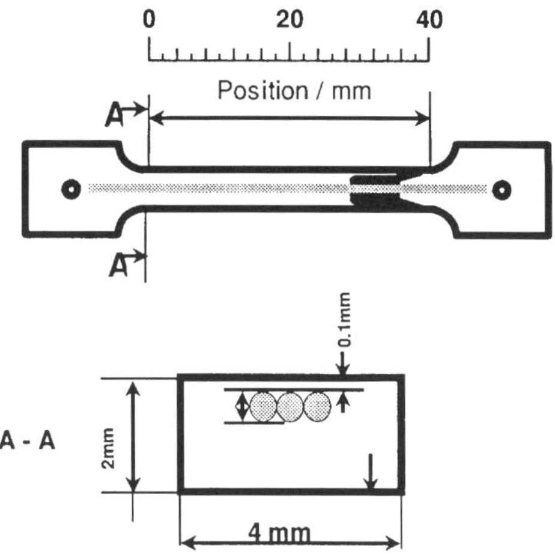

Applied composite strain= 0.75%

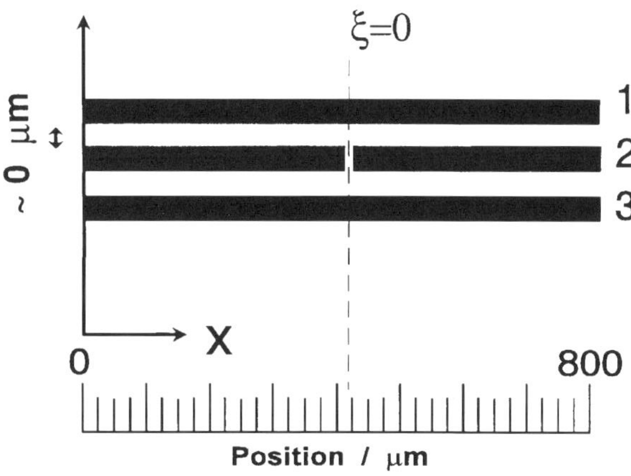

8.15 Schematic illustration of a three-fibre microcomposite specimen and of the corresponding area of stress measurements at 0.75% applied strain.

incrementally in tension up to fracture. The procedure involves the identification of a single fibre fracture at a certain increment of applied load and the point-by-point measurement of fibre stress within a 'window' of seven fibres (Fig. 8.19) around the locus of the first fibre failure. A representative axial stress profile along the fractured fibre at an applied composite strain of 0.4% is shown in Fig. 8.20. As expected, the fibre stress drops to zero at the fibre fracture and then reaches a maximum on either side of the fibre break. As in the case of the 2D microcomposites, the transfer length on either side of the fracture point is of the order of 200 μm. The maximum ISS developed in the case of the full composite on either side of fibre fracture at 0.4% applied strain, was of the order of 30

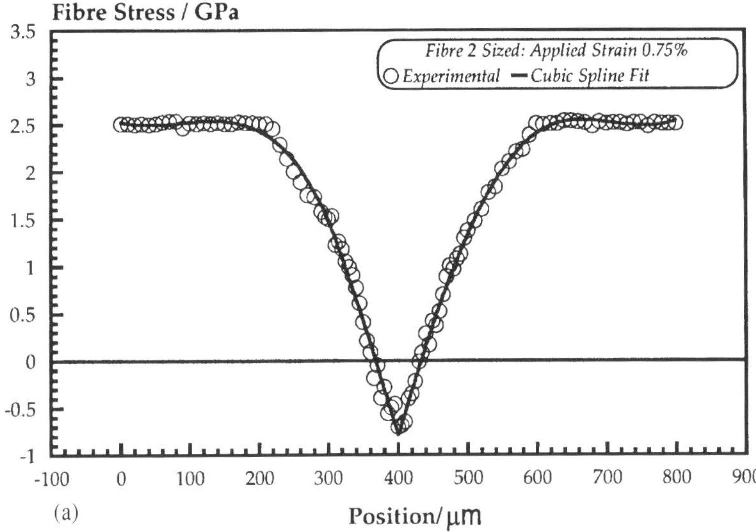

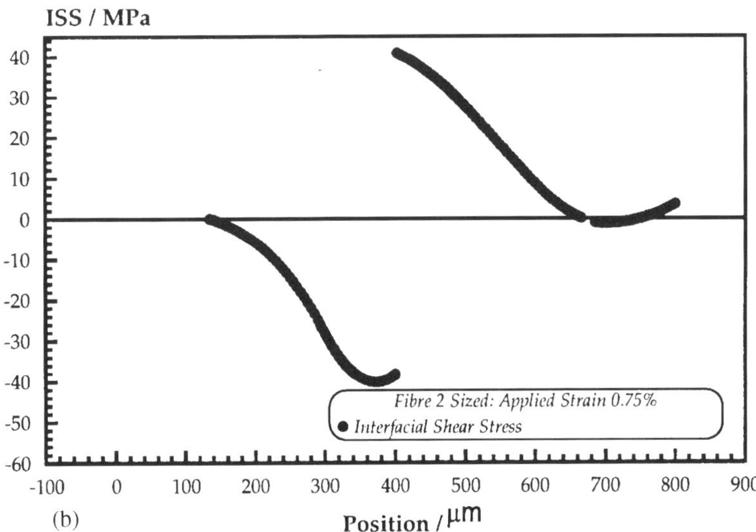

8.16 (a) Stress profile and corresponding cubic spline fit for fibre 2 at 0.75% applied composite strain. (b) Interfacial shear stress (ISS) profile for fibre 2 at 0.75% applied composite strain.

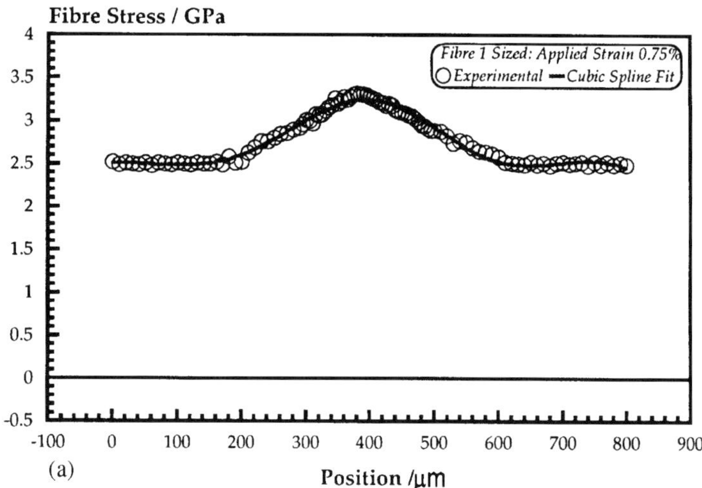

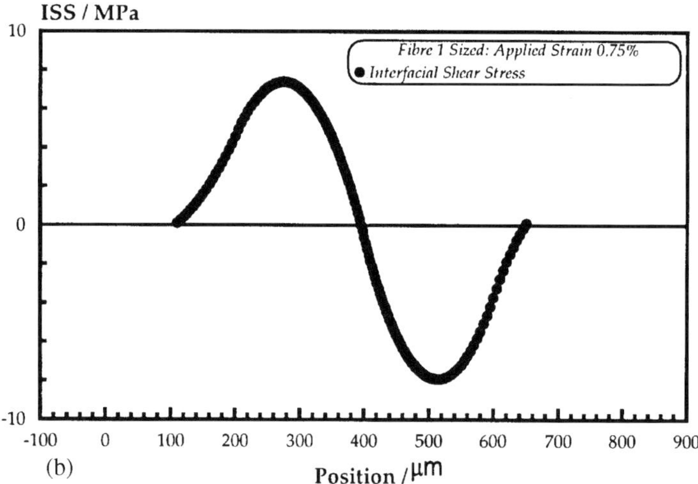

8.17 (a) Stress profile and corresponding cubic spline fit for fibre 1 at 0.75% applied composite strain. (b) Interfacial shear stress (ISS) profile for fibre 1 at 0.75% applied composite strain.

MPa. Finally, the observed significant 'knee' of the ISS distribution on either side of the fibre fracture indicates that considerable interfacial damage was initiated in tandem with the fibre fracture process. The size of this zone, as defined by the separation of the two ISS maxima, was of the order of 150 μm (Fig. 8.20).

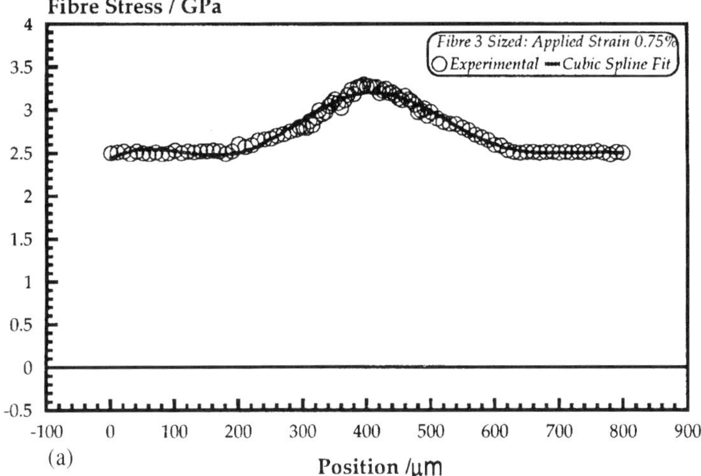

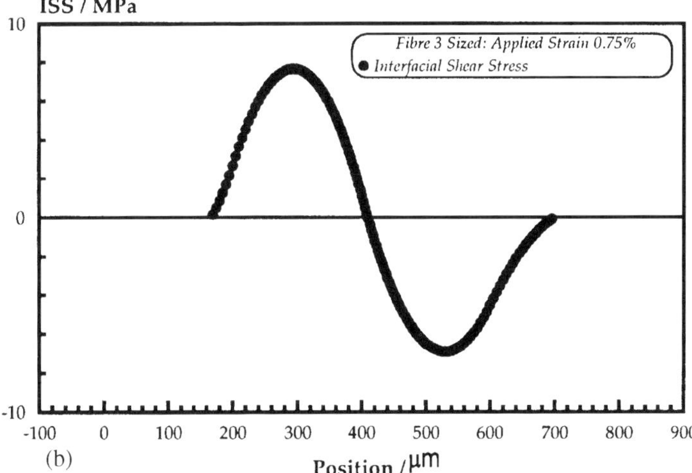

8.18 (a) Stress profile and corresponding cubic spline fit for fibre 3 at 0.75% applied composite strain. (b) Interfacial shear stress (ISS) profile for fibre 3 at 0.75% applied composite strain.

In Fig. 8.21 and 8.22 the fibre stress distribution in the remaining six fibres of the 'window' of measurements is given. The considerable scatter in the data points is not surprising, since in a full composite the axial stress values along any individual fibre are affected by shear field perturbations present not only in the plane of Raman measurements but also in planes immediately beneath it [4]. As can be seen in Fig. 8.21 and 8.22, the stress magnification was particularly

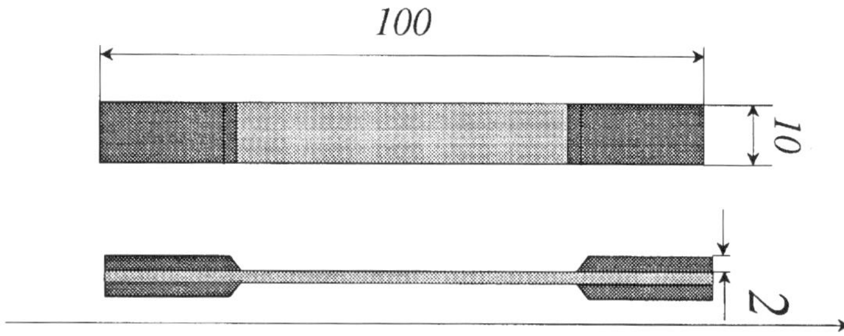

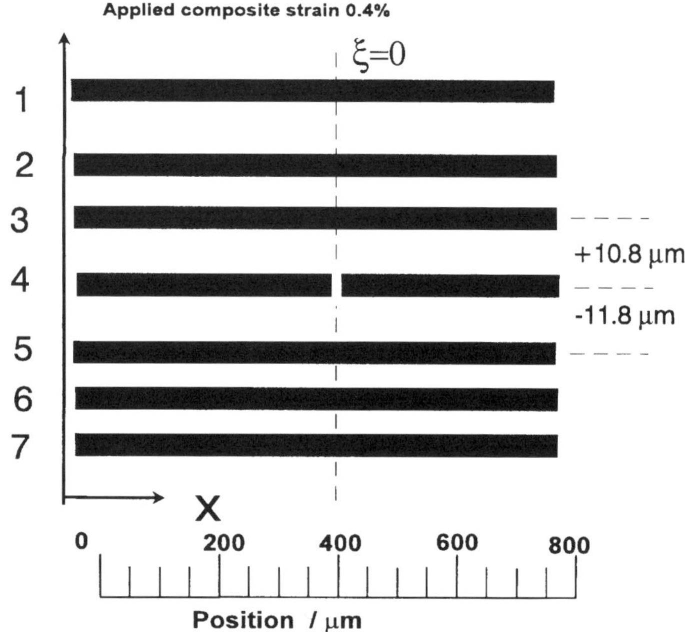

8.19 Schematic illustration of the 4-ply unidirectional composite and of the corresponding area of stress measurements. The dimensions of the ASTM standard coupon are given in mm. The interfibre distance of the near-surface fibres was variable.

evident in fibres 3 and 5, located at interfibre distances of +10.8 and − 11.8 μm, respectively. As expected, the axial stress profile of fibre 3, which was closest to the fractured fibre 4, showed a more intense stress magnification effect. Farther away from the locus of fibre failure the stress concentration clearly diminished (Fig. 8.21 and 8.22, fibres 1, 2, 6, 7). Values of stress concentration factor, K_q, of 1.18 and 1.11 were measured for fibres 3 and 5, respectively.

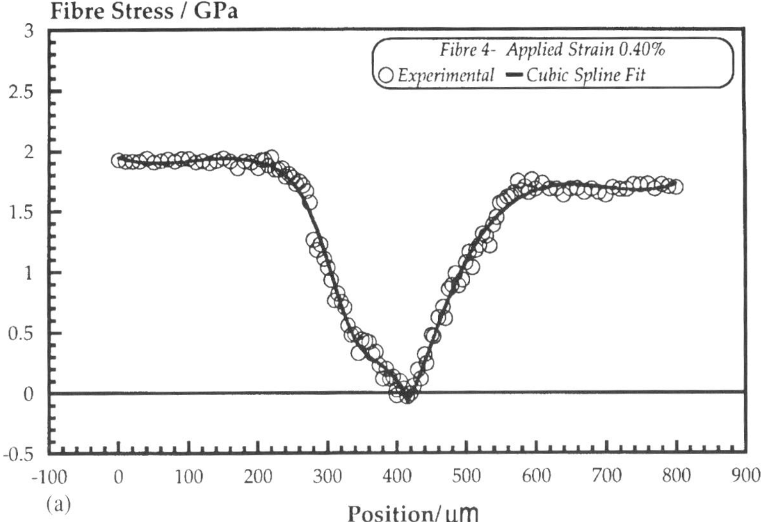

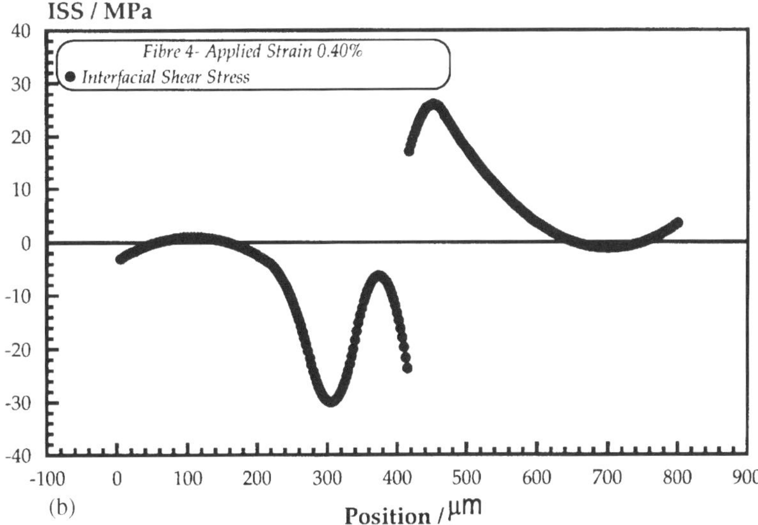

8.20 (a) Stress profile and corresponding cubic spline fit for fibre 4 at 0.40% applied composite strain. (b) Interfacial shear stress (ISS) profile for fibre 4 at 0.4% applied composite strain for the full composite coupon.

The two sets of data for K_1 (single fibre fracture, $q = 1$) as a function of the normalised interfibre distance R/r, where R is half the centre-to-centre distance and r is the radius of the fibre, are presented in Fig. 8.23. As can be seen, two distinct families of data are obtained for each respective specimen geometry. The data derived from planar fibre arrays lie higher than those obtained from the

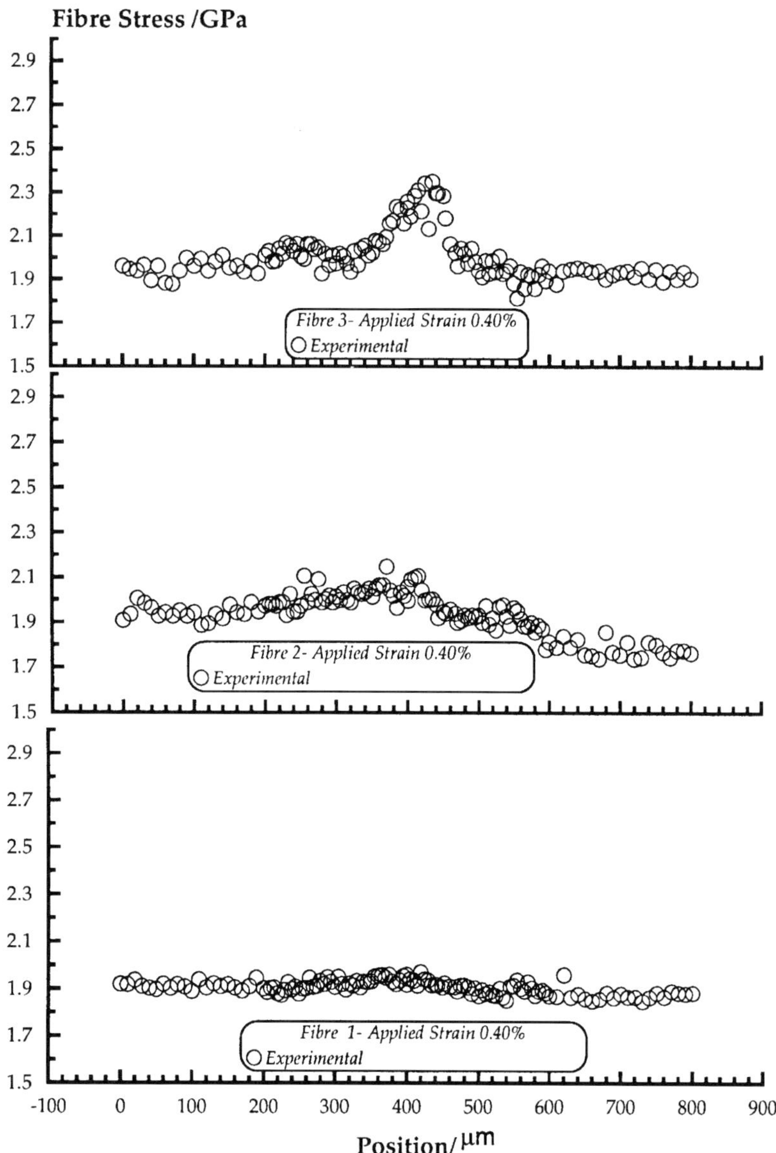

8.21 Stress profiles for fibres 1, 2 and 3 at 0.40% applied composite strain.

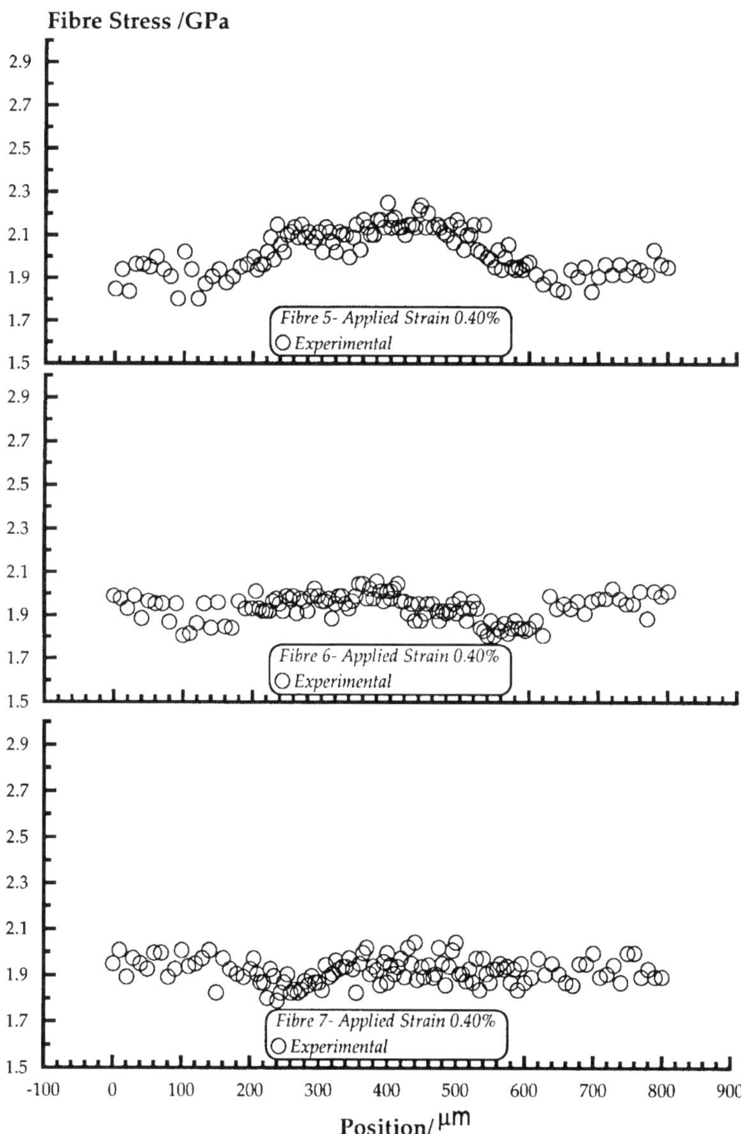

8.22 Stress profiles for fibres 5, 6 and 7 at 0.40% applied composite strain.

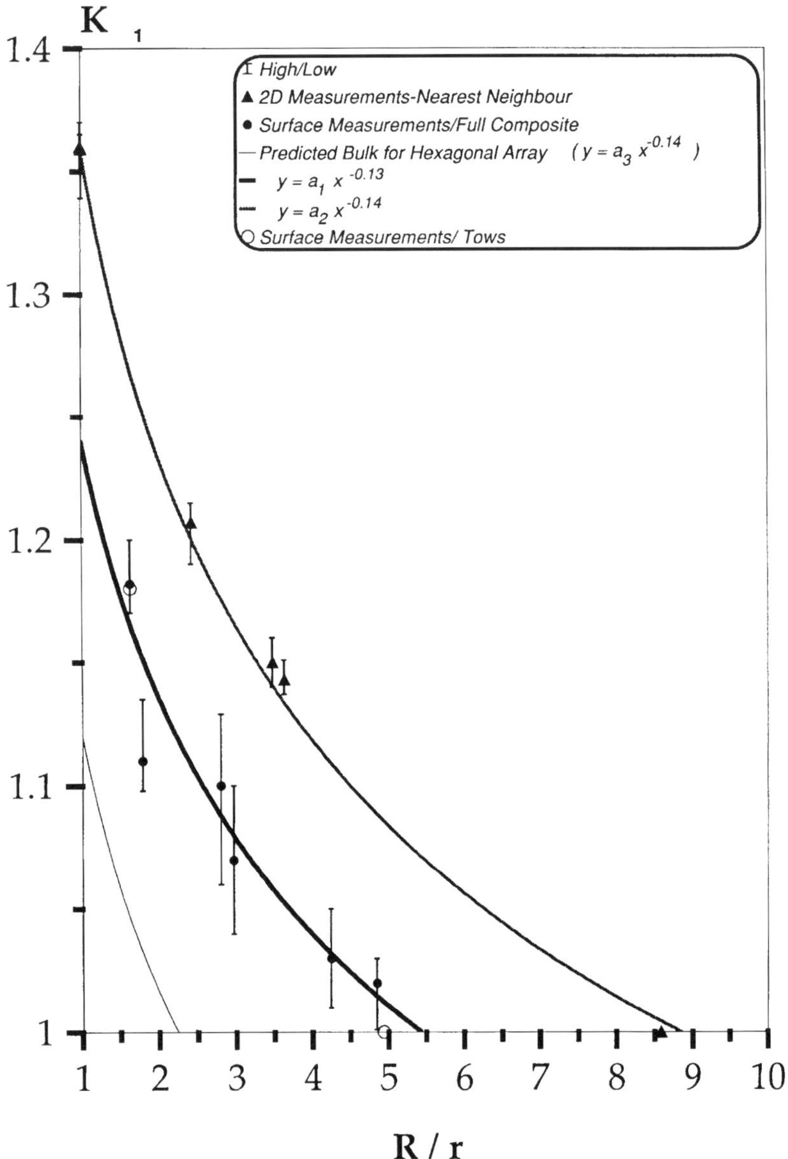

8.23 Graph of stress concentration for a single fracture ($q = 1$) for all geometries as a function of normalised interfibre distance R/r, where R is half the centre-to-centre distance and r is the radius of the fibre.

Table 8.3 Interfacial characteristics and maximum stress concentration of planar and full composite geometries

Composite system	Geometry	Maximum interfacial shear stress at fracture (MPa)	Maximum stress concentration [$K_1^{r=R}$]	Nearest neighbours
M40-3k-40B®[a] /LY-HY5052®[,b]	Planar	35 ± 5	1.36	2
	Surface/full composites	30 ± 5	1.24	4
	Bulk/full composites	–	1.12*	6

[a]Trademark of Toray Industries, Soficar (France). The fibre is supplied with a commercial sizing.
[b]Trademark of Ciba-Geigy plc (UK).
*Predicted value.

surface of full composites. For either set of values the stress concentration value K_1 seems to decay exponentially with R/r in agreement with previous predictions [22]. Regression analysis performed on this set of data [18] has shown that the stress concentration value, K_1, relates to R/r via the equation:

$$K_1 = K_1^{r=R}\left[\frac{R}{r}\right]^{-0.14} \tag{3}$$

where $K_1^{r=R}$ is the maximum stress concentration at close proximity to the broken fibre. For the case of the 2D array the value of $K_1^{r=R}$ has been measured experimentally (Fig. 8.17 and 8.18), whereas for the full composite it has been obtained by extrapolation [18]. The relationship between K_1 and R/r for the bulk of the sized M40/LY-HY5052 composite has also been predicted assuming hexagonal geometry (Fig. 8.23). The values of stress concentration for all geometries are listed in Table 8.3. As can be clearly seen, the 'zone of influence' over which a broken fibre has no measurable effect upon adjacent fibres in full composites, is about 11 and 5 fibre radii for the surface and bulk of the laminate, respectively. Such a dramatic reduction in the stress concentration with interfibre distance is a consequence of the dramatic decrease of the interfacial shear stress that the system can accommodate in the radial direction [18].

8.4 Conclusions

Laser Raman microscopy can be employed to determine the stress transfer profiles, as well as the interfacial shear stress distribution in single fibre composites, in planar fibre arrays and, finally, full composites. The advent of remote laser Raman microscopy allows measurements of stresses and strains in composite structures which are located at larger distances from the Raman detector.

For carbon fibre/epoxy composites, it has been shown that the exact nature of stress transfer and interfacial damage development depends upon fibre treatment, fibre sizing and associated fabrication conditions which are geometry specific. Stress mapping of full composites provided information about the stress concentrations in these materials resulting from fibre fracture(s). A strong dependence between stress concentration and interfibre distance has been found for all geometries. The stress concentration versus R/r data were described via an equation of the type $y=ax^b$, where the exponent b is an interfacial material parameter unaffected by specimen geometry and the coefficient a is the maximum attained stress concentration for close fibre contact and is geometry (or volume fraction) dependent.

Acknowledgements

The author would like to thank the Engineering and Physical Sciences Research Council (EPSRC), the Defence Research Agency (DRA) at Farnborough, the Commission of European Communities (BRITE-EurAM program), British Aerospace (Sowerby, Filton), the British Council (British-German ARC Programme) and the Advisory Group for Aerospace Research and Development (AGARD) for financial support. Dr. N. Melanitis and Messrs V. Chohan, A. Paipetis and C. Marston are thanked for performing the experiments mentioned in this chapter. Drs. M. Pitkethly (DRA) and P. Marshall (BAe) are thanked for advice and encouragement during the accomplishment of this work.

References

1. P C Powell, *Engineering with Fibre-Polymer Laminates*, Chapman and Hall, 1994, ISBN 0-412-49610-0.
2. A M Sastry and S L Phoenix, Shielding and magnification of loads in elastic, unidirectional composites, *Soc. Adv. Mater. Process Eng. (SAMPE) J.* 1994 **30**(4), 61–67.
3. C Zweben, Advanced composites for aerospace applications: a review of current status and future-prospects, *Composites* 1981 **12**(4) 235–240.
4. C Galiotis, V Chohan, A Paipetis and C Vlattas, *Interfacial Measurements in Single and Multi-fibre Composites using the Technique of Laser Raman Spectroscopy*, eds J C Spragg and L T Drzal, ASTM-STP 1290, American Society for Testing and Materials, 1996, pp 19–33.
5. C Galiotis, Laser Raman spectroscopy; a new stress/strain measurement technique for the remote and on-line non-destructive inspection of fibre-reinforced polymer composites, *Mater. Tech.* 1993 **8** 203–209.

6. B Arjyal and C Galiotis, Localised stress measurements in composite laminates using a Raman stress sensor, *Adv. Composites Lett.* 1995 **4** 47–52.
7. L S Schadler and C Galiotis, A review of the fundamentals and applications of LRS microprobe strain measurements, *Internat. Mater. Rev.* 1995 **40**(3) 116–134.
8. A Paipetis, C Vlattas and C Galiotis, Remote laser Raman microscopy (ReRaM); Part 1: design and testing of a confocal microprobe, *J. Raman Spectros.* 1996 **27** 519–526.
9. C Vlattas and C Galiotis, Deformation behaviour of liquid crystal polymer fibres: Part 1: converting spectroscopic data into mechanical stress–strain curves in tension and compression, *Polymer* 1994 **35** 2335–2347.
10. C C Chamis, Mechanics of load transfer at the interface, *Composite Materials*, Volume 6, Academic Press, New York, 1974, pp 31–37.
11. L T Drzal, M J Rich and P F Lloyd, Adhesion of graphite fibers to epoxy matrices; 1: the role of fiber surface treatment, *J. Adhesion* 1982 **16** 1–30.
12. H L Cox, The elasticity and strength of paper and other fibrous materials, *Br. J. Appl. Phys.* 1952 **3** 72–79.
13. T F McLaughlin, A photoelastic analysis of fiber discontinuities in composite materials, *J. Composite Mater.* 1968 **2**(1) 44–55.
14. N Melanitis and C Galiotis, Interfacial micromechanics using laser Raman spectroscopy, *Proc. Roy. Soc. London* 1993 **440** 379–398.
15. N Melanitis, C Galiotis, P L Tetlow and C K L Davies, Interfacial shear stress distribution in model composites: Part 2: fragmentation studies on carbon fibre/ epoxy systems, *J. Composite Mater.* 1992 **26**(4) 574–610.
16. A Paipetis and C Galiotis, Effect of sizing on the stress transfer characteristics and interface failure modes of model carbon fibre/epoxy composites, *Composites* 1996 **27**A(9) 755–767.
17. N Melanitis, C Galiotis, P L Tetlow and C K L Davies, Monitoring the micromechanics of reinforcement in carbon fibre/epoxy resin systems, *J. Mater. Sci.* 1993 **28** 1648–1654.
18. V Chohan and C Galiotis, Effects of interface, volume fraction and geometry upon stress redistribution in polymer composites under tension, *Composite Sci. Technol.*, 1997, *57*(8), 1089–1101.
19. H D Wagner and L W Steenbakkers, Microdamage analysis of fibrous composite monolayers under tensile-stress, *J. Mater. Sci.* 1989 **24** 3956–3975.
20. V Chohan and C Galiotis, Interfacial measurements and fracture characteristics of 2D microcomposites using remote laser Raman microscopy, *Composites* 1996 **27**A(9) 881–888.
21. C Marston, B Gabbitas, J Adams, S Nutt, P Marshall and C Galiotis, Failure characteristics in carbon/epoxy composite tows, *Composites* 1996 **27**A(12) 1183–1194.
22. H D Wagner and A Eitan, Stress-concentration factors in 2-dimensional composites: effects of material and geometrical parameters, *Composite Sci. Technol.* 1993 **46**(4) 353–362.

9 Acoustic microscopy of ceramic fibre composites

CHARLES W LAWRENCE AND
G ANDREW D BRIGGS

9.1 Introduction

In scanning acoustic microscopy a high frequency acoustic wave is excited by a transducer and focused on the axis of a sapphire lens. A fluid (usually hot water) is used to couple the acoustic waves generated in the lens into a sample. By mechanically scanning the lens in a raster parallel to the surface of the sample, and using the strength of the reflected signal to modulate the brightness of a monitor, an acoustic image of the sample can be produced. As the propagation of acoustic waves is governed by the elastic properties of the material through which they pass, the images produced by a scanning acoustic microscope (SAM) can reveal information about the microstructure and elastic properties of the sample.

For many samples, the ability of the SAM to image below the surface of opaque materials is an important factor in its use. An example of this is its use to image the damage at successive interfaces between the different plies of a laminated composite that has experienced impact damage. Between the plies nearest the surface the damage is slight, but between plies at a greater depth the damage becomes greater, and the delaminated areas take on different orientations depending on the ply orientations. The SAM has an enhanced depth discrimination which derives from its confocal nature so that contrast is obtained only from the plane in focus. This makes acoustic microscopy particularly powerful for high resolution, non-destructive inspection of subsurface damage in composites used in critical applications. Typically, frequencies of the order of 50 MHz are used for this kind of application, which is intermediate between traditional non-destructive testing (NDT) frequencies (5 MHz) and those used for high resolution imaging (1–2 GHz).

SAM is highly sensitive to the presence of cracks and elastic discontinuities in monolithic ceramics and metals [1–3]. The contrast in SAM is directly related to the elastic properties of the surface being examined. Contrast also arises from anything that scatters Rayleigh waves, such as cracks and boundaries between materials. This allows the discrimination of optically transparent phases and the imaging of extremely fine surface cracks and pores via the mechanism of

9.1 The ELSAM scanning acoustic microscope used in the work described here.

Rayleigh wave fringe formation [4], with a theoretical detection limit of about 10 nm [5]. It is possible not only to image these features, but also to make quantitative measurements of elastic properties as well. In this chapter we will illustrate this by observations of composites consisting of various ceramic fibres in matrices of glasses, glass-ceramics and metals. All the images presented here were obtained using an Leica scanning acoustic microscope (ELSAM) (Fig. 9.1) operating at 1.9 GHz, with a spatial resolution of 0.7 μm.

9.2 Glass-matrix composites

Engineering ceramics and glasses possess excellent thermal properties coupled with low density, high elastic modulus and reasonable mechanical strength. Unfortunately, though, they exhibit poor fracture toughness. Considerable effort has been expended on overcoming this Achilles heel. In the early 1970s it was found that the addition of 0.40 volume fraction (V_f) of carbon fibres to borosilicate (Pyrex) glass increased its work of fracture from $10\,J\,m^{-2}$ to $3.7\,kJ\,m^{-2}$ [6]. The subsequent development of small diameter (< 20 μm) silicon oxycarbide fibres such as Nicalon and Tyranno has permitted the fabrication of ceramic matrix composites (CMCs) with fracture toughnesses approaching those of metals. For instance the addition of 0.49 volume fraction of Nicalon fibres to borosilicate glass produces a CMC with a work of fracture of $50\,kJ\,m^{-2}$ [7]. This increase in work of fracture is primarily due to the low interfacial frictional shear strength (IFSS) between the fibres and matrix, allowing broken fibres to be pulled out of the matrix [8].

The full potential of this type of material is not often achieved because of loss in fibre properties due to aggressive processing conditions, the generation of inappropriate microstructures during composite fabrication and matrix microcracking around fibres at relatively low stresses which allows environmental ingress and loss in composite properties. Careful optimisation of microstructures during fabrication in order to tailor fibre, matrix and fibre/matrix interfacial properties helps avoid the first two problems. Matrix microcracking under stress is the principal obstacle to the use of CMCs. At high temperatures access of the environment to the fibres [9] causes their fracture, and the usable strength of the material is therefore limited to the matrix microcracking stress.

Defects such as matrix microcracks generated during fabrication or use are difficult to detect in CMCs using optical microscopy (OM) due to their small dimensions. In back-scattered scanning electron microscopy (SEM), the low atomic numbers of the composite constituents gives rise to poor contrast. In addition, since most ceramics are poor electrical conductors, CMCs normally have to be carbon or gold coated to obtain a clear image.

9.2.1 Experimental procedure

The material described in this section was borosilicate glass reinforced with Nicalon silicon oxycarbide fibres, fabricated at the Harwell Laboratory [7]. The material was fabricated by hot pressing at 950°C and 10 MPa for 10 min, to give optimum mechanical properties. The reinforcing fibres were arranged uniaxially, with a nominal fibre volume fraction of 0.45. In order to investigate the effect of fibre volume fraction on composite microstructure, a specimen containing a single reinforcing fibre (i.e. effectively zero fibre volume fraction) was fabricated under identical conditions. Typical mechanical properties which have been measured for this composite system are shown in Table 9.1, along with the equivalent properties for the borosilicate glass matrix and Nicalon fibres.

Acoustic microscopy requires samples to be flatter than the acoustic wavelength in water at the frequency used (0.8 μm at 1.9 GHz for the current work) in order to minimise spurious topographical contrast so that the unique contrast due to elastic properties and structure can be seen clearly and unambiguously. To ensure an adequate surface for examination of the composite, samples (typical dimensions 2 × 10 × 5 mm) were cut from the as-pressed composite coupon using a thin-bladed annular diamond saw. The samples were then vacuum embedded in an epoxy resin so that the composite presented a transverse section to the surface. Once the epoxy was fully cured, the sample was attached to a precision polishing jig and progressively lapped and polished using in turn 6, 1, 0.25 and 0.1 μm diamond pastes with a silicone oil

Table 9.1 Mechanical properties of SiC fibres and ceramic matrices

Property	Nicalon SiC fibre	Tyranno SiC fibre	Textron SCS-6 SiC monofilament	Sigma SiC monofilament	7740 borosilicate glass	CAS glass-ceramic	MAS glass-ceramic	0.49 V_f 7740/SiC
Young's modulus (GPa)	196	206	400–415	420	63	106	110	120
Shear modulus (GPa)	77	—	—	—	28	36	—	51
Density ($kg\,m^{-3}$)	2550	2300–2400	3045	3040	2230	2860	2700	2390
Tensile strength (GPa)	2.75	2.74	3.95	3.60	0.05	—	0.15	580
Thermal expansion coefficient ($10^{-6}\,°C^{-1}$)	3.5	3.1	1.5	1.5	3.25	5.3	4.6	—
Poisson ratio	—	—	0.2	—	0.2	0.26	0.25	—
Mean diameter (μm)	16	8–10	140	100	—	—	—	—

lubricant on scrolled brass platens, at low loads. Napless cloths were found to give undesirable relief between fibres and matrix, giving rise to spurious contrast in the acoustic microscope, although these cloths give an adequate finish for optical microscopy. Indeed, for examination in the SEM using secondary electron imaging, it was found necessary to include an extra polishing stage (0.25 μm diamond paste on a napless cloth) in order to emphasise relief, and hence contrast, between microstructural constituents. The final polishing stage for acoustic microscopy involved the use of Syton (colloidal silica solution with a 0.1 μm particle size) on a polyurethane-coated platen for 10 min to remove fine scratches and provide a mild etch of the glass matrix.

Specimens were examined using an Olympus BHM optical microscope (OM) a Leica SAM (hereafter ELSAM), an Hitachi S-530 SEM with secondary electron detector, a JEOL JSM 35X in back-scattered mode and a Cameca microprobe. Samples for SEM examination were carbon coated to prevent charge buildup on the specimen surface and to minimise beam damage to the composite. In addition, a non-imaging acoustic microscope, a line focus acoustic microscope (LFAM), was used to measure Rayleigh wave velocities in bulk borosilicate glass and hot pressed silicon carbide.

9.2.2 Composite microstructure

OM was used to examine the basic composite microstructure. Figure 9.2 shows an optical micrograph of Nicalon reinforced borosilicate glass (7740/SiC). The fibre distribution can be seen to be comparatively poor, with clusters of fibres occurring. There is no indication of the arrangement of the fibres in the original tows. Only sporadic instances of open surface porosity (about 0.001 area fraction) were visible which were invariably associated with matrix-denuded regions between close packed fibres. The matrix consisted of a crystalline phase in an amorphous glass matrix, with microcracks just visible in the matrix linking adjacent fibres, the contrast from both being poor.

More detailed microstructural information was obtained on examining the composite with a SAM. Figure 9.3 shows a low magnification acoustic micrograph of the 0.45 V_f composite. The crystalline matrix phase has high contrast with respect to both the glassy matrix phase and the silicon carbide fibres. Cracks, delineated by fringes, occur in both the glass and crystalline components of the matrix. The volume fraction of the latter was measured as 0.40 ± 0.05. The fibres deviate from their nominal circular cross-section and diameter. Mean fibre diameter as measured from electron micrographs was 15.7 μm, with an approximately normal distribution (Fig. 9.4) and a standard deviation of 2.6 μm.

Acoustic micrographs such as Fig. 9.3 all showed interference fringes within both the silicon carbide fibres and the glass matrix component. These fringes

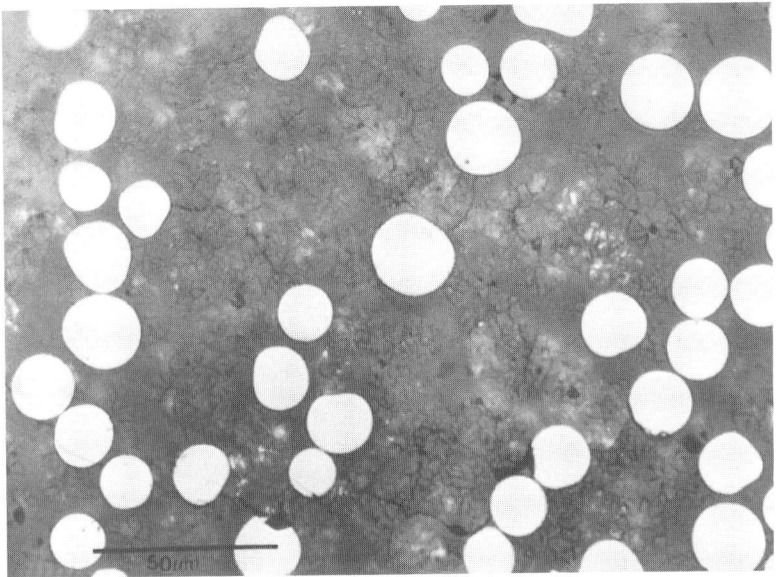

9.2 Optical micrograph showing the microstructure of 7740/SiC.

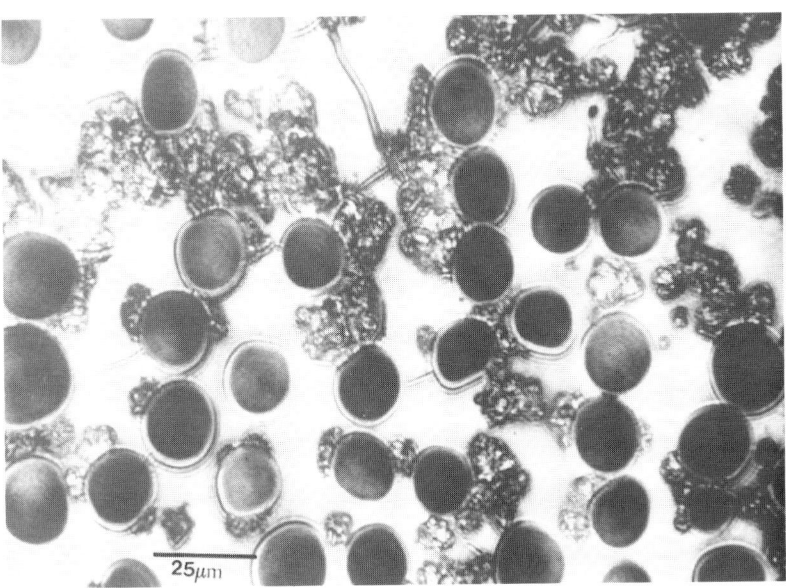

9.3 Acoustic micrograph of 7740/SiC showing the general microstructure of the composite. $z = -2.8\,\mu m$.

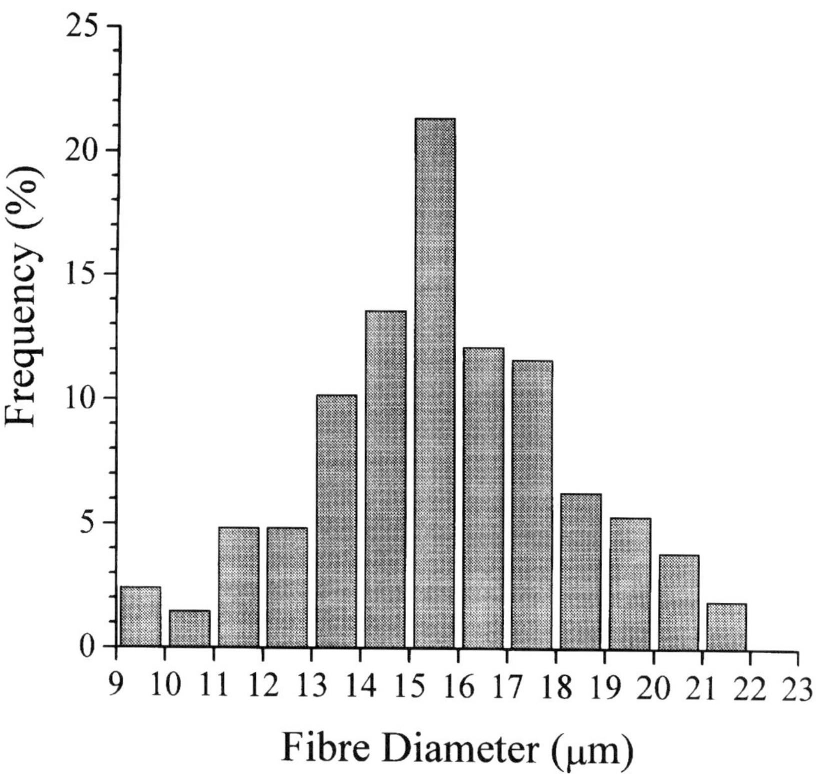

9.4 Histogram showing the distribution of fibre diameters in 7740/SiC as measured from electron micrographs.

were faint at focus but their contrast increased at negative defocus (i.e. the sample moved towards the lens). These fringes are a manifestation of the scattering of Rayleigh waves at the interface between two materials of different acoustic properties [5]. As the lens travels parallel to the surface in the course of its scan, the path length of Rayleigh waves that are reflected from the fibre/matrix interface changes, but the path of rays that are geometrically reflected from the surface remains constant (Fig. 9.5). The interference of these two components at the transducer leads to the fringes that are seen here. This is essentially the same mechanism as that which gives rise to fringe formation from surface breaking cracks oriented normally to a surface [4].

The fringes in the fibres have an appearance similar to growth rings in trees. Just as the growth rings in trees reveal information on the age of the tree, we can use Rayleigh wave fringes to reveal the elastic properties of the fibres. The fringe spacing, Δx, is given by

$$\Delta x = \frac{\lambda_r}{2} \tag{1}$$

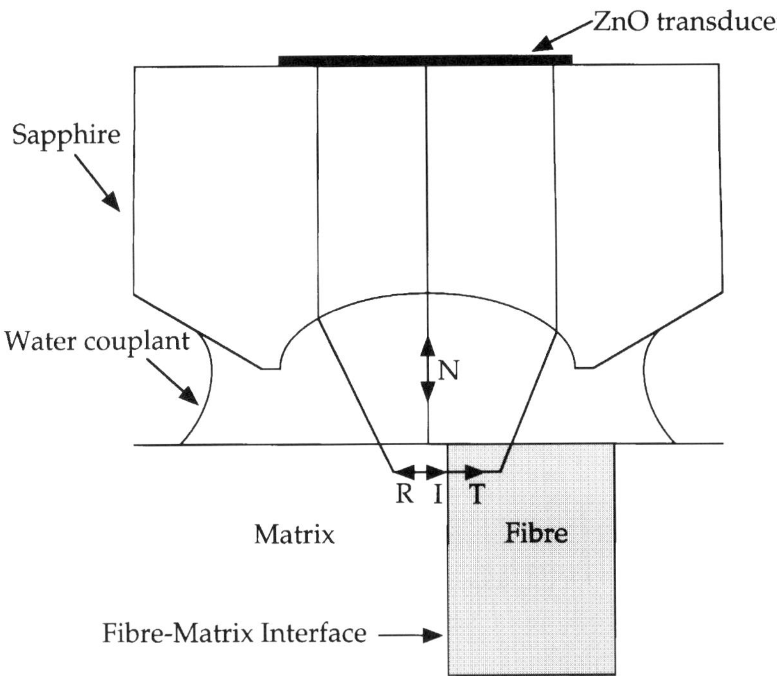

9.5 Schematic diagram of the mechanism of fringe formation in the fibres and matrix. I is the Rayleigh wave incident on the fibre/matrix interface, R is the component reflected from the interface and T is the transmitted wave component. N is the normal component of the longitudinal wave in the water couplant incident on the surface of the composite. Note that the incident and reflected components follow the same path.

where λ_r is the Rayleigh wavelength for the materials in which they form [4]. Therefore if the acoustic frequency is f, then the Rayleigh velocity is

$$v_r = 2f\Delta x \tag{2}$$

Hence by carefully measuring the spacing of the Rayleigh fringes from high magnification acoustic micrographs, the Rayleigh velocity for the fibres can be calculated for the known imaging frequency. A mean Rayleigh velocity of 5320 $m\,s^{-1}$ was calculated for the Nicalon silicon carbide fibres. This value is lower than the value of 6729 $m\,s^{-1}$ for bulk hot-pressed silicon carbide as measured using a LFAM. The measured Rayleigh wave velocity can then be used to calculate the shear modulus and Young's modulus of the material in which the fringes form (see Section 9.3).

Inserting the appropriate values of the density (2550 $kg\,m^{-3}$) yields an *in situ* shear modulus of 89 GPa and a Young's modulus of 205 GPa for Nicalon silicon carbide fibres incorporated in a glass matrix (Poisson ratio of the fibre was assumed to be 0.19). These values are in good agreement with the manufacturer's quoted values [10] shown in Table 9.1, but are slightly higher than values published in the literature [11].

The Rayleigh wave interference fringes formed in the glass component of the matrix can, by the same process, be used to calculate the corresponding elastic properties of the glass. The measured Rayleigh wave velocity of 3800 m s^{-1} yields a shear modulus of 39 GPa and a Young's modulus of 93 GPa. In comparison the Rayleigh wave velocity of bulk borosilicate glass measured using the LFAM was 3324.2 m s^{-1}, giving a shear modulus of 29.7 GPa and a Young's modulus of 62.3 GPa, in good agreement with the data shown in Table 9.1. As the borosilicate glass matrix has undergone significant devitrification, we would expect some change in its elastic properties. Table 9.2 summarises the elastic properties determined for the composite systems discussed in this chapter.

9.2.3 Matrix crystalline phase

The strong contrast between the fibres, glass and crystalline phases arises primarily as a result of differences in Rayleigh wave propagation in the various phases and in the different crystallographic orientations. The crystalline phase in the glass matrix has a granular 'flowery' appearance with an individual grain (or 'petal') size of 3–5 μm (Fig. 9.6). This phase resulted from devitrification of the borosilicate glass matrix during fabrication of the composite. Chemical analysis of this phase using both wavelength and energy dispersive spectrometers on a Camebax electron microprobe indicated that it was silicon rich (46 wt%), with no other heavy elements detected. Owing to the requirement of a carbon coating for the specimen, the presence of elements lighter than carbon could not be determined. Analysis of the peaks in a X-ray diffraction pattern obtained from the single fibre specimen identified the crystalline phase as α-cristobalite, a tetragonal allotrope of silica which is metastable at room temperature. This agrees with previous results reported in the literature [12,13].

It has long been known that, despite being the high temperature (1470–1713°C) silica phase, cristobalite can precipitate in borosilicate glass and remain metastable down to room temperature. Cox and Kirby [14] have linked the diffusion process controlling cristobalite crystal growth to the removal of sodium and boron ions from the silicon–oxygen structure of the borosilicate glass, resulting in the formation of a crystalline region without any major positional changes in silicon and oxygen ions. More recently Clarke *et al.* [15] have investigated the formation of cristobalite in bulk hot-pressed borosilicate glass. The volume fraction of cristobalite in samples of borosilicate glass hot-pressed at temperatures in the range 750–950°C was found to decrease with temperature but increase with applied pressure.

Table 9.2 *In situ* measured properties of composite constituents

Material	Nicalon SiC fibre	Tyranno SiC fibre	Textron SCS-6 SiC monofilament	Bulk hot-pressed SiC[a]	7740 glass matrix	Bulk 7740 glass	CAS	MAS	Bulk MAS
Rayleigh wave velocity ($m\,s^{-1}$)	5320	5150	6325	6729	3800	3324.2	3600	3520	3550
Young's modulus (GPa)	205	177	352	—	93	62.3	101	93	—
Shear modulus (GPa)	89	77	147	—	39	29.7	—	37	—

[a] measured by LFAM.

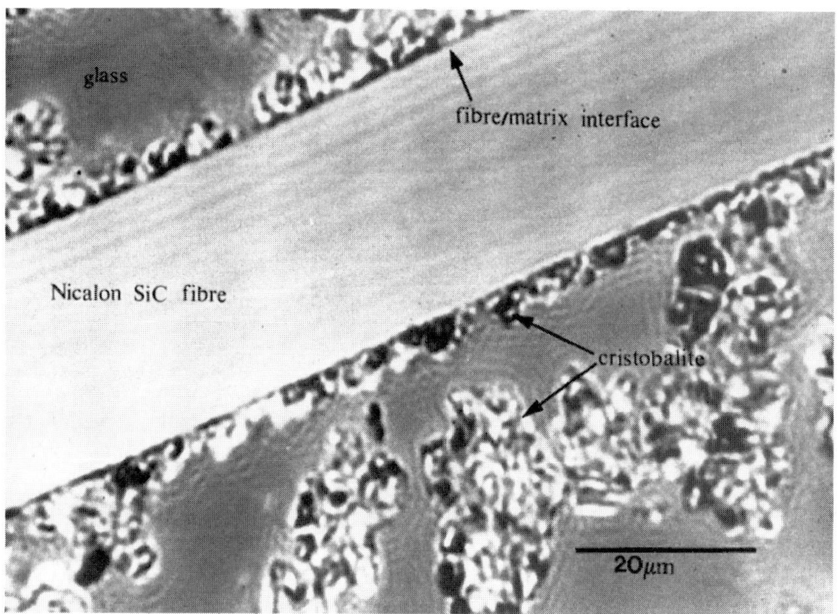

9.6 Acoustic micrograph showing cristobalite formation at the fibre/matrix interface. $z = -1.0\,\mu m$.

As can be seen from the 0.45 V_f composite specimen (Fig. 9.6), cristobalite has formed at the fibre/matrix interface and also randomly throughout the borosilicate glass matrix. Note the fringes formed in the glass phase surrounding the cristobalite grains. These fringes have varying spacing, which in some instances corresponds to that of Rayleigh waves. The variation is due to the shape of the grains. The surfaces of the grains are inclined slightly with respect to the specimen surface, hence they act as oblique cracks and form fringes by interference between longitudinal waves excited in the specimen and the wave directly reflected from the specimen surface [16]. The fringe spacing, Δx, will depend on the angle θ, which the cristobalite grain makes with the surface as follows:

$$\Delta x = \frac{\lambda}{2 \sin \theta} \tag{3}$$

where λ is the longitudinal wavelength in the glass phase. For grains oriented at 90° fringes formed will be due to Rayleigh waves and hence λ will be the Rayleigh wavelength in the glass phase.

The volume fraction of cristobalite in the matrix was estimated at 0.4. In Fig. 9.6, cristobalite has intermittently precipitated at the fibre/matrix interface along its entire length. This has the effect of forming an *in situ* coating about 3–5 μm thick along the fibre length. Comparing the volume fractions of cristobalite in the single fibre and the high fibre volume fraction composites we see that the presence of fibres has given a comparatively small increase in the amount of cristobalite formed. In agreement with Bleay and Scott [17], these results indicate that nucleation of cristobalite at fibre surfaces is not the primary mechanism of matrix devitrification in 7740/SiC. Devitrification is most likely to occur in silica rich regions of the glass [13,15] or at the surfaces of the powdered glass during hot pressing.

The fabrication of carbon fibre reinforced borosilicate is performed at temperatures of the order of 1200°C [18–20]. Cristobalite is not observed to form in the matrix of the composite, although Bleay and Scott [17] report the formation of small quantities of cristobalite at the fibre/matrix interface. We have also observed the presence of a thin layer of cristobalite grains at the fibre/matrix interface in Corning Code 1723 glass reinforced with Nicalon fibres. Murty *et al.* [21] have observed the same behaviour in 7740/SiC (incorporating both Nicalon and Tyranno fibres) fabricated at temperatures above the liquidus of borosilicate glass (about 1100°C). They concluded that at the more common hot-pressing temperature of about 950°C matrix devitrification initiates at the surfaces of the powdered glass. As processing above the liquidus reduces these surfaces, the only mechanism enabling devitrification is precipitation at fibre/matrix interfaces.

The presence of cristobalite is detrimental to composite strength as cristobalite grains provide multiple stress concentrators. Hence the effective composite matrix microcracking stress will be lowered. However the profusion of partially debonded cristobalite/glass and cristobalite/fibre interfaces introduces a high volume fraction of interfaces along which cracks preferentially propagate. Hence we expect the composite to be comparatively tough. This is supported by the high work of fracture which has been measured for this CMC system [7].

9.2.4 Matrix microcracking

The acoustic microscope reveals the presence of extensive matrix microcracking (Fig. 9.3). Matrix cracks are delineated by means of fringes which run parallel to the crack on either side (Fig. 9.7). The fringes have a spacing of one half of the Rayleigh wavelength and are due to Rayleigh wave interference in the same manner as the Rayleigh wave fringes around the fibres. The fringes end at a definite distance along the crack, thus giving a good indication of the crack length [5]. These microcracks can be divided into three categories, those which (1) connect neighbouring fibres, (2) form around cristobalite grains and (3) radiate out into the devitrified borosilicate glass from the fibre/matrix interface. If large residual tensile radial thermal stresses were generated in the composite and the fibre/matrix interface had a comparatively low fracture toughness (an essential

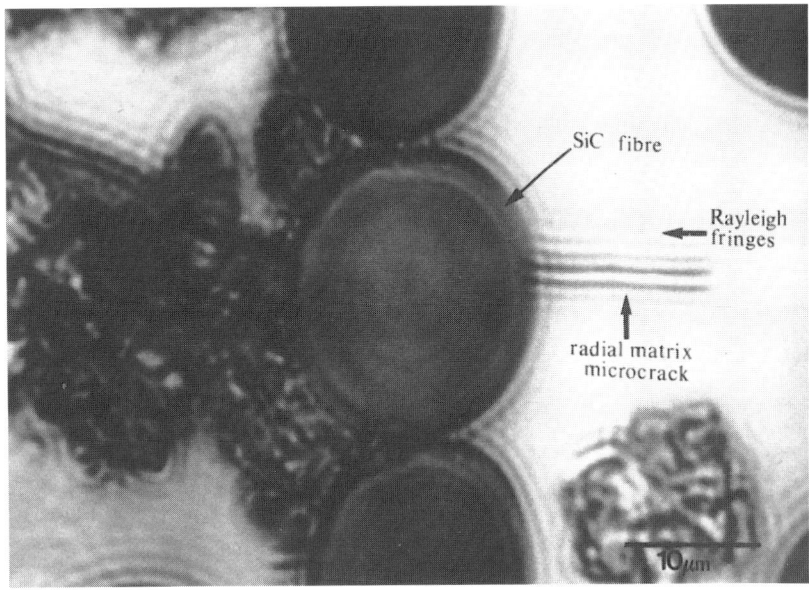

9.7 Acoustic micrograph of 7740/SiC showing a microcrack in the amorphous matrix phase. $z = -2.0\,\mu m$.

requirement for high composite fracture toughness) then we would expect to observe fibre/matrix debonding using the SAM.

Examination of the composite using SEM (secondary electron and back-scattered modes) also enabled images of matrix microcracking to be obtained (Fig. 9.8 and 9.9). Whilst the contrast between fibres, glass and cristobalite is

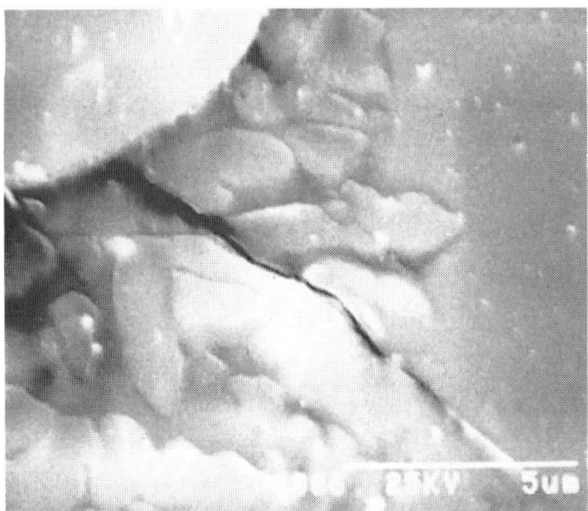

9.8 Secondary electron micrograph of 7740/SiC (30° specimen tilt, 25 kV).

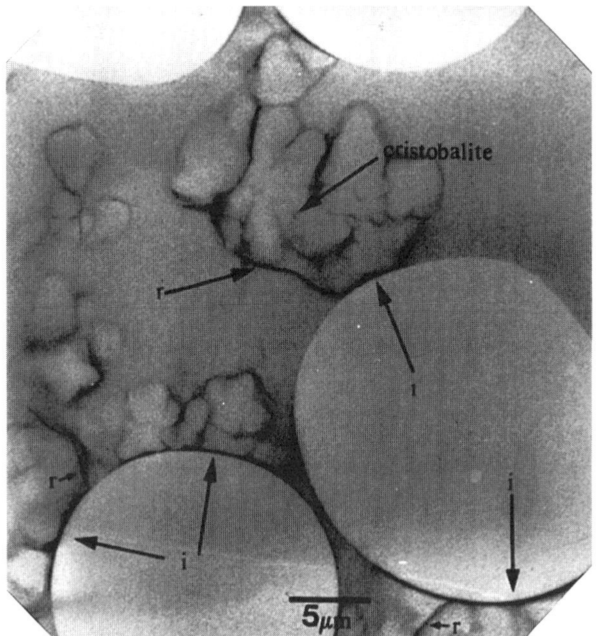

9.9 Back-scattered electron micrograph of 7740/SiC: i indicates interfacial/circumferential cracks, r indicates radial microcracks.

inferior to that obtained from SAM, the contrast from cracks is excellent. Of particular note is propagation of cracks along the fibre/matrix interface (arrowed in Fig. 9.9). Fibre/matrix debonding (decohesion) is only seen where cristobalite grains are present at the interface. Thus fibre/matrix debonding appears to be associated with the precipitation and cooling of cristobalite to room temperature.

In acoustic images such as Fig. 9.3 and 9.7 it is not possible to observe directly the fibre/matrix debonding seen in the SEM, yet matrix microcracks are easily resolved. Owing to the low ratio of fibre to matrix elastic modulus, the fibre/matrix interface in CMCs has a much lower impedance to acoustic waves than is the case for polymer matrix composites (PMCs). Hence only a small fraction (about 0.23 for 7740/SiC) of the acoustic power of Rayleigh waves incident on the fibre/matrix interface will be reflected back, i.e. the interface has a low reflection coefficient. Thus the Rayleigh wave fringes formed on either side of the fibre/matrix interface are comparatively weak with respect to the background contrast. Debonding at the fibre/matrix interface results in the formation of a circumferential crack. As cracks have a comparatively high reflection coefficient we would expect stronger fringes either side of a debonded interface*. As fibre/matrix debonding is primarily associated with the presence of

* Reflection of Rayleigh waves from a crack has an associated phase change which will result in a shift of the fringe pattern by a fraction of the wavelength in the matrix.

cristobalite in the matrix, it is difficult to observe fringe formation in the matrix side of the debond. In Fig. 9.7, variations in the fringe pattern around and within the fibres are just visible indicating partial fibre/matrix debonding has occurred around some of the fibres. Contrast across fibre/matrix interfaces in acoustic micrographs will be discussed more fully in Section 9.3.

Matrix microcracking has been observed in several other CMC systems (both glass and glass-ceramic [22]), and was attributed to the high residual thermal stresses generated in the matrices during cooling down from the processing temperatures (of the order of 1000°C). These CMCs, however, had matrices which were essentially single phase, whereas for 7740/SiC the presence of cristobalite introduces a complicating factor. This leads us to conclude that the radial microcrack indicated in Fig. 9.7 is a product of residual thermal stresses, whereas the microcracks labelled i and r in Fig. 9.9 and associated with cristobalite grains are due to matrix devitrification. We conclude that at least two mechanisms contribute to matrix cracking in 7740/SiC: residual thermal stresses and matrix transformations.

If the coefficient of thermal expansion (CTE) of the reinforcing fibres differs from the matrix, then cooling the composite from hot pressing to room temperature will result in thermal stresses being generated in both fibres and matrix. If the axial CTE of the fibres is lower than that of the matrix then the fibres will contract less than the matrix, hence placing the matrix in a state of tension which, if sufficiently great, may lead to a reduction in the matrix cracking stress of the composite [23], or even spontaneous matrix microcracking. Conversely, if the axial CTE of the matrix is less than that of the fibres, the fibres will compress the matrix, effectively increasing the matrix microcracking strength of the composite when it is subjected to tensile loading. Residual radial and hoop thermal stresses in the composite can also affect composite tensile strength. Tensile radial thermal stresses across the fibre/matrix interface can result in fibre/matrix debonding. Such behaviour has been observed in lithium aluminosilicate reinforced with Nicalon silicon carbide fibres [24] which can lead to a lower IFSS and hence enhanced composite toughness [25]. High compressive thermal stresses across the fibre/matrix interface give rise to an increased IFSS resulting in a brittle material.

Above 230°C*, cristobalite exists in the β form; however, on cooling through 230°C it undergoes a martensitic phase change from β to α cristobalite with an associated volume decrease of 5.7%. As the α cristobalite shrinks away from the glass matrix, tensile stresses will be generated across the cristobalite/borosilicate glass interface. If these stresses are sufficiently high, they will lead to particle debonding and possibly crack growth in the glass phase. We note from micrographs such as Fig. 9.7–9.9 that complete particle debonding seldom

*The precise temperature at which β cristobalite transforms to α cristobalite is dependent on the initial purity of the β cristobalite, ranging from 220 to 270°C [40].

occurs and that some cracks terminate in the devitrified glass matrix. The irregular shape of the cristobalite particles may well be the cause of the few instances of observed crack growth into the devitrified glass. Micrographs such as Fig. 9.8 and 9.9 show that crack surface separation can be quite large around cristobalite particles, supporting the calculation of large matrix strains. Cracks in the matrices of other CMC systems due to fibre/matrix CTE mismatch were found to have very little crack opening [22].

9.3 Glass-ceramic matrix composites

Glass ceramics are polycrystalline solids produced by the controlled crystallisation of glasses [26]. They have the advantage over engineering ceramics such as alumina and silicon nitride that at high temperatures before they are cerammed they have a low viscosity, so that good penetration and densification are possible in the fabrication of the composite [27, 28]. By adding appropriate nucleating agents and by controlled heat treatment of the glass, stronger crystalline phases can be produced. Over 90% of the parent phase can be crystallised in this way, and the higher temperature performance limit of the glass ceramic compared with the glass comes from this polycrystalline structure. The composition of the parent glass can be chosen to give optimum compatibility with the fibres. In particular the thermal expansion characteristics of the matrix can in principle be tailored to match those of the fibres, in order to prevent, or at least reduce, residual thermal stresses.

Two glass-ceramic matrix composites will be discussed here, magnesium aluminosilicate reinforced with silicon carbide fibres and calcium aluminosilicate reinforced with silicon carbide fibres.

9.3.1 Magnesium aluminosilicate matrix composites

Composites with a magnesium aluminosilicate (MAS) matrix were studied with two different kinds of fibre reinforcement. The first contained small Tyranno silicon carbide fibres, denoted by subscript t. The second had much larger Sigma silicon carbide monofilaments, denoted by subscript m. The manufacture of the composites was proprietary, but from published literature [29] it may be expected that the fabrication temperature was in the range 1200–1400°C, with pressures of 7–14 MPa applied for up to 20 min. The matrix powder was magnesium aluminosilicate glass with a small amount of glass ceramic added to initiate crystallisation, together with ZrO_2 as a nucleating agent and Nb_2O_5 to give barrier protection to the SiC fibres. Mechanical properties of the two kinds of fibre are summarised in Table 9.1, together with the corresponding properties of the glass-ceramic matrices. X-ray diffraction (XRD) indicated that the matrix predominantly consisted of cordierite.

9.3.1.1 MAS/SiC$_t$

The microstructure of the matrix of the MAS/SiC$_t$ composite can be seen in the acoustic micrograph in Fig. 9.10. The predominant contrast mechanisms are variation of Rayleigh wave propagation in the different phases and variation in reflection of acoustic waves due the different acoustic impedances of these phases. There are two phases visible in Fig. 9.10. The majority phase is darker than the minority phase at this defocus of $z = -1\,\mu m$ (and also at focus); the minority phase comprising about 0.15 volume fraction of the matrix. There is also a fine dispersion of porosity of diameter 1–2 μm, comprising about 0.02 volume fraction of the matrix. This volume fraction was confirmed by reflected polarised light microscopy, which also confirmed that the majority phase was crystalline and the minority phase was glassy. Polarised light microscopy was not able to elucidate further the nature of the fine dispersion of porosity. Both optical and acoustic microscopy of the composite indicated a fibre volume fraction of 0.4, with excellent distribution and almost no fibre bunching.

The nature of the composite in cross-section is shown in Fig. 9.11, which is a series of images of the same area at increasing defocus (i.e. the acoustic lens is moved towards the specimen surface so that the lens–specimen separation is smaller than the focal length of the lens). At focus, in Fig. 9.11a, the fibres appear white; they have an average diameter of 9 μm, consistent with Table 9.1. The matrix shows two phases, just as it did in the lower magnification acoustic micrograph of Fig. 9.10. The minority phase is lighter than the rest of the matrix, with an average grain size of about 6 μm. As with the 7740/SiC composite

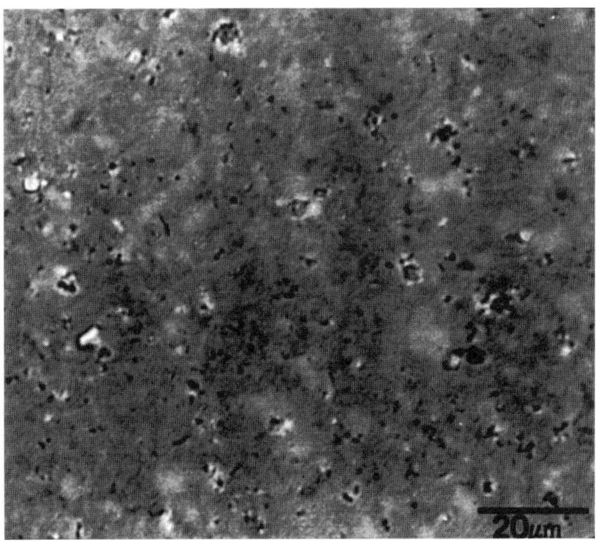

9.10 Acoustic micrograph of matrix in MAS/SiC$_t$. $z = -1.00\,\mu m$.

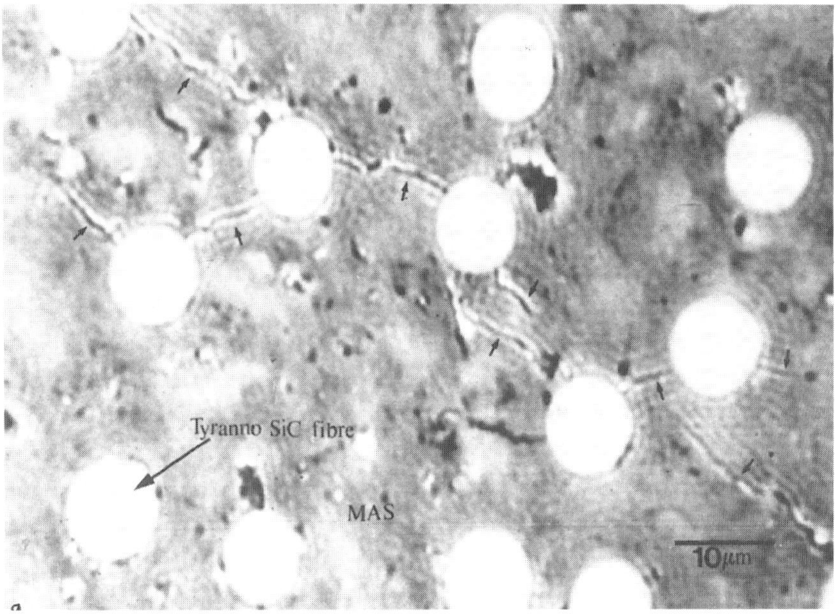

Tyranno SiC fibre
MAS
10μm

(a)

10μm

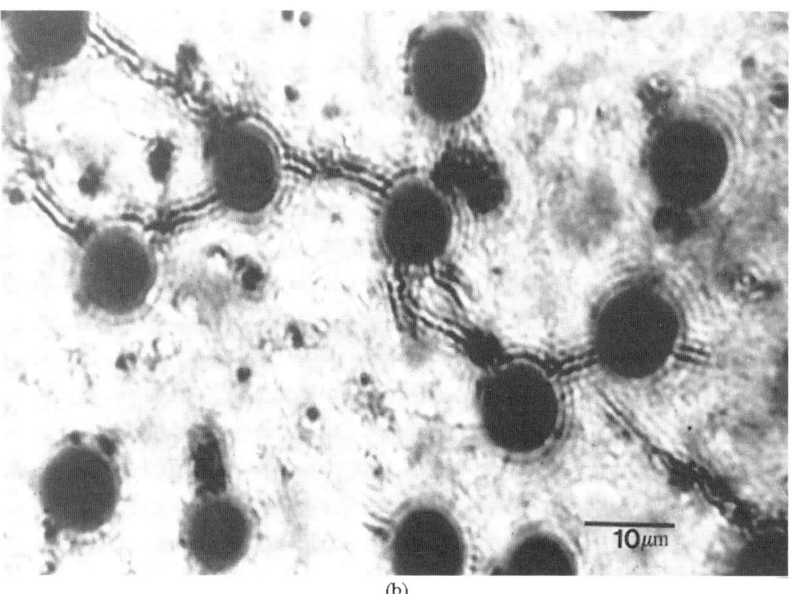

(b)

9.11 Acoustic micrographs of MAS/SiC_t. Arrows indicate microcracks. (a) $z = 0\,\mu m$, (b) $z = -1.0\,\mu m$.

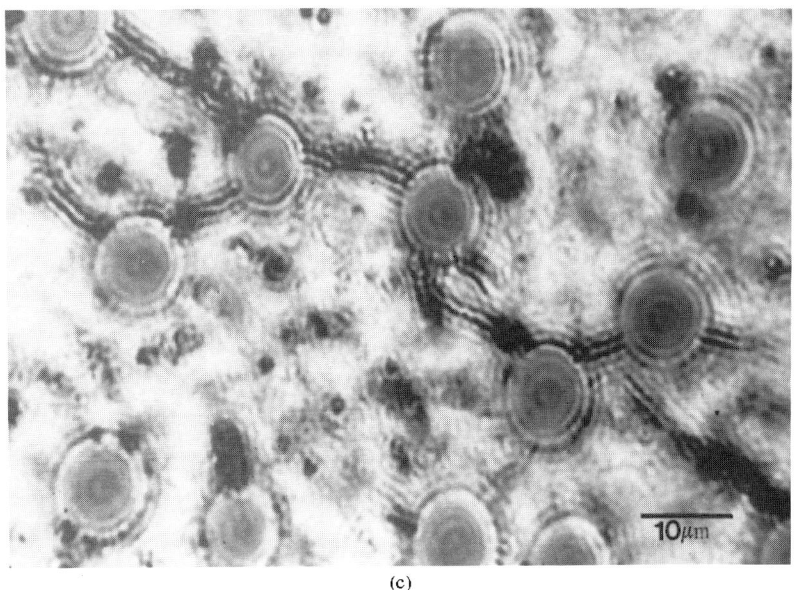

(c)

9.11 (c) $z = -2.0\,\mu\text{m}$.

discussed in Section 9.2, fringes can be seen around the fibres. At a defocus of 1 μm, in Fig. 9.11b, the contrast is considerably enhanced.

First, it can be seen that a change of only 1 μm in the defocus has reversed the contrast of the fibres from light to dark. This is a consequence of the well known $V(z)$ effect [30]; it is a caution against interpreting contrast in acoustic images naïvely. Bright contrast cannot simply be associated with high density or high elastic stiffness, though both of these affect the contrast. Detailed interpretation must be performed in the light of the behaviour of the oscillations in $V(z)$ and their dependence on parameters that determine Rayleigh wave propagation [31].

Secondly, the fringes around the fibres have become stronger with respect to the background matrix contrast. Finally, and perhaps most striking of all, there is remarkably strong contrast from cracks that run through the matrix between the fibres. The cracks can also be seen in Fig. 9.11a, but their contrast is greatly enhanced at the defocus of Fig. 9.11b. The sensitivity to cracks comes from the fact that even when the width of the cracks is much less than the nominal spatial resolution of the microscope, Rayleigh waves propagating in the surface can nevertheless be strongly scattered. Two kinds of scattering, as perceived by the nature of the fringe contrast, can be seen in Fig. 9.11b. In the fringes along the cracks in the matrix the contrast extends from almost full black to full white, i.e. the full range of the grey scale; whereas around most of the circumference of the fibres the contrast is somewhat more subdued. But around some parts of the fibre/matrix interface the contrast is more characteristic of a crack. This is noticeable, for example, around the top of the three most central fibres. Following the course of the various segments of crack suggests that these are places where a crack has been deflected by (or has nucleated at) a fibre, and that what is being seen are regions of fibre/matrix interface where debonding has occurred, and where Rayleigh waves are being scattered just as they would be by a crack. This impression is confirmed by Fig. 9.11c, which is at twice the defocus of Fig. 9.11b. Here the cracks in the matrix are characterised by bright contrast at the crack itself, with at least one dark fringe on either side. It would therefore be expected that a fibre–matrix debond would be characterised by a bright fringe with a dark fringe on the matrix side (the fibre side of the debond might be quite different, because it is a different material, but there are no cracks inside the fibres here to compare with): this is exactly what is seen, with very pleasing correspondence between where the cracks appear to go around the fibres in Fig. 9.11b and Fig. 9.11c.

Measurement of the fringes in Fig. 9.11 (using equation (2)) gave Rayleigh velocities of 3520 m s^{-1} for the MAS and 5150 m s^{-1} for the Tyranno SiC. It is possible to make independent measurements of the Rayleigh velocity by analysing the oscillations in $V(z)$ in a line-focus-beam acoustic microscope [32]; measurements in this way yielded a Rayleigh velocity of 3550 m s^{-1} for MAS. The Rayleigh velocity is related to the density and the elastic constants by a sextic equation involving both the longitudinal and the shear bulk velocities [31]. It is therefore not possible to deduce the elastic constants from the Rayleigh velocity alone, even when the specimen is isotropic and the density is known. However, it is possible to write the Rayleigh velocity in terms of the density ρ, the shear modulus G and an approximate polynomial expansion in the Poisson ratio ν (Scruby *et al.* [33]),

$$v_r = \frac{1}{R_\nu}\left(\frac{G}{\rho}\right)^{1/2} \tag{4}$$

where $R_\nu \approx 1.14418 - 0.25771\nu + 0.126617\nu^2$.

As can be seen from Table 9.3, R_ν varies only slowly with ν, for example when $\nu = 0.15$, $\partial R_\nu/\partial\nu = -0.207$. Thus if the Poisson ratio is known to be, say, 0.15 ± 0.01 (an uncertainty of 7%), the shear modulus can be deduced from the Rayleigh velocity to an accuracy of 0.4% (plus any error in the measurement and in the value of the density). The Poisson ratio of Tyranno silicon carbide fibres has not been measured, though a value of 0.15 has been quoted for the similar Nicalon fibre. For a Poisson ratio $\nu = 0.15$, $R_\nu = 1.108$. For the Tyranno fibre, $\rho = 2350$ kg m^{-3} (Table 9.1), and so from the measured Rayleigh velocity of 5150 m s^{-1}, a shear modulus $G = 77$ GPa can be deduced. Similarly for the MAS, with $\rho = 2500$ kg m^{-3} and $\nu = 0.25$, a mean Rayleigh velocity of 3535 m s^{-1} yields a shear modulus of 37 GPa.

Table 9.3 Effect of Poisson ratio (ν) on the calculated shear (G) and Young's moduli (E) for an error in ν of 0.01

ν	$1/R_\nu$	$\Delta G/G(\%)$	$\Delta E/E(\%)$
0.0	0.8740	−0.45	0.55
0.1	0.8931	−0.42	0.49
0.2	0.9110	−0.38	0.46
0.3	0.9274	−0.34	0.43
0.4	0.9422	−0.29	0.42
0.5	0.9551	−0.25	0.42

The Young's modulus of an isotropic material is related to the shear modulus and the Poisson ratio by:

$$E = 2G(1 + \nu) \tag{5}$$

Hence a determination of E can be more sensitive to an error in ν and an estimate of $\nu = 0.2 \pm 0.01$ would give an uncertainty of 0.46% in E (including the uncertainty in G, which is of opposite sign). The Young's modulus deduced for the Tyranno fibres in this way is $E = 177$ GPa, and for the MAS $E = 93$ GPa. For comparison, the *in situ* Young's elastic modulus of the Tyranno fibres was measured using a Nanoindenter. Using the simple analysis of Doerner and Nix [34] the elastic modulus of the fibres was determined to be 166 GPa with a standard deviation of 13 GPa (from 25 measurements), in satisfactory agreement with the acoustically derived value. Whilst more precise analysis has recently become available [35] for load/displacement data from Nanoindention machines, the measured value is consistent with values determined by Pysher *et al.* [36] and Fischbach *et al.* [37], 150 GPa and 171 GPa, respectively, from tensile testing. Unlike the tensile testing measurements, neither the Nanoindenter nor acoustic microscope measurements are gauge length dependent. These figures are not intended to be definitive, but they do illustrate the kind of measurement that can be made using the acoustic microscope.

The mean linear thermal expansion coefficient of magnesium aluminosilicate is $4.6 \times 10^{-6}\,^{\circ}\mathrm{C}^{-1}$ (20–1000°C), compared with $3.1 \times 10^{-6}\,^{\circ}\mathrm{C}^{-1}$ for the Tyranno SiC fibres. Hence the fibre–matrix interface will be in radial compression, while the matrix will be in longitudinal and circumferential tension. The matrix monolithic tensile strength is 138 MPa [38], so it is quite plausible that the microcracking is a direct consequence of the failure to match the thermal expansion of the fibre and the matrix. The way that the cracks follow the interface between fibre and matrix suggests that the interface is weaker than either the matrix or the fibre, and so fails preferentially: this in turn suggests that the matrix is indeed essentially crystalline with a comparatively fine grain size.

9.3.1.2 MAS/SiC$_m$

An acoustic picture of a transverse section through a Sigma SiC monofilament in a MAS matrix is shown in Fig. 9.12. There is one long radial crack in the eight o'clock position, which was also seen without difficulty using a light microscope. There are two smaller radial cracks between four o'clock and six o'clock, and there may even be another crack between three and four o'clock. These cracks were not detected in the light microscope: they are identified in the acoustic micrograph by their characteristic fringe patterns. As with the Tyranno reinforced composites, it is believed that the cracks form due to large thermal stresses generated in the matrix during the cooling of the composite from its fabrication temperature [39].

Not all ring patterns are due to Rayleigh wave scattering and interference. Fig. 9.13 shows the microstructure of a Sigma monofilament, at somewhat higher magnification than Fig. 9.12 (and different defocus). The central core is 14 μm diameter tungsten fibre. Onto this core β-SiC is deposited to give radial columnar grains. The concentric rings in the fibre in Fig. 9.13 are not due to interference. A simple analysis of their spacing via equation (2) would give a surface wave velocity of 24 000 m s^{-1}, which is faster than any known elastic wave! Besides, the rings do not have the appearance of interference fringes. It therefore seems that the rings reveal regions of different microstructure due to fluctuating SiC deposition rates during the fabrication of the monofilament. The fringes in the matrix next to the interface are Rayleigh interference fringes, and

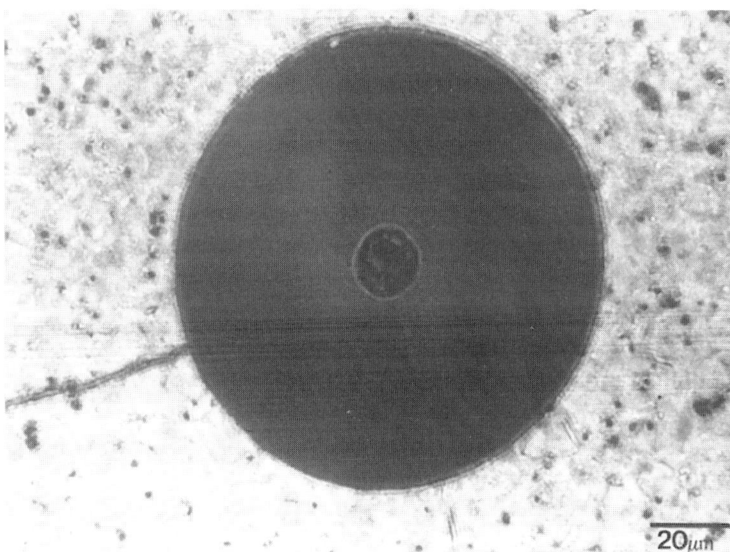

9.12 Acoustic micrograph of MAS/SiC$_m$. $z = -4.6\,\mu$m.

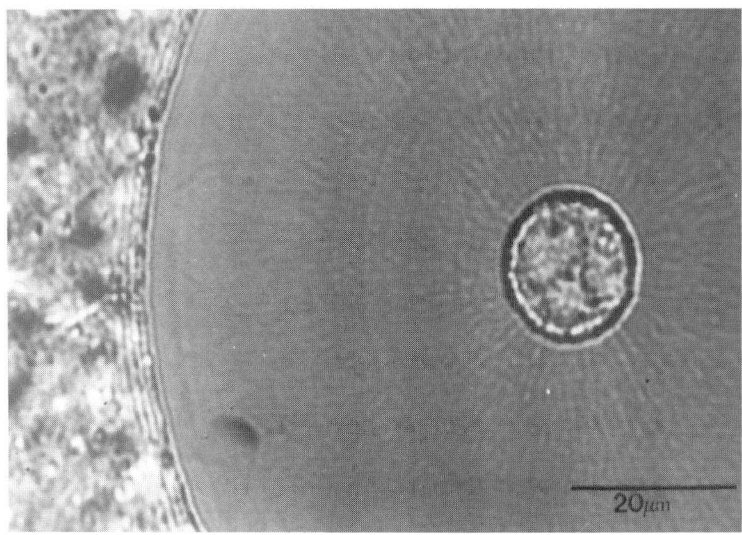

9.13 Acoustic micrograph of MAS/SiC_m showing the microstructure of the monofilament. $z = -2.0\,\mu m$.

they can be analysed in exactly the same way as they were in the MAS/SiC_t composite.

9.3.2 Calcium aluminosilicate CAS/SiC

Calcium aluminosilicate is a relatively simple glass ceramic, forming stoichiometric anorthite, $CaAl_2Si_2O_8$. The proprietary specimens were made by hot pressing at a temperature in the range 1300–1400°C. One percent by weight of As_2O_3 can be added as a fining agent to thicken the interfacial carbon layer and to improve its mechanical properties; a good interface consists of a layer of graphitic carbon 100 nm thick [40]. For low temperature applications the performance of a CAS/SiC composite is comparable with a glass matrix composite such as the 7740/SiC composite discussed in Section 9.2, but because the matrix is crystalline it can be used up to temperatures of 1000°C, approaching the composite fabrication temperature and the usable temperature limit of the fibres. Nicalon silicon carbide fibres were used as the reinforcing phase (see Table 9.1 for properties). The microstructure of a transverse section of a CAS/SiC composite (Fig. 9.14) shows the fibres and their distribution, a reasonably uniform matrix and some porosity. The volume fraction of the fibres is not completely uniform; an average value is 0.4. The uniformity of the matrix, as characterised by density and elastic homogeneity, could not be determined so

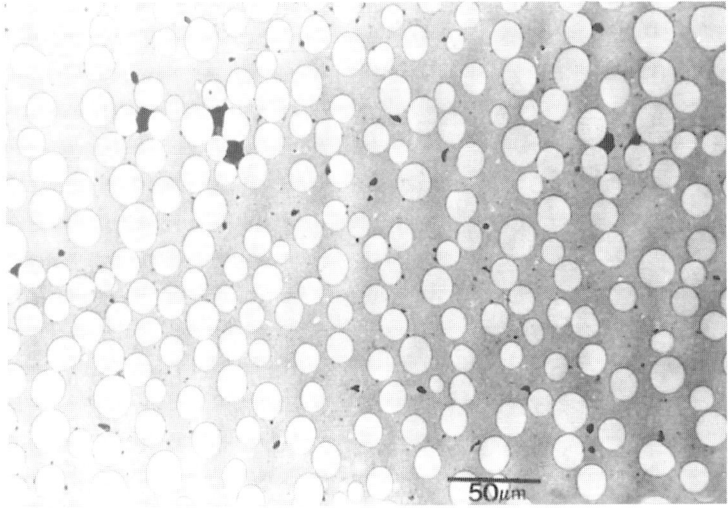

9.14 Optical micrograph of CAS/SiC showing the general microstructure of the composite.

directly optically; but this is something that the acoustic microscope can reveal readily.

Figure 9.15 shows an individual fibre in the composite at higher magnification, at focus and at a defocus $z = -3.2\,\mu m$. At focus (Fig. 9.15a), the fibre and the matrix can be seen, with the interface between them quite well delineated. Two radial cracks are present and fringes can be seen either side of the fibre/matrix interface. But the contrast is relatively flat. It is much more dramatic in the defocused image (Fig. 9.15b). There is more contrast from the microstructure of the matrix and the contrast from the cracks is greatly enhanced, in the same way that it was in Fig. 9.11. However, there seems to be a problem: in the defocused image the fibre seems to have shrunk. This is a perfectly genuine effect, there has been no change in the magnification. The dark region in the centre, especially towards the bottom of the picture, does not indicate the area of the fibre. The fibre actually extends two fringes further out, to the dark fringe indicated by I. A clue to this is given by the radial crack in the bottom of Fig. 9.15b. In Fig. 9.15a it appears to stop at the fibre/matrix interface but in Fig. 9.15b it appears to stop short of the dark region (though the contrast at the end of the crack shows some further complications). Moreover the circular fringes associated with the interface show curious effects: they have one combination of spacing and contrast outside the dark region and then different spacing and contrast in the centre of the fibre; certainly not an abrupt change at the fibre/matrix interface itself. Quantitative modelling adds confidence to the interpretation that can be made of such micrographs and also to the kind of

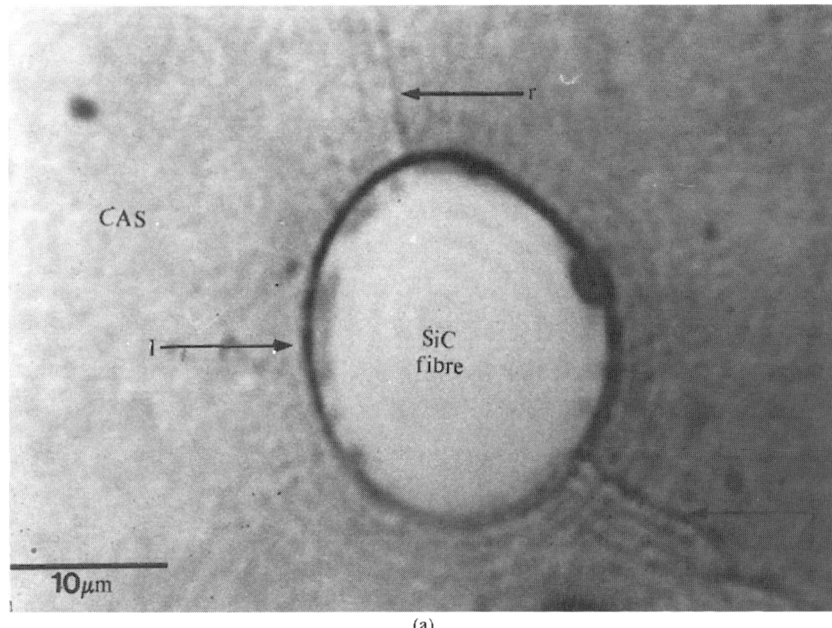

(a)

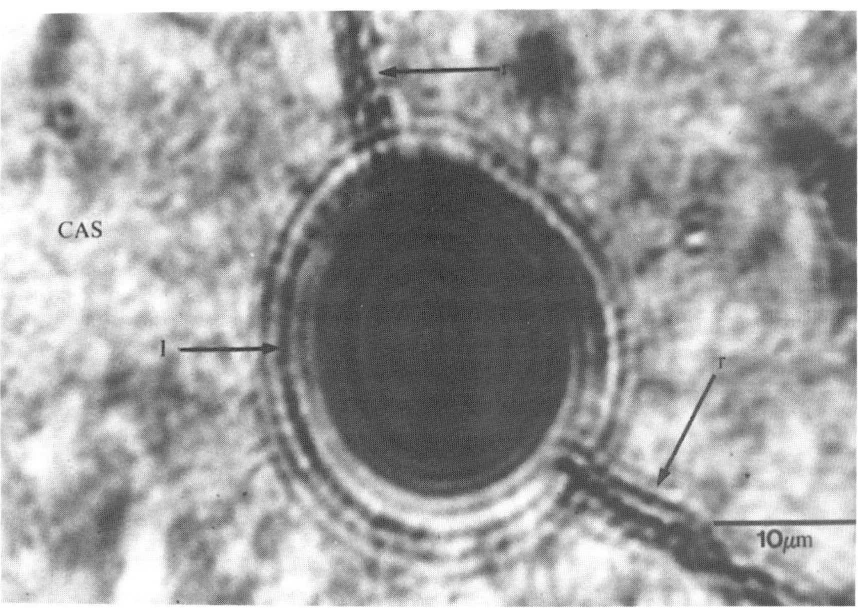

(b)

9.15 (a) and (b) Acoustic micrographs of CAS/SiC. Arrows indicate radial microcracks (r) and fibre/matrix interface (i), $z = 0\,\mu m$ and $z = -3.2\,\mu m$, respectively.

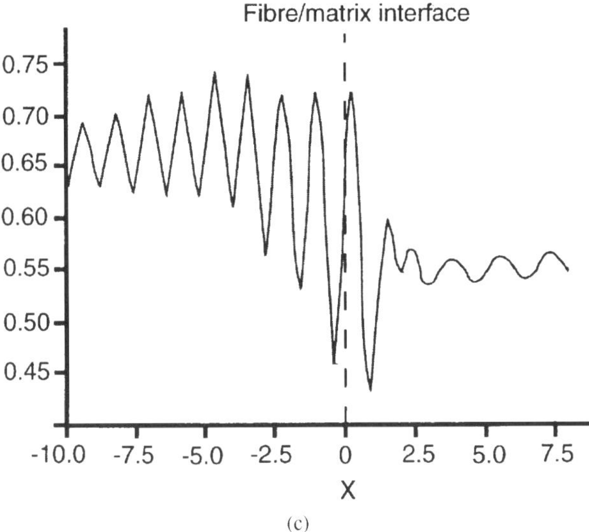

(c)

9.15 (c) shows a calculated linescan for (b) where $x < 0$ is the matrix and $x > 0$ is the fibre. x is in units of water wavelength, where one unit is 0.82 μm.

distinction that was made in the discussion of Fig. 9.11 between the contrast from a crack and the contrast from an interface.

The contrast from the interface between two media has been analysed using a combination of ray theory and diffraction theory [41]. The materials on either side of the interface are characterised by their density and elastic properties, and the interface is characterised by transmission and reflection coefficients for Rayleigh waves. This has the great advantage that the results of calculations (often lengthy) from the literature for scattering of Rayleigh waves by different configurations can be immediately incorporated. The theory was originally developed in two dimensions: it has subsequently been extended to three dimensions, but it is found that the two-dimensional theory gives an adequate account of the behaviour of the contrast over a crack. A calculation has been performed, with no free parameters, for the situation in Fig. 9.15b, and the result is shown in Fig. 9.15c. The frequency and defocus used were the same as in Fig. 9.15b, and the materials parameters were taken from Table 9.1. The calculated curve follows the changes reasonably well, not only in the level of the contrast, but also in the period of the oscillations. In particular, it exhibits an apparent sideways displacement of the interface in the same way that the experimental image does. This gives considerable confidence in the use of contrast theory in the interpretation of images of fibre–matrix interfaces.

Two specimens of CAS/SiC were examined that have been deformed in a flexural test. In each case the specimen was held at a known strain then a quick setting resin was poured into a mould that was placed around it, with the aim of preventing the closure of any matrix microcracks generated during the deformation. The strain was parallel to the fibre direction: the first specimen was strained close to the elastic limit of the matrix, $\varepsilon = 0.235\%$; the second was strained beyond it, $\varepsilon = 1.25\%$. Longitudinal sections were then prepared for microscopy.

Acoustic micrographs of two areas of a section through the first specimen are shown in Fig. 9.16. In Fig. 9.16a there are a number of microcracks, indicated with arrows, that had not been detected when this section was examined in a light microscope. As usual, these cracks are characterised by Rayleigh fringes running alongside them. In many ways these cracks look similar to those seen in the as-fabricated specimens in Fig. 9.15 (although of course this is a different section) and the relative abundance is similar. But the cracks in the strained specimen tend to be longer and in many cases they bridge closely spaced fibres. The strain may have caused inherent microcracks to propagate through the matrix until they were deflected by the fibres along the fibre–matrix interface. Cracks were also found parallel to the fibre axis throughout the specimen; an example is given in Fig. 9.16. These cracks are the same orientation as the ones seen in a transverse section, but since they invariably started where a fibre emerged from the section, it is feared that they are probably an artefact of section preparation. The difficulty of preparing these sections should not be underestimated and it is remarkable that the other specimens are relatively free from such artefacts.

The specimen that had been strained beyond the elastic limit was much more interesting. Two acoustic micrographs of different areas at different magnifications are shown in Fig. 9.17. The fuzzy variations in contrast are due to the $V(z)$ effect where the surface is not quite flat. Figure 9.17a shows a part of the beam that was in compression. The cracks are again similar to what was seen in the as-fabricated material, though the orientation is different. Rayleigh fringes were seldom seen beside these cracks, suggesting that they may be subject to closure stresses. In a tensile region in Fig. 9.17b, the appearance is quite different. Once again there are fine cracks perpendicular to the fibres (labelled D), which again may have been there before the deformation. There are also cracks (labelled C) that were wide enough open to show that the fracture surfaces had separated. Like the other cracks, these were probably intergranular, but they often bifurcated, with the subsidiary cracks propagating parallel to the main crack, and occasionally rejoining it.

The crack spacing can be related to the composite stress at which cracking first occurs and the fibre–matrix interfacial shear strength τ_f [42]. Taking the measured fibre and matrix properties into the appropriate equations (see Lawrence *et al.* [39]) yields a value for the interfacial shear strength of $\tau_f = 6.3\,\text{MPa}$. If the *in situ* measurements of the elastic properties of the fibre and matrix ($E_m = 76\,\text{GPa}$ and $E_f = 184\,\text{GPa}$ measured using a Nanoindenter) are used, the deduced IFSS rises slightly to 8.6 MPa. Both values lie within the range of reported values [43].

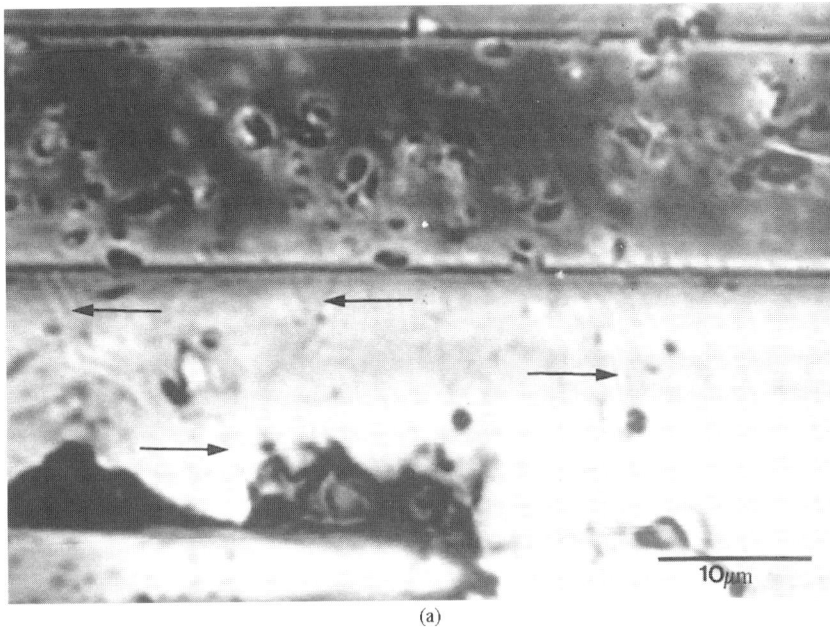

(a)

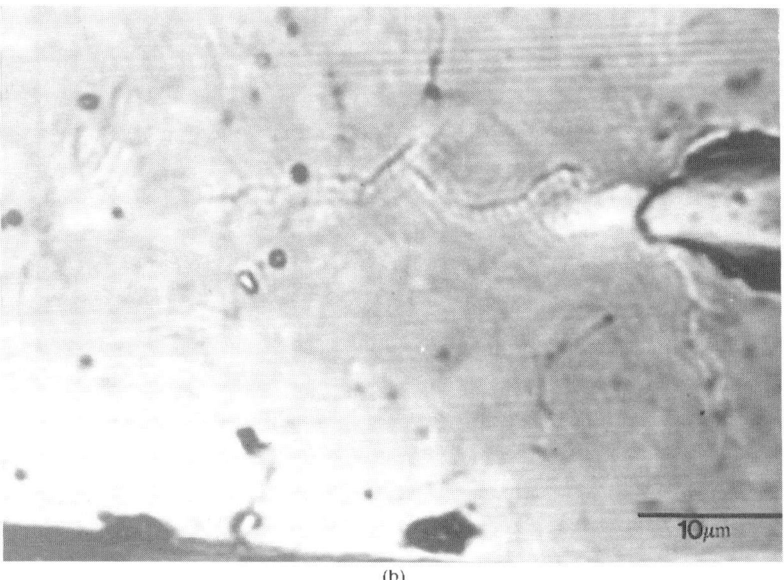

(b)

9.16 Acoustic micrographs of CAS/SiC bend specimen (strain below elastic limit). (a) $z = -1.0\,\mu m$, arrows indicate microcracking between fibres. (b) Different area from (a), with $z = -0.8\,\mu m$.

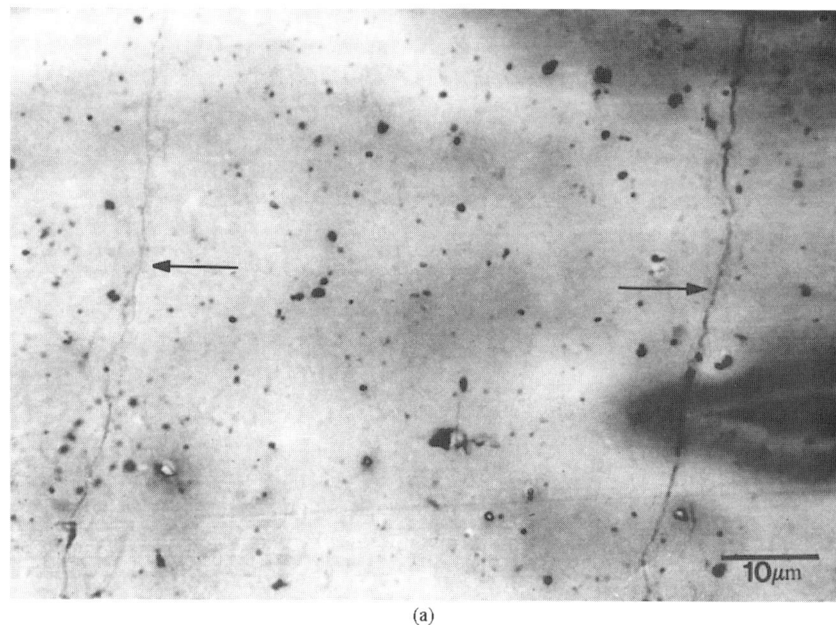

(a)

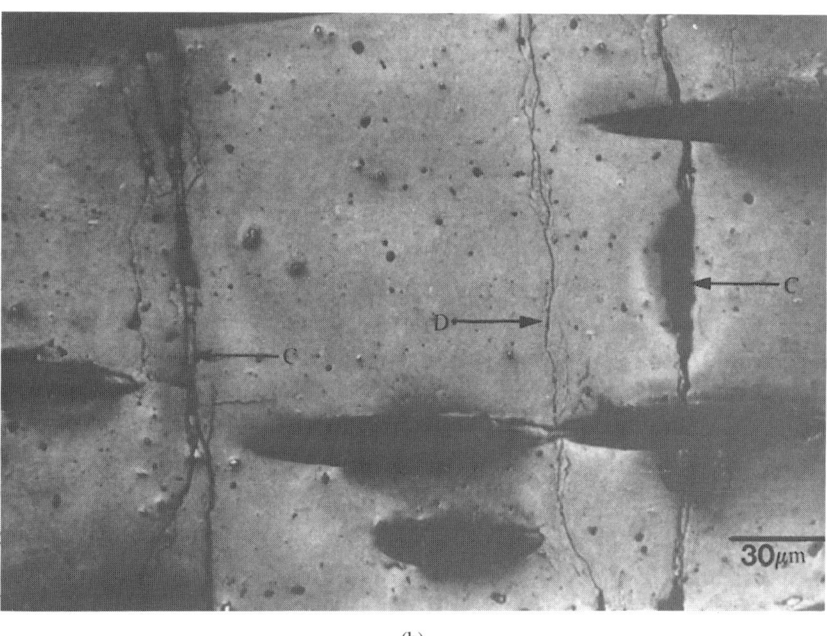

(b)

9.17 Acoustic micrographs of CAS/SiC bend specimen (strained to the elastic limit). (a) Compressive region of specimen, $z = 0\,\mu m$. (b) Tensile region, $z = 0\,\mu m$.

9.4 Metal-matrix composites

Metal-matrix composites (MMCs) have been developed for superior specific mechanical properties compared with unreinforced metals for the medium temperature range 300–600°C [44–46]. They are thus intended for use at lower temperatures than the glass and ceramic matrix composites (CMCs) which were discussed in Sections 9.2 and 9.3.

One MMC with potential for high temperature applications, e.g. gas turbine components, where weight, strength and stiffness are all important factors uses titanium alloyed with 6% aluminium and 4% vanadium (Ti-6Al-4V) as the matrix and silicon carbide monofilaments as reinforcement. In this composite the function of the matrix is to provide ductility and toughness while the fibres give extra strength, rigidity and creep resistance. A good interface between matrix and fibres is essential to ensure that tensile stresses are adequately transferred to the fibres. It is of crucial importance to be able to determine how the composite performs during typical service conditions. These can readily be simulated by various thermal ageing treatments. While there is interest in being able to monitor all aspects of material degradation, the condition of the fibre–matrix interface is of paramount importance to the strength and integrity of the composite. The degradation of the SiC-reinforced Ti-6Al-4V composite as a function of thermal ageing can be studied by acoustic microscopy because of its sensitivity to changes in elastic properties caused by variations in microstructure and its ability to image very small defects such as cracks [4,5]. Quantitative measurements of ultrasonic velocity can be made using both point-focus and line-focus beam [32,47] microscopy.

9.4.1 Experimental procedure

The main material studied was manufactured by Textron and supplied by Rolls Royce plc, Derby. The matrix was Ti-6Al-4V titanium alloy (6% aluminium, 4% vanadium, balance titanium) reinforced with Textron SCS-6 silicon carbide monofilaments, designed Ti-6Al-4V/SiC_m hereafter. This material is manufactured from layers of a prepreg of parallel silicon carbide monofilaments held together in a thin sheet by polymer binder, alternating with thin foils (50–100 μm) of the titanium alloy. After removing the binder by vacuum heating to 400–450°C, the composite is consolidated by hot pressing to 105 MPa at 925°C for 45 min. The specimens used for this study were all 8-ply material. Optical and acoustic microscopic examination of the as-fabricated material revealed a reasonably even distribution of monofilaments and an average monofilament volume fraction of 0.36 [48].

One specimen was examined in the as-fabricated condition, while the remaining four specimens were studied after a range of heat treatments carried out by Rolls Royce plc. Their purpose was to simulate potential service

Table 9.4 Details of heat treatment of Ti-6Al-4V/SiC_m specimens

Specimen	Temperature (°C)	Duration (h)
1	—	—
2	450	500
3	450	1000
4	600	500
5	600	1000

conditions and investigate degradation of the material. Ageing temperatures of 450°C and 600°C were chosen, and the duration was either 500 or 1000 h (Table 9.4).

All the specimens were mounted in transverse section and polished for optical metallography. As they had been prepared using polishing cloths, there was a certain amount of relief polishing around the monofilaments. These tended to stand proud, the matrix sloping away from the interface. The specimens were not repolished for fear of damaging the fibre–matrix interface. The size of the monofilaments ensured that they themselves were flat enough not to perturb the acoustic image. However, extra care was needed to avoid topological effects when interpreting acoustic contrast from matrix regions adjacent to the fibres.

A specimen of α-titanium aluminide (Ti_3Al) reinforced with Textron SCS-6 SiC monofilaments was also examined in the as-fabricated condition by acoustic microscopy. The specimen was supplied with the same surface condition as specimens 1–5 above, and there was no need for further polishing.

9.4.2 Monofilament microstructure

Figure 9.18a shows an acoustic micrograph of a single monofilament which has been mounted in epoxy resin. It reveals the carbon core, pyrolytic graphite (PG) layer, silicon carbide and carbon rich coatings (CRC). The diameter was measured as 143 μm, in agreement with the manufacturer's figure. Imaging the same monofilament at a higher magnification (Fig. 9.18b) revealed the carbon rich coating to consist of at least three clearly defined layers. The middle layer (lightest contrast) was 1 μm thick, and the inner and outer layers each 1.7 μm, to give a total thickness of 4.5 μm.

9.4.3 As-fabricated Ti-6Al-4V/SiC_m

Acoustic microscopy of the as-fabricated material revealed an approximately even distribution of SiC monofilaments in a Ti-6Al-4V matrix with no obvious signs of

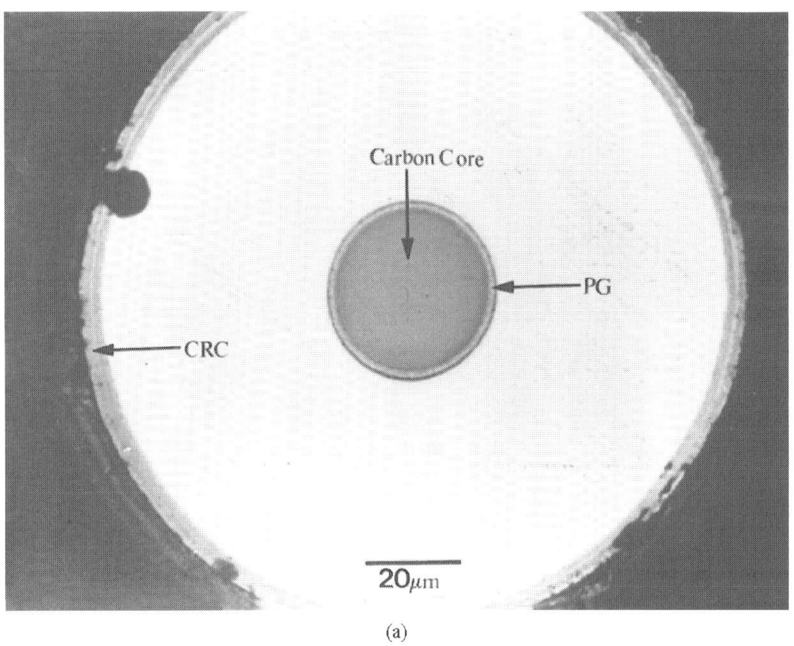

(a)

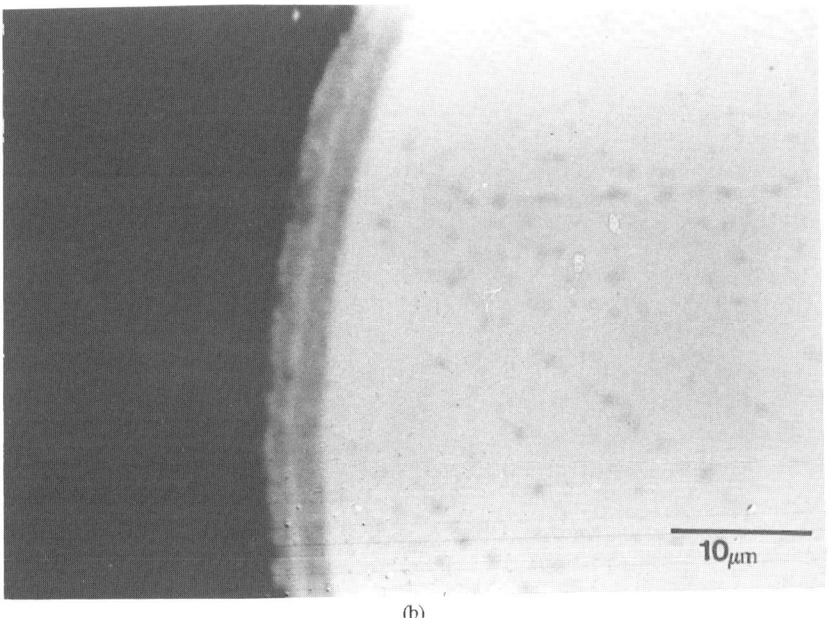

(b)

9.18 Acoustic micrographs of Textron SCS-6 monofilament, $z = 0\,\mu m$. (a) Cross-section of monofilament, (b) magnified image of coatings. PG, pyrolytic graphite; CRC, carbon rich coatings.

fabrication damage. Figure 9.19a shows an acoustic image of a typical monofilament. There are a number of features in common with the virgin monofilament of Fig. 9.18a. Thus a 33 μm carbon core (CC) is still visible together with a surrounding 1.3 μm layer of pyrolytic graphite. The irregularities in the graphite–SiC interface are indicative of chemical reaction between the monofilament and the matrix metal.

Moving further outwards, there are microstructural features in the silicon carbide that were not visible in Fig. 9.18. There is a band extending out to the mid-radius boundary (MR) at 22 μm. The microstructure of SCS-6 silicon carbide monofilaments is known to consist of columnar grains radiating out from the centre of the monofilament. When negative defocus is used to image the same monofilament (Fig. 9.19b) Rayleigh waves are generated which are sensitive to this microstructure, so that the contrast difference between the two regions is enhanced. Within the band fine-grained silicon carbide is observed (f), whereas outside MR the silicon carbide has a coarser grain (g). This observation is consistent with changes in concentration of the reactants in the chemical vapour deposition chamber. The resolution in the coarser region is sufficient to estimate the grain length as 3–4 μm, which is in good agreement with the literature [49].

The carbon rich coating surrounding the SiC is partially replaced by a reaction layer (RL, Fig. 9.19a). Higher magnification acoustic micrographs (Fig. 9.20) show at least three thin regions of varying contrast that separate the fibre from the matrix. These data suggest that certainly the inner layer and possibly also the middle layer of the carbon rich coating prior to fabrication (Fig. 9.18) remain intact, whereas the outer carbon layer has reacted with the titanium of the matrix. The structure of this layer as deduced from acoustic microscopy is shown schematically in Fig. 9.21. The reaction layer is likely to consist mainly of TiC; it is also observed (Fig. 9.19 and 9.20) to vary somewhat in thickness. The titanium alloy matrix consists of two phases α and β; the latter is known to be more reactive with the fibre material than the former. A microstructure of alternating α and β grains would thus be expected to react unevenly with the coated monofilament.

Figures 9.19b and 9.20b both show low contrast fringes on the SiC side of the fibre–matrix interface. These are consistent with interference between Rayleigh waves excited in the SiC at negative defocus and partial reflections of these waves from the interface. Measurement of the fringe spacing implies a Rayleigh wave velocity of 6325 m s^{-1}, which is less than the 6729 m s^{-1} that was measured for monolithic SiC. It is however appreciably larger than the value (5150 m s^{-1}) for Nicalon and Tyranno SiC fibres reported in Sections 9.2 and 9.3, respectively (see also Table 9.2). That the value for the SCS-6 monofilament should lie between the values for the much smaller Tyranno fibre (diameter 9 μm) and for bulk material is reasonable. Using the same reasoning as in Section 9.2, and assuming a density of 3045 kg m^{-3} and Poisson's ratio of 0.2, we deduce a shear modulus of

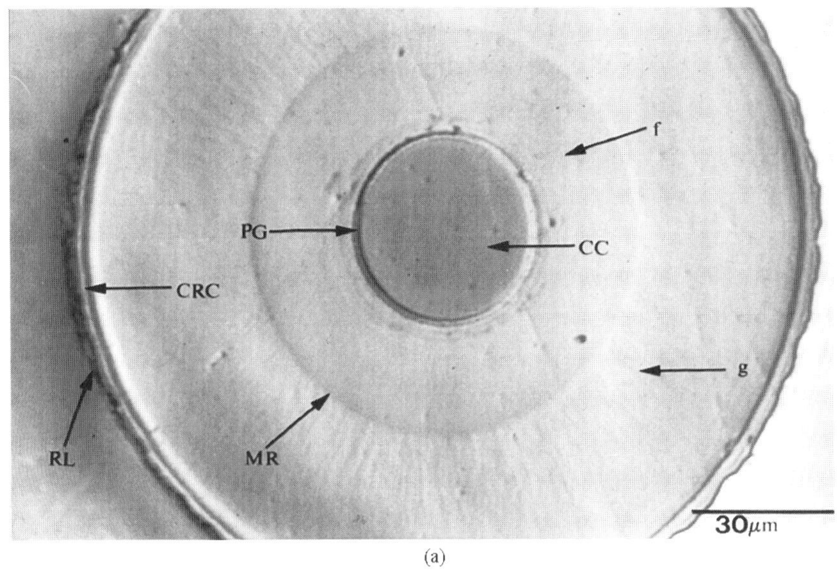

(a)

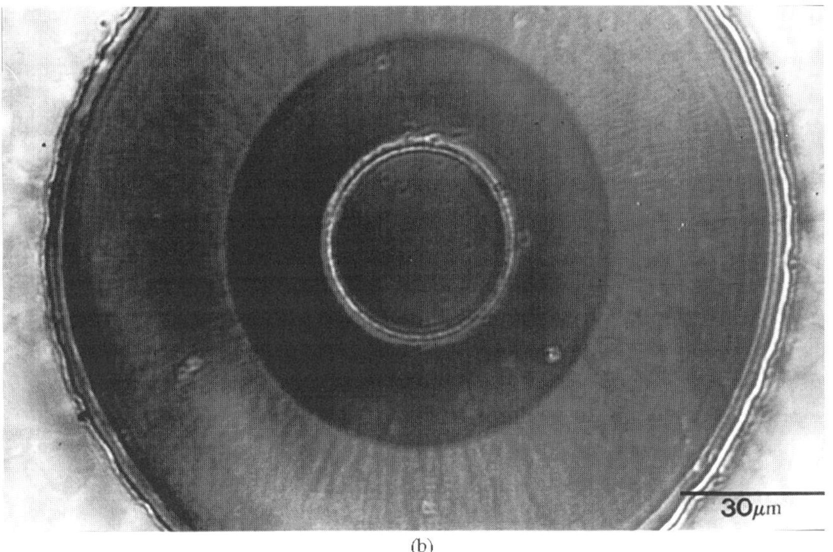

(b)

9.19 Acoustic micrographs of as-fabricated Ti-6Al-4V/SiC_m composite. (a) $z = 0\,\mu m$, (b) $z = -0.5\,\mu m$. CC carbon core; PG, pyrolytic graphite; CRC, carbon rich coatings; RL, reaction layer; MR, mid-radius boundary between fine grained, f, and coarse grained, g, β-SiC.

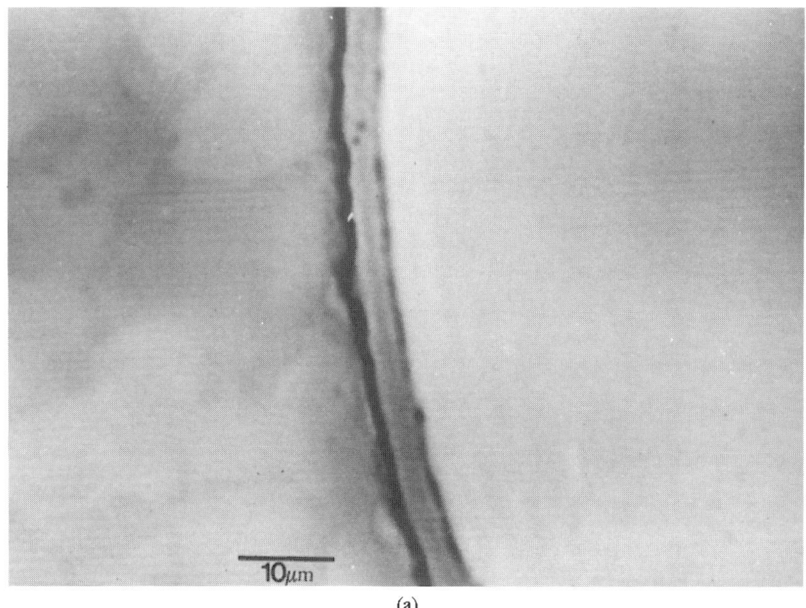

(a)

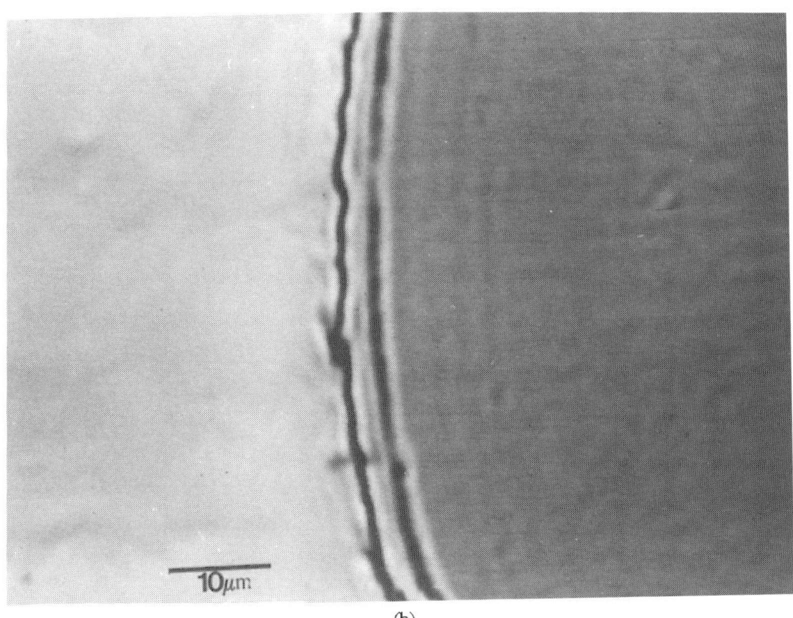

(b)

9.20 Acoustic micrographs of monofilament/matrix interface in as fabricated Ti-6Al-4V/SiC_m. (a) $z = 0\,\mu m$, (b) $z = -1.0\,\mu m$.

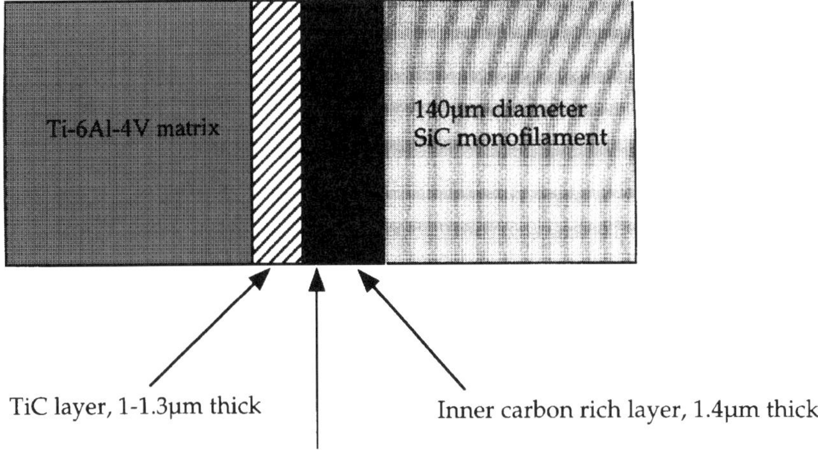

9.21 Schematic diagram of observed interfacial structure in the as-fabricated Ti-6Al-4V/SiC_m.

147 GPa and Young's modulus of 352 GPa for Textron SCS-6 monofilament SiC. The latter is lower than the quoted value for Young's modulus of 400–415 GPa for Textron SiC monofilaments by about 12% – a similar variation to that reported for Sigma SiC monofilaments and Tyranno fibres in Section 9.3.

9.4.4 Ti_3Al/SiC_m

Turning briefly to the material with the Ti_3Al matrix, it was found that the general appearance of the SiC monofilaments in the acoustic microscope was similar to those in specimen 1 examined above. The same features of carbon core, pyrolytic graphite layer and mid-radius boundary were all clearly observed. However, the fibre/matrix interface (Fig. 9.22) is noticeably different in appearance from Fig. 9.20. The interfacial region is more uniform in this material than in the composite with the Ti-6Al-4V matrix. Furthermore, differences in contrast are evident, suggesting differences in the composition of the various layers identified. This observation is consistent with the findings of Yang and Jeng [50] who examined the reaction layer and found it to consist of two sublayers of complex titanium carbides and silicides. As with some of the Ti-6Al-4V matrix specimens, cracks (i.e. diffusion debonds) were observed between closely spaced monofilaments. Certain cracks appeared to initiate at the interface between the inner and outer carbon rich layers. Other shorter cracks were observed to traverse the reaction layer. These latter are believed to be due to thermal contraction stresses following fabrication.

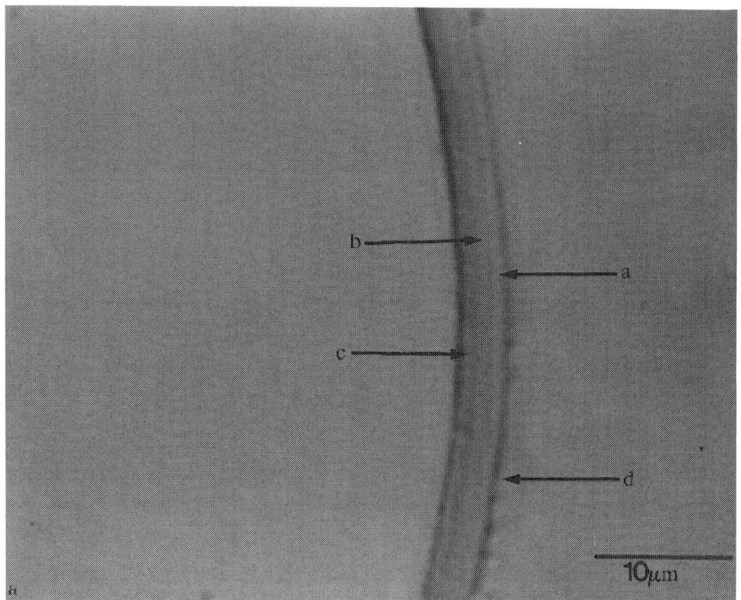

9.22 Acoustic micrograph of monofilament/matrix interface in Ti_3Al/SiC_m. Arrows indicate a, outer carbon rich layer, b, silicon rich layer, c, inner carbon rich layer, and d, monofilament/matrix reaction layer.

9.4.5 Heat treated Ti-6Al-4V/SiC_m

Low power optical and acoustic microscopy reveal two main effects of heat treatment: loss of carbon cores and fibre bunching [48]. While there are no missing cores in either specimens 1 (as-fabricated) or specimen 2 (500 h at 450°C), specimen 3 has lost some of its cores, and specimens 4 and 5 (the highest temperatures) have lost all their cores due to oxidation of the carbon. This oxidation of carbon cores (and the carbon rich monofilament coatings) may place a restriction on the use of this type of material unless oxygen diffusion inhibiting coatings are used on the outer surface. Bunching of the fibres is observed in all the heat treated specimens, which appears to be associated with the formation of defects as discussed below.

Figure 9.23 enables the general features of the monofilaments and interfaces to be compared as a function of heat treatment. Each specimen still exhibits the same change in microstructure at the mid-radius (MR) from finer to coarser SiC grains in the as-fabricated material. However, all four heat treated materials show a new feature in the form of an annular band at the mid-radius that was not observed in the as-fabricated material (Fig. 9.19). The annulus was consistently 6–7 μm wide, and can be seen most clearly in Fig. 9.23. This annulus corresponds to the change in monofilament structure and composition observed

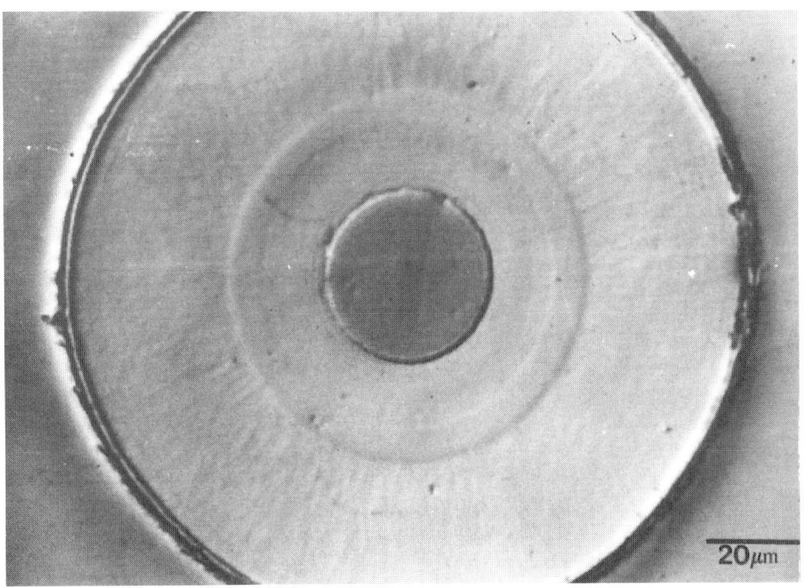

9.23 Acoustic micrograph of a typical monofilament in Ti-6Al-4V/SiC_m specimen 2. $z = 0\,\mu m$.

in as-fabricated monofilaments by Ning and Pirouz [51]. They found two SiC sublayers, each about 4.5 µm wide, to have formed at the point at which we observe the annular structure. Ning and Pirouz found that the inner of the two SiC sublayers showed an elongation and alignment of SiC grains, and the SiC grains in the outer sublayer had lengths of the order of micrometres and were heavily faulted. On crossing the sublayers the composition of the monofilament changed from about 10–20% excess carbon to stoichiometric SiC. Ning and Pirouz speculated that the excess carbon was present at the SiC grain boundaries. As the annulus was observable by the acoustic microscope only in heat treated specimens, we conclude that the heat treatment resulted in preferential growth of the SiC grains in this region.

Increasingly harsh heat treatments result in progressive deterioration of the interface between the fibres and the titanium alloy matrix (Fig. 9.24). The reaction layer in the as-fabricated specimen 1 varied only slightly in thickness as a result of differences between α and β titanium grains (Fig. 9.19). Specimen 2 is markedly different, the reaction layer between the fibre and matrix being variable in thickness. Figure 9.24a is taken at focus so that the contrast variations in the interfacial region are mainly due to variations in acoustic impedance rather than to interference fringes. Thus the contrast variations observed must be due to a more complex structure than in the as-fabricated material. The inner continuous layer that appears grey in Fig. 9.24a has a

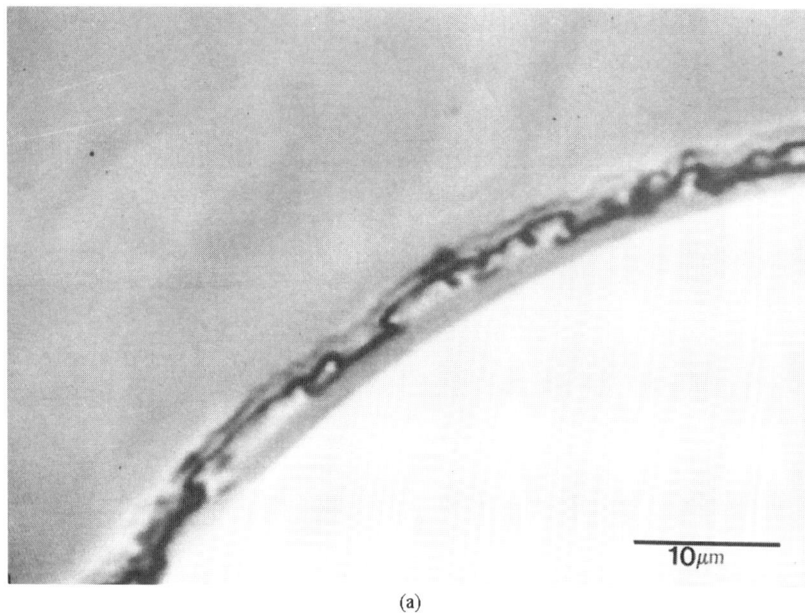

(a)

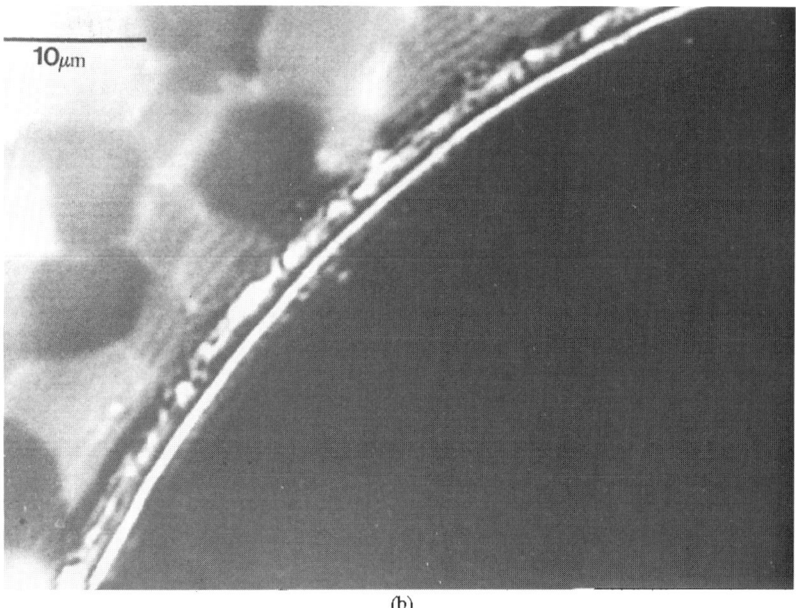

(b)

9.24 Acoustic micrographs of monofilament/matrix interfaces in thermally aged specimens. (a) Specimen 2, $z = 0\,\mu m$, (b) specimen 2, $z = -2.1\,\mu m$.

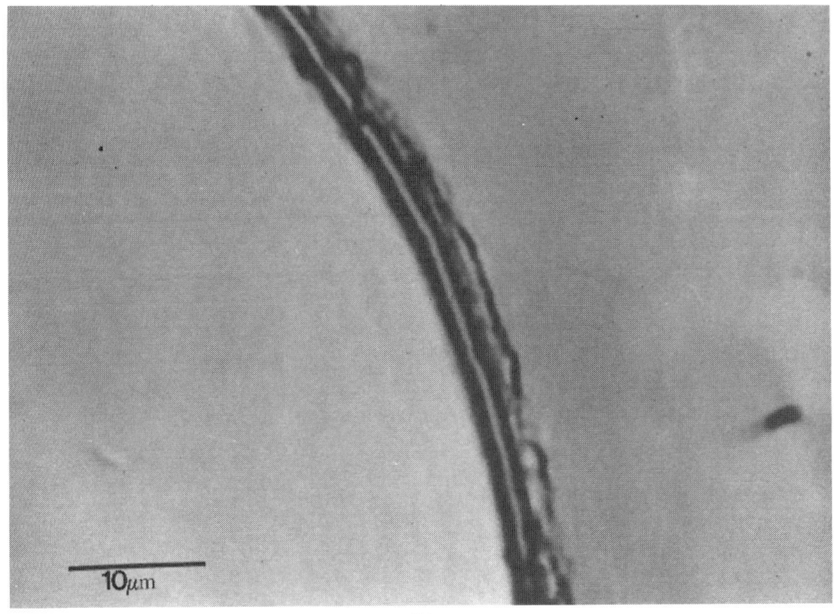

(c)

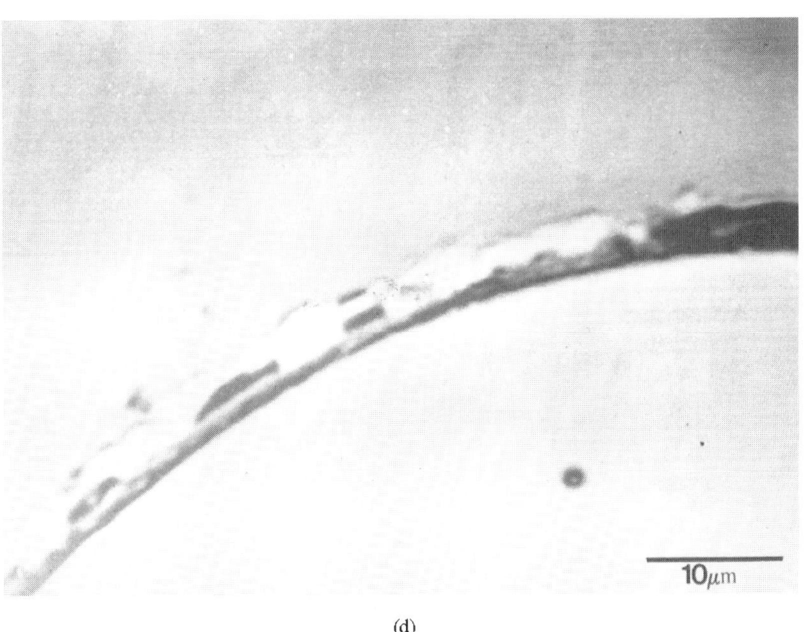

(d)

9.24 (c) Specimen 3, $z = 0\,\mu m$ and (d) specimen 4, $z = 0\,\mu m$.

thickness of 1.4 μm and therefore corresponds to the inner layer of the original carbon rich coating.

Outside this layer there is a discontinuous 'crenellated' layer up to 1.5 μm thick which is lighter in contrast (Fig. 9.24a). Since Fig. 9.24a was taken at focus this implies higher acoustic impedance, i.e. greater density and/or higher elastic modulus. The length of the crenellations corresponds approximately to the size of the matrix grains, which are revealed at negative defocus in Fig. 9.24b. This suggests that parts of this outside layer have been consumed in a reaction with certain grains in the matrix. It would thus be consistent with the higher reactivity of the β-phase Ti grains. Between the crenellated layer and the matrix is a dark line, but the resolution is insufficient to determine whether this signifies reduced bonding or the presence of remnant carbon. There are further contrast variations in the matrix: this could possibly be due to the presence of TiO_2 or TiC.

Further changes to the fibre–matrix interface are observed (Fig. 9.24c) in specimen 3, indicative of further degradation compared with specimen 2. Thus the inner remaining carbon rich layer is no longer present, having been replaced by a narrow dark region indicating a much lower acoustic impedance such as would be caused by a narrow gap. It is possible that the inner carbon rich layer had deteriorated to such an extent that it was easily damaged and removed during the polishing of the specimen. Some form of light-coloured reaction layer is still observed, varying in thickness from 1 to 1.5 μm. However, Fig. 9.24c shows that in specimen 3 there is also discrete damage to the SiC fibre material itself, which takes the form of notches or pores.

More extensive degradation of this interface was observed in specimens 4 and 5. In Fig. 9.24c a light-contrast reaction layer is present, but here it is both intermittent and thicker (3 μm) than previously. The reaction layer is believed to be mainly composed of TiO_2. There are also dark regions which indicate absence of material and porosity. Some of the porosity could have developed as the reaction layer degraded at 600°C. Alternatively, the reaction layer was very weak so that it was easily damaged during specimen preparation.

The changes that occur to the layers on the fibre–matrix interface are summarised in Table 9.5. The outer carbon rich layer is rapidly consumed during

Table 9.5 Interfacial structures in Ti-6Al-4V/SiC_m as a function of thermal ageing

Specimen	Inner carbon layer (μm)	Outer carbon layer (μm)	Reaction layer (μm)	Reaction layer composition
1	1.4	1.0–1.3	0.2–0.7	TiC + Ti silicide
2	1.5	0.0–1.5	0.2–1.3	TiO_2 + TiC + Ti silicide
3	1.5	0*	1.5	TiO_2 + some TiC
4	1.5	0*	1.0–2.0	TiO_2 + trace TiC
5	1.5	0*	2.0–2.5	TiO_2

* Denotes that the layer has been totally consumed.

prolonged thermal ageing, as a progressively thicker reaction layer is formed simultaneously. The composition of the reaction layer, as determined by an energy dispersive spectrometer (EDS) on an electron microprobe, changes progressively from one which is mainly TiC (with Ti_5Si_3 and/or Ti_3SiC_2) to one which is mainly TiO_2. The inner carbon layer mainly remains intact, continuing to protect the SiC monofilament itself, although as noted above, some discrete defects do form to penetrate this layer in, for instance, specimen 3 (Fig. 9.24c).

The bunching of the fibres that is observed in the heat treated materials is accompanied by defect formation in the matrix, which are a result of the composite fabrication process [52]. The first type of defect (Fig. 9.25a) consists of a fine crack which joins up regions of porosity at the matrix/fibre interface in adjacent fibres. The crack was invisible in the optical microscope but made visible acoustically by the presence of Rayleigh wave fringes on either side. The nature and geometry of this type of defect suggests its formation during fabrication, the close proximity (less than half a diameter) of two monofilaments preventing adjacent titanium foils from making a good diffusion bond. The second type of defect (Fig. 9.25b) consists of a porous region between two monofilaments that are too close for the titanium foils even to make contact. Whereas the second type of defect was observed in all the heat treated materials, the first type was relatively rare, occurring only in specimen 2. An unusual type of defect was observed in specimen 5, where the interfacial TiO_2 reaction layer formed during the heat treatment extends between adjacent monofilaments. Figure 9.25c shows that this region is traversed by two distinct cracks. They do not however appear to cross the inner interfacial layer of the right monofilament, suggesting that it is the residual carbon rich coating layer.

All heat treatments except the lowest were observed to induce radial microcracks in the annular banded region of the SiC monofilaments (Fig. 9.26a and 9.26b). In specimens 3, 4 and 5, 23%, 9% and 31%, respectively, of the monofilaments were microcracked. Specimens 3 and 5 (the longest heat treatments) also had the largest numbers of the second type of matrix defect discussed above. The microcracks in the monofilaments could not be observed by optical microscopy, indicating them to be very tightly closed, their acoustic contrast being enhanced by Rayleigh wave fringes. They were often associated with bunched groups of monofilaments and porosity, which must cause concentrations of stress and strain, and it is postulated that these cracks formed as a result of stress relief during or immediately following the heat treatments.

9.5 Conclusions

The scanning acoustic microscope generates images whose contrast is governed by the variations in acoustic impedance of the material being imaged. As the acoustic impedance of materials is a function of their elastic properties, the

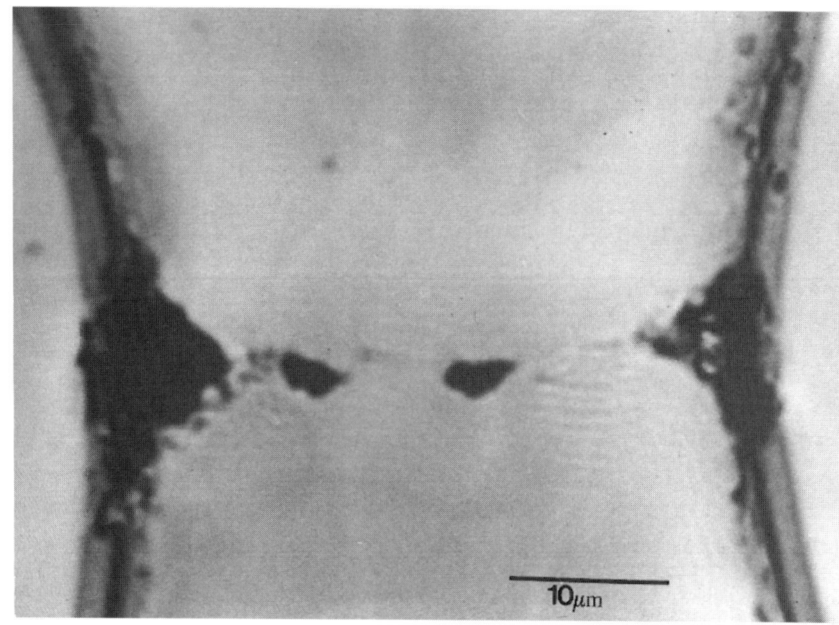

(a)

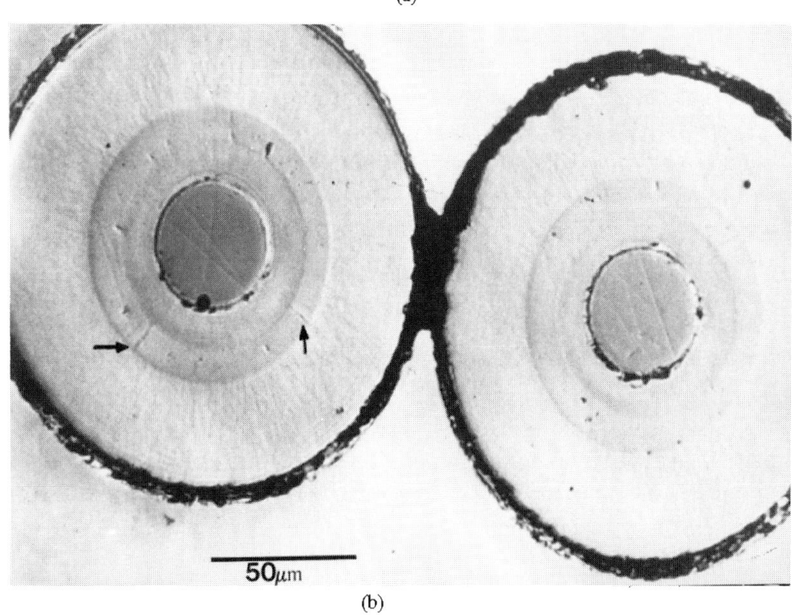

(b)

9.25 Acoustic micrographs of fabrication defects in the thermally aged specimens. (a) Crack due to poor diffusion bonding in specimen 2 ($z = 0\,\mu m$), (b) porous void between bunched monofilaments in specimen 4 ($z = 0\,\mu m$), the arrows indicate radial microcracking of the monofilament.

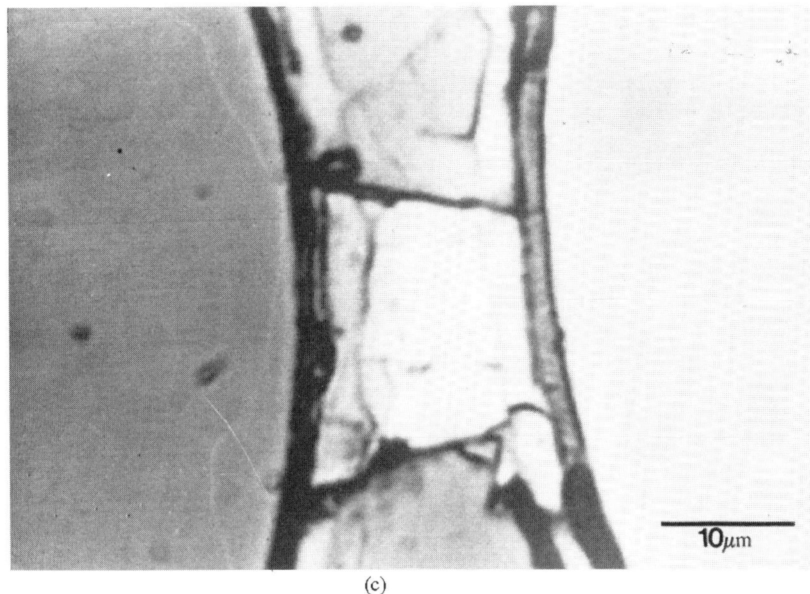

(c)

9.25 (c) Cracks in the reaction layer joining adjacent monofilaments.

scanning acoustic microscope enables the production of images which are effectively maps of the variations in elastic properties of a particular material. The images presented here indicate how strong the contrast between various material constituents can be. In addition, microstructural features can perturb the Rayleigh waves excited in the material by the microscope, giving enhanced contrast by the process of Rayleigh wave fringe formation. This enables features much smaller than the spatial resolution of the microscope to be clearly imaged. Features such as microcracks, porosity, differences in composition, second phases and fibre/matrix boundaries can be imaged.

In addition to providing images of materials, the acoustic microscope can be used to make quantitative measurements of elastic properties. A simple way to do this would be to calibrate the contrast at focus in terms of known impedances. In the case of relatively homogeneous materials whose properties are constant over comparatively large distances, accurate (better than 1 part in 10^3) measurements of Rayleigh wave velocity can be made over a region of size a millimetre or so. Under high resolution imaging conditions the range of $V(z)$ data is small, which severely limits the accuracy. In these circumstances the local elastic properties of composite materials can be determined from the spacing of Rayleigh wave fringes in the vicinity of boundaries and cracks.

The acoustic microscope can give information on the microstructure and elastic properties of composite materials which is difficult or impossible to obtain by more traditional techniques. Future work is expected to focus on making more

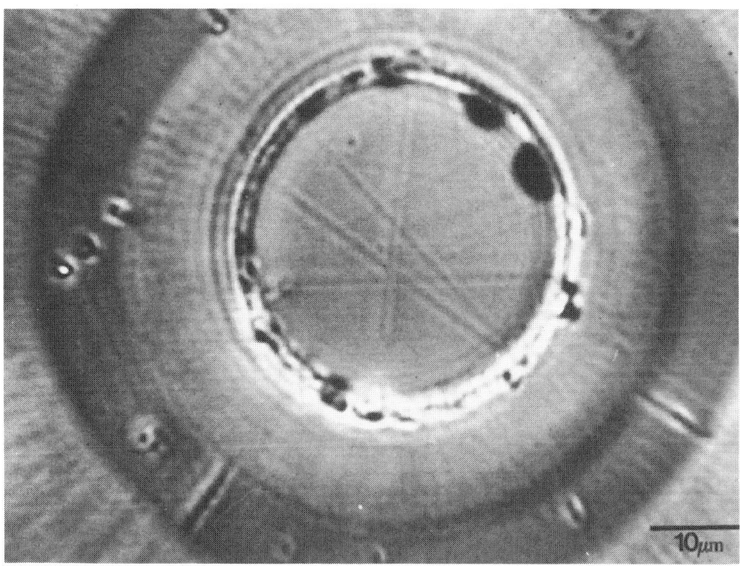

(a)

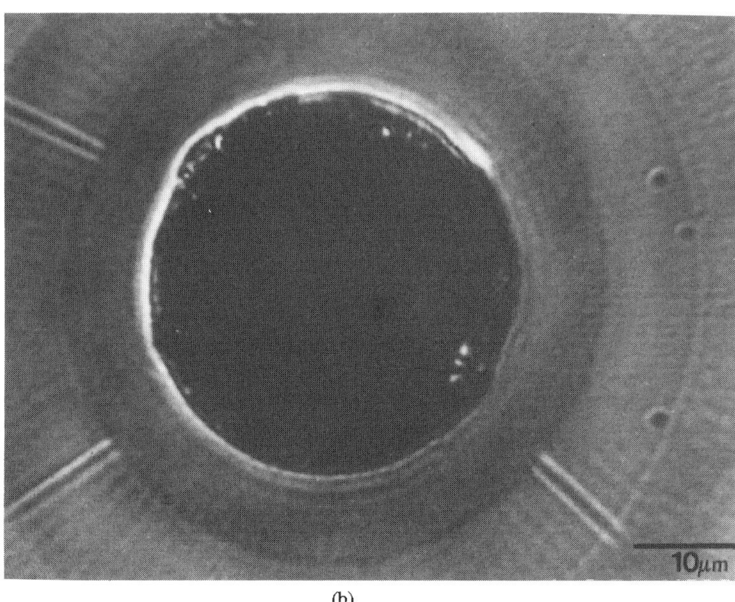

(b)

9.26 (a) Acoustic micrograph of radial microcracking within a SCS-6 monofilament in specimen 3, $z = -2.0\,\mu m$. (b) Acoustic micrograph showing radial microcracking in specimen 5, $z = -2.5\,\mu m$.

accurate Rayleigh wave velocity measurements (and hence elastic property measurements) over much finer scales.

Acknowledgements

This work was undertaken whilst CWL was supported by a SERC research studentship sponsored by AEA Technology. The authors thank AEA Industrial Technology, Rolls Royce plc and Textron Speciality Materials for provision of material, and Dr C B Scruby for support and encouragement throughout the work described here.

References

1. A F Fagan, G A D Briggs, J T Czernuszka and C B Scruby, Microstructural observations of two partially stabilized zirconia ceramics using acoustic microscopy, *J. Mater. Sci.* 1992 **27**(5) 1202–1206.
2. D G P Fatkin, C B Scruby and G A D Briggs, Acoustic microscopy of low ductility materials, *J. Mater. Sci.* 1989 **24**(1) 23–40.
3. C B Scruby, C W Lawrence, D G P Fatkin, G A D Briggs, A Dunhill, A E Gee and C-L Chao, Non-destructive testing of ceramics by acoustic microscopy, *Brit. Ceram. Trans. J.* 1989 **88**(4) 127–132.
4. K Yamanaka and Y Enomoto, Observation of surface cracks with scanning acoustic microscope, *J. Appl. Phys.* 1982 **53**(2) 846–850.
5. G A D Briggs, P J Jenkins and M Hoppe, How fine a surface crack can you see in a scanning acoustic microscope? *J. Microscopy* 1990 **159**(1) 15–32.
6. D C Philips, The fracture energy of carbon-fibre reinforced glass, *J. Mater. Sci.* 1972 **7** 1175–1191.
7. D M Dawson, R F Preston and A Purser, Fabrication and materials evaluation of high performance aligned ceramic fiber-reinforced glass-matrix composite, *Brit. Ceram. Proc.* 1987 **39** 221–226.
8. R W Davidge and J J R Davies, Ceramic matrix fibre composites: mechanical testing and performance, *Internat. J. High. Tech. Ceram.* 1988 **4** 341–358.
9. T Mah, M Mendiratta, A Katz, R Ruh and K Maz Digasni, High temperature mechanical behavior of fiber-reinforced glass-ceramic matrix composites, *J. Amer. Ceram. Soc.* 1985 **68**(9) C248–C251.
10. *Nicalon Silicon Carbide Continuous Fiber Data Sheet*, Nippon Carbon Company, 6-1, Hatchabori 2-Chome, Chuo-K, Tokyo, Japan.
11. R Simon and A R Bunsell, Mechanical and structural characterization of the Nicalon silicon carbide fibre, *J. Mater. Sci.* 1984 **19**(11) 3649–3659.
12. V S R Murty and M H Lewis, Interface structure and matrix crystallization in SiC (Nicalon)-Pyrex composites, *J. Mater. Sci. Lett.* 1989 **8**(5) 571–572.
13. S M Bleay and V D Scott, Microstructure property relationship in Pyrex glass composites reinforced with Nicalon fibres, *J. Mater. Sci.* 1991 **26**(8) 2229–2239.
14. S M Cox and P L Kirby, Rate of crystal growth in glass, *Nature* 1947 **159** 162–163.

15. S G Clarke, S M Bleay and V D Scott, An investigation into the formation of second-phase crystals in amorphous Pyrex glass, *J. Mater. Sci. Lett.* 1991 **10**(3) 149–153.
16. G A D Briggs, Cracks and discontinuities, *An Introduction to Scanning Acoustic Microscopy*, Oxford University Press, Oxford, 1985, chapter 7, pp 49–60.
17. S M Bleay and V D Scott, Microstructure and micromechanics of the interface in carbon fibre reinforced Pyrex glass, *J. Mater. Sci.* 1991 **26**(13) 3544–3552.
18. D C Philips, Interfacial bonding and the toughness of carbon fibre reinforced glass and glass-ceramic, *J. Mater. Sci.* 1974 **9**(11) 1847–1854.
19. K M Prewo, Carbon fibre reinforced glass matrix composite tension and flexure properties, *J. Mater. Sci.* 1988 **23**(8) 2745–2752.
20. D C Philips, R A J Sambell and D H Bowden, The mechanical properties of carbon fibre reinforced Pyrex glass, *J. Mater. Sci.* 1972 **7**(12) 1454–1464.
21. V S R Murty, M W Pharoah and M H Lewis, Interface microstructure and matrix crystallization in SiC-borosilicate (Pyrex) composites, *Mater. Lett.* 1990 **10**(4/5) 161–164.
22. C W Lawrence, C B Scruby and G A D Briggs, A study of ceramic matrix composites by acoustic microscopy, *Inst. Phys. Conf. Ser.* 1990 **98** 139–142.
23. B Budiansky, J W Hutchinson and A G Evans, Matrix fracture in fiber-reinforced ceramics, *J. Mech. Phys. Solids* 1986 **34**(2) 167–189.
24. R Chaim and A H Heuer, The interface between (Nicalon) SiC fibers and a glass-ceramic matrix, *Adv. Ceram. Mats.* 1987 **2**(2) 154–158.
25. T P Weihs, O Sbaizero, E Y Luh and W D Nix, Correlating the mechanical properties of a continuous fiber-reinforced ceramic matrix composite to the sliding resistance of the fibers, *J. Amer. Ceram. Soc.* 1991 **74**(3) 535–540.
26. P W McMillan, *Glass-ceramics*, Academic Press, London, 1964.
27. Y Imanaka, K Yamazaki, S Aoki, N Kamehara and K Niwa, *J. Ceram. Soc. Jpn.* 1989 **97** 309–313.
28. L M Sheppard, Toughening glass with fiber reinforcements, *Amer. Ceram. Soc. Bull.* 1988 **67**(1) 1779–1782.
29. B L Metcalfe, I W Donald and D J Bradley, Development and properties of a SiC fibre-reinforced magnesium aluminosilicate glass-ceramic matrix composite, *J. Mater. Sci.* 1992 **27**(11) 3075–3081.
30. A Atalar, C F Quate and H K Wickramasinghe, Phase imaging in reflection with the acoustic microscope, *Appl. Phys. Lett.* 1977 **31**(12) 791–793.
31. G A D Briggs, *Acoustic Microscopy*, Oxford University Press, Oxford, 1992.
32. J Kushibiki and N Chubachi, Material characterization by line-focus-beam acoustic microscopes, *IEEE Trans. Sonics Ultrasonics* 1985 **SU-32**(2) 189–212.
33. C B Scruby, K R Jones and L Antoniazzi, Diffraction of elastic waves by defects in plates, *J. NDE* 1987 **5** 145–156.
34. M F Doerner and W D Nix, A method of interpreting the data from depth sensing instruments, *J. Mater. Res.* 1986 **1**(4) 601–609.
35. W C Oliver and G M Pharr, An improved technique for determining hardness and elastic modulus using load and displacement sensing indentation experiments, *J. Mater. Res.* 1992 **7**(6) 1564–1583.
36. D J Pysher, K G Goretta, R S Hodder and R E Tressler, Strengths of ceramic fibers at elevated temperatures, *J. Amer. Ceram. Soc.* 1989 **72**(2) 284–288.

37. D B Fischbach, P M Lemoine and G V Yen, Mechanical properties and structure of a new commercial SiC-type fibre (Tyranno), *J. Mater. Sci.* 1988 **23**(3) 987–993.
38. *Ceramic Source Volume 5*, The American Ceramic Society, OH, 1989.
39. C W Lawrence, G A D Briggs, C B Scruby and J J R Davies, Acoustic microscopy of ceramic-fibre composites. Part 1, Glass-matrix composites, *J. Mater. Sci.* 1993 **28**(13) 3635–3644.
40. R F Cooper and K Chyung, Structure and chemistry of fibre–matrix interfaces in silicon carbide fibre-reinforced glass-ceramic composites: an electron microscopy study, *J. Mater. Sci.* 1987 **22**(9) 3148–3160.
41. M G Somekh, H L Bertoni, G A D Briggs and N J Burton, A two-dimensional imaging theory of surface discontinuities with the scanning acoustic microscope, *Proc. Roy. Soc. London A* 1985 **401** 29–51.
42. J Aveston, G A Cooper and A Kelly, Single and multiple fracture, *Proceedings of the National Physical Laboratory Conference on The Properties of Fibre Composites*, Leuven, Belgium, 1971, pp 15–26.
43. C W Lawrence, S Kooner, T P Weihs and B Derby, *Interfacial Phenomena in Composite Materials '91*, Butterworth-Heinemann, Oxford, 1991, pp 208–211.
44. M Taya and R I Arsenault, *Metal Matrix Composites: Thermo-Mechanical Behaviour*, Pergamon Press, Oxford, 1989.
45. J A McElman, Continuous silicon carbide fiber MMCs, *Engineering Materials Handbook, Volume 1: Composites*, American Society for Metals, Ohio, 1987, chapter 13C, pp 858–866.
46. J C Romine, Continuous aluminium oxide fibre MMCs, *Engineering Materials Handbook, Volume 1: Composites*, American Society for Metals, Ohio, 1987, chapter 13E, pp 874–877.
47. J Kushibiki, A Ohkubo and N Chubachi, Linearly focused acoustic beams for acoustic microscopy, *Electron. Lett.* 1981 **17**(15) 520–522.
48. C W Lawrence, *Acoustic Microscopy of Ceramic Fibre Composites*, D.Phil Thesis, University of Oxford, 1990.
49. C Jones, C J Kiely and S S Wang, The characterization of an SCS6/Ti-6Al-4V MMC interphase *J. Mater. Res.* 1989 **4**(2) 327–335.
50. J M Yang and S W Yeng, Interfacial reaction kinetics of SiC fiber-reinforced Ti_3Al matrix composites, *Scripta Metall.* 1989 **23**(9) 1559–1564.
51. X J Ning and P Pirouz, The microstructure of SCS-6 SiC fiber, *J. Mater. Res.* 1991 **6**(10) 2234–2248.
52. Z X Guo and B Derby, Microstructural characterization in diffusion-bonded SiC Ti-6Al-4V composites, *J. Microscopy* 1993 **169**(2) 269–277.

Index